Sustainable Engineering Forum 2018

Core Programming Area at the 2018 AIChE Annual Meeting

Pittsburgh, Pennsylvania, USA
28 October - 2 November 2018

ISBN: 978-1-5108-7613-2

Printed from e-media with permission by:

Curran Associates, Inc.
57 Morehouse Lane
Red Hook, NY 12571

Some format issues inherent in the e-media version may also appear in this print version.

Printed by Curran Associates, Inc. (2019)

For permission requests, please contact AIChE
at the address below.

AIChE
120 Wall Street, FL 23
New York, NY 10005-4020

Phone: (800) 242-4363
Fax: (203) 775-5177

www.aiche.org

Additional copies of this publication are available from:

Curran Associates, Inc.
57 Morehouse Lane
Red Hook, NY 12571 USA
Phone: 845-758-0400
Fax: 845-758-2633
Email: curran@proceedings.com
Web: www.proceedings.com

TABLE OF CONTENTS

Some abstracts in this book refer to images that were not made available by the publisher

526233 Effects of Solvent Selection on Efficient Furfural Production

Abstract:

Conversion of 5-carbon (C_5) sugars in biomass to furfural is an early step in the value chain to producing renewable value-added chemicals such as furfural alcohol, methyl tetrahydrofuran, tetrahydrofuran, 1,4 butanediol, and other four and five-carbon chemical intermediates. Molar yields of furfural from C_5 sugars are limited by competitive reactions to form humins. The commercially practiced technology for furfural production involves the simultaneous conversion of C_5 sugars in solid biomass and steam stripping of the resulting furfural. The process has a molar yield between 45-55 mol%.

Recently, researchers have examined more efficient methods of converting aqueous C_5 sugars, generally available as byproduct streams from pulp and paper manufacturing, into furfural in a two-liquid phase process. In this process, a water-immiscible organic solvent and acid catalyst are added to the aqueous C_5 stream. The organic solvent is used to selectively partition furfural away from the acid catalyst in the aqueous phase, thereby reducing the formation of humins and increasing furfural selectivity.

In this talk we will discuss the overall process considerations used to select an optimized solvent and the effect of the solvent on the energy load on distillation column used to separate furfural from the organic solvent. We will discuss the how varying the 4-tert-butyl phenol content in solvent and the organic to aqueous phase ratio affected the xylose conversion rate. We observed a 3x increase in reaction rate over the parameter space explored. The likely reasons and the implications of this finding will be discussed.

526657 Synthesis & Characterization of Molecularly Hybrid Bisphenols Derived from Lignin & Cashew Nutshell Liquid: Resin and Polymer Properties

Abstract:

Vinyl Ester (VE) resins can be used to create polymers with favorable properties for high performance applications. Standard industrial VE resins are comprised of monomers with a bisphenolic structure, which result in polymers with high strengths, moduli, thermal stabilities, and chemical resistances. Most of the commercially available VE resins are derived from bisphenol A (BPA). Unfortunately, BPA is considered unsustainable because it is synthesized from petrochemical feedstocks. It has also been scrutinized for its toxicity to the human endocrine system. Bio-based alternatives are being investigated to mitigate the negative aspects of BPA.

Lignin, a bio-based resource, is a favorable feedstock for the synthesis of VE resins due to its extensive aromaticity. Lignin is the second most abundant natural resource and a major byproduct of the pulp and paper industry. Vanillin can be derived from lignin and further reduced to vanillyl alcohol, a viable platform chemical for the production of bisphenols. Cashew nutshell liquid (CNSL) is an aromatic, renewable resource produced from the inedible shells of cashew plants. CNSL contains many phenolic lipids, the primary of which is cardanol. Cardanolâ€™s C15 alkyl chain, which can be naturally saturated or unsaturated, may impart water and thermal resistant properties to the resultant polymer network.

This study shows the synthesis and properties of lignin-derived vanillyl alcohol coupled with CNSL-derived cardanol (VAC). The VAC is an asymmetric bisphenol that is then functionalized to produce a dimethacrylate (VACDM). All dimethacrylates used in this work were blended with a reactive diluent, analyzed for their rheological properties, and cured. The resulting polymers were characterized via FT-IR, TGA, DMA, and fracture toughness to quantify the thermomechanical, thermogravimetric, and mechanical properties. The bio-based dimethacrylates were compared to BPA-derived industrial standard Vinyl Ester 828 (VE828) and the structurally similar bisphenol A dimethacrylate (BPADM).

535374 Synthesis, Design and Thermodynamic Analysis of Hybrid Processes Gasifying Biomass and Smelting Iron

Abstract:

Emerging research has revealed tremendous benefits in hybridizing the processes of iron ore reduction and biomass gasification. The iron ore acts as both an oxidant and a catalyst for the biomass gasification, while the biomass in turn replaces coking coal as the reducing agent for iron ore. The result is a carbon neutral iron ore reduction process co-producing synthesis gas.

This research presents detailed thermodynamic analysis of this system of chemical reactions to develop process targets and offer a starting point for design and optimization, as well as a basis for economic evaluation. Also presented is a sensitivity analysis to determine the effect of silica content on process performance.

It has been found that there is an inverse linear proportionality between iron reduction capacity and synthesis gas production and that the best economic performance under present market conditions is achieved by maximizing iron smelting and producing little or no synthesis gas.

However, the capacity for co-production allows for the operational flexibility to operate at maximum capacity during periods of reduced iron throughput, by transitioning to the production of synthesis gas. Once an iron deposit is depleted, a smelter capable of biomass gasification need not be scrapped; it can instead transition to full-time fuel production. In this way, iron/synthesis gas co-production can be thought of as a stepping stone to long-term renewable fuel production.

534805 F D M E (2,5-Furandicarboxylate, DiMethyl Ester) Process Development: Scale-up through Pilot

Abstract:

DuPont Industrial Biosciences and the Archer Daniels Midland (ADM) company are in a strategic partnership to develop and commercialize a new process to produce renewable FDME from corn-derived fructose. This newly commercial monomer is the dimethyl ester of 2,5-furan dicarboxylic acid (FDCA), which along with existing renewable diols such as BioPDO (1,3-propanediol) and ethylene glycol, can be polymerized to make a new family of bio-based furanic polymers (DuPont Everact^{TM}) including polytrimethylene furandicarboxylate (PTF) and polyethylene furandicarboxylate (PEF). These polyesters are of interest because of their enhanced gas barrier properties relative to incumbent polymers used in the beverage, food, and industrial packaging markets such as poly(ethylene terephthalate) (PET), and they provide opportunity for improved shelf life and decreased overall material use in packaging applications where O_2 and CO_2 permeation is critical.

This presentation will provide an overview of the process as well as the scale-up methodology being employed to develop an economically-viable FDME manufacturing process. The presentation will include a summary of best practices incorporated into the design phase such as internal third-party technical reviews of conceptual process designs and development of a dynamic process model to develop control strategies, tune process controllers, and train operators. Finally, weâ€™ll summarize the structured risk assessment and mitigation methodology employed to identify and address key technical risks to commercialization.

531142 Reuse and Valorization of Used Cooking Oils By Transformation into Epoxidized Oils

Abstract:

Currently, sustainability and circular economy are fundamental elements that are increasingly incorporated as design principles of novel chemical processes. In this direction, the use of waste residues as raw materials for the manufacture of chemical products is a clear example of these ecofriendly principles. Among different biobased residues suitable for chemical exploitation, used cooking oils (UCOs) are of importance as they represent a major problem for modern large cities. In several urban centers worldwide, UCOs are incorrectly disposed through de sewage systems, generating a cascade of problems (sewage clocking, floodings, vectors generation, bad odors, etc.) with huge environmental, social and economic impacts. Alternatively, UCOs can be used as second generation raw materials for the manufacture of different derivatives (e. g. fuels and chemicals), with the advantage of being renewable, low-cost, high-volume, readily-available, and suitable to exploit as they already delivered their nutritional value.

After pretreatment and purification processes, the UCOs can be considered as an oleochemical feedstock. It can be transformed into a large variety of byproducts such as resins, surfactants, monomers, additive for polymers, lubricants, epoxidized oils, etc. Generally, a major limitation identified when dealing with UCOs is their chemical heterogeneity due to large variety of impurities generated during the cooking process (such as aldehydes, ketones, peroxides, cylindrical and polymeric compounds). However, even after the cooking process, the UCOs maintain around 90% of its original triglycerides content, making them useful in the oleochemical industry.

In this direction, this work developed a process for the reuse and upgrade of UCOs into epoxidized oils. Samples of UCOs collected from larger generators (fast food restaurants) in the city of Bogotá were used as raw material for the process. Initially, the heterogeneity of UCOs was assessed by collecting and characterizing UCOs from different fried chicken and hamburgers restaurants along the city. The recollected samples were characterized by evaluation of acid value, content of volatile matter, density, peroxide value, color, and content of total polar compounds. This allowed to stablish the variability of the properties, and the required pretreatment process to use the UCOs as oleochemical feedstock. The range of iodine values and saponification index indicated that the collected UCOs can be used as raw materials for epoxidation. After performing filtration, degumming and bleaching processes, the UCOs were adjusted for further epoxidation.

Finally, the epoxidation of the pretreated UCOs was studied by using a traditional *in-situ* peroxyacetic acid method, with a homogeneous acid catalyst. The impact of main process variables of the epoxidation process such as temperature (50, 65, 80ºC), hydrogen peroxide excess respect acetic acid (15, 30, 45%p/p), ratio of acetic acid to UCO (1.0, 3.5, 6.0 %p/p) were assessed by a three-level factorial design, using 2%molar relation of catalyst concentration. The reaction was analyzed by measuring changes in iodine value and oxirane value. The obtained epoxidized oil exhibited oxirane oxygen content of around 3.2% with conversions above 60%. The observed epoxidation efficiency and the product specifications were similar to those obtained with virgin vegetable oils, turning the developed process into a feasible alternative for UCOs upgrading and valorization.

542555 Important Hard and Soft Skills Difficult to Teach in the Classroom

Abstract:

There are many important skills and experiences imparted through working on international humanitarian engineering projects. Understanding your customer, dealing with other cultures, safety, dealing with uncertainty, and thinking on your feet are a few of them. Students can benefit greatly from taking on this kind of work either in implementing full projects or doing the conceptual work in a capstone design class.

542581 A Student Perspective: What I Learned from Working on an International Project

Abstract:

Students volunteer to work on international projects for many different reasons. They take on many different roles in planning, design, execution, communications, project management, and research; and they each develop different skills, have different experiences and are impacted differently. Undergraduate students will discuss why *they* volunteered, what *they* have done, and what *they* have gotten out of this work professionally and personally.

542554 What Does Sustainability Really Mean in These Projects?

Abstract:

To make a project successful partnering with a small community in the developing world requires many important attributes besides technical accuracy. In order to sustain the benefits over the long term, the project team must consider and master many other aspects of sustainability in addition to environmental sustainability. Financial and cultural are two of the many types of sustainability that will be discussed.

542556 What Is EWB-USA and Why Might You Want to Work with Them?

Abstract:

EWB-USA is organization of volunteer students and professionals. Student and professional chapters can be found on campuses and cities across the country. EWB-USA has very specific standards in terms of how partnerships with local communities are maintained and projects executed. The EWB-USA philosophy is tightly focused on executing projects that will actually deliver value over the long-haul, and the mandatory work processes take into account the unique characteristics of student volunteers.

542585 How International Projects Might Help with ABET Accreditation

Abstract:

Team humanitarian engineering projects align with many ABET Criterion 3 Student Outcomes. Incorporating the projects into a programâ€™s assessment structure is difficult when the projects are extra-curricular. We will present solutions from different schools that allow the projects to assist accreditation.

542580 Panel Discussion -- Making Successful International Projects Happen

Abstract:

There are many challenges to implementing humanitarian engineering projects in a university setting. Where do the ideas / projects come from? How can these projects be turned into meaningful capstone design classes? How can they be fit into existing curriculum and academic environment? How do you motivate students, faculty, administration? How can faculty motivated to make these projects happen collaborate across different universities? This panel is intended to be a forum for all those in the room to define challenges and offer solutions.

541124 Controlling Particle Size in 2 Stage Grinding Processes

Abstract:

Grinding raw biomass into a format suitable for conversion to a fuel, whether a dry or liquid product for thermal, chemical, or biological conversion, requires a specific particle size which is critical for the conversion pathway. Particle size also factors into the flowability of the biomass during the conversion process. Controlling the particle size of biomass through the grinding and milling process is difficult because on the inherent variability of most feedstocks. Crop type, harvest method, moisture content, age of biomass and processing method all affect the particle size of the processed biomass. Through research at the Idaho National Laboratoryâ€™s Biomass Feedstock National User Facility, several processing parameters have been identified when can be used to alter the biomass particle size during processing to meet the requirements of the conversion process. The factors tested were screen size, feed rate, mill speed, and moisture content. This research is aimed at creating a smart, dynamic control system that reacts to feedstock characteristics during processing to create a feedstock that is on spec and a control system that will ensure that is stays on spec despite the inherent variability of the feedstock.

540151 Impact of Parcel Size, Field Shape, Crop Yield, Storage Location and Collection Equipment on the Performance of Harvest System in Shrub Willow Fields

Abstract:

In this study, an integrated system of harvesting, collecting and transporting willow biomass crop to a storage site is modeled and evaluated using the IBSAL simulation model. A scenario analysis is used to quantify the impacts of five major input parameters on the performance of the integrated system. These parameters include parcel size, field shape, willow yield, distance to the storage site, and type of the collection equipment. Multiple performance indicators are identified to quantify the impacts on the system such as size of the equipment fleet, effective material and field capacity of the harvester, operating costs, and waiting times. The input data are collected from 36 commercial, short-rotation, shrub willow fields in northern New York State. The simulation results indicate that crop yield and type of collection equipment have the highest impact on operating costs and the equipment fleet size. As the size of equipment fleet increases in the system, variability in the system performance tends to increase. Field shape has the lowest impact on the overall system performance compared to the other four input parameters. The simulation results suggest that a combination of performance indicators need to be considered to evaluate the overall performance of a dynamic and complex system of harvesting, collection and transportation. This study enhances the understanding of the dynamics and complexity of this system. The developed IBSAL model and scenario analysis approach can assist in planning this system based on the characteristics of field, crop and logistical equipment to reach a high system performance.

534030 Investigating the Distributed and Centralized Preprocessing Depot in the Supply Chain Network Design of a Biorefinery Designed for Biochemical Conversion

Abstract:

This research develops an optimization model to investigate distributed and centralized preprocessing depot in the supply chain design of a biorefinery designed for biochemical conversion. The distributed depot in the supply chain network of a biorefinery deploys distributed preprocessing depots near biomass resources whereas centralized depot in the supply chain network deploys a primary centralized depot co-located with biorefinery to meet feedstock demand. The optimization model determines optimal supply chain network (e.g. optimal depot locations, required number of depots and size) while identifying the least cost feedstock blend of a biorefinery based on resource availability, quality requirements, and logistics cost for a biochemical conversion. A mixed-integer linear programming model is developed to determine optimal supply chain structure of a biorefinery sourcing multiple feedstocks. A case studyâ €"based on a biorefinery located in western Kansas, USA that uses three pass corn stover, two-pass corn stover, switchgrass, and municipal solid waste fraction to meet biochemical carbohydrate specifications and feedstock demandâ€"shows that utilizing a distributed depot in the supply chain network is more economical than the centralized depot in the supply chain network.

539718 Willow Biomass As a Feedstock for Biorefinery: Evaluation of Bark Effect on Hot Water Extraction Output, and Lifecycle Assessment of Cellulosic Ethanol Production

Abstract:

The use of cellulosic biofuels is projected to increase in order to offset greenhouse gas emissions from the combustion of non-renewable fuels. Shrub willows are being developed as a short rotation woody crop (SRWC), and have a high net energy ratio (energy produced/ fossil fuel energy consumed), low greenhouse gas footprint and high carbohydrate production potential. Willow biomass can be combined with woody biomass from forests but willow often has higher bark levels that creates challenges because it increases the ash content and lowers the melting point.

This study evaluated how the amount of bark (0%, 33%, 66% and 100%) on three willow cultivars impact the output of hot water extraction in term of mass removal and extract composition, as well as its influence on the heating value and ash. Additionally, a cradle-to-grave life cycle assessment (LCA) was performed to assess the greenhouse gas emissions associated with the production of bioethanol from the hot water extract of willow biomass.

The hot water extraction process reduced ash content up to 50% for sugar maple and willow, but there was variation among the varieties of willow. After hot water extraction the heating value of biomass increased about 5% with removal of mostly hemicelluloses that have relatively low heating value. The hot water extraction provided a fermentable sugar stream for biofuels and bioproducts production, and improved the characteristics of the hot water processed willow biomass that can be used to produce high quality pellet. The lifecycle assessment revealed the greenhouse gas saving associated with ethanol production from willow biomass via fermentation of sugars extracted by the hot water extraction process. The most sensitive parameters to greenhouse gas emissions included fermentation yield, mass removal by hot water extraction, and amount of belowground carbon sequestration, in addition to transportation distance.

535771 Techno-Economic Analysis of Supplying Forest Biomass Feedstock for Biopower Applications

Abstract:

Forest biomass residues in the Northeast U.S. are low-carbon fuels that can reduce fossil fuel consumption. Forest biomass utilization for the bioenergy feedstock production, however, demands capital-intensive forestry equipment for felling, skidding, processing and chipping felled trees. Techno-economic analysis (TEA) is used to evaluate the economic feasibility of forest biomass feedstock supply chains. This research uses two methodsâ€"machine rate and net present value (NPV) analysis methodsâ€"to build up the deterministic TEA model with baseline cost and revenue components of a whole tree harvesting system over an eight year project.

The current findings from sensitivity analysis indicate that the prices of logging products such as sawlog, pulpwood, firewood, and wood chips most affect the economic viability of the logging activities, followed by capital costs invested. We found that the stumpage price of sawlog is closely associated with the minimum selling price of wood chips. Higher profits from sawlog sales would give loggers a buffer against the current, low price of wood chip production.

Further research will focus on uncertainty analysis of logging economic viability to develop financial risk management tools for the forest biomass feedstock supply chain. The nature of forest operations involves uncertainty due to variations in the price of logging products, equipment costs, fuel price, and other factors. To incorporate the uncertainty of cost and revenue components, this research applies stochastic analysis with Monte Carlos simulations and obtains probability distributions of NPVs of harvest system models.

Final findings from this TEA study will contribute to biopower applications through integrated analysis using life cycle assessment (LCA) of climate change impacts of the equivalent feedstock production. We will incorporate variations of the modeled feedstock price into the fuel cost of biopower model. For the integrated analysis, one implication is that using biofuels for equipment operations can mitigate climate change impacts compared to using conventional diesel fuels; however, it also results in higher financial operating costs when not subsidized. As such, the integration of TEA and LCA modeling provides insights into the development of forest biomass feedstocks to supply affordable and low carbon energy. The biopower model integrated with this feedstock model will provide robust findings for economic viability of electricity generation with bio-feedstock.

540135 Performance and Logistics of a Forage Harvester and Collection System in Short Rotation Willow Crops

Abstract:

Demand for renewable energy in the United States is at its highest and is projected to increase. Bioenergy sourced from perennial energy crops is expected to be an important resource and short rotation woody crops (SRWC) such as shrub willow and hybrid poplar have shown great potential in their respective regions in terms of biomass production and job creation while also providing environmental benefits. However, ambitious cost targets have been set for bioenergy crops in order for them to compete with other sources of energy. Harvesting and logistics continue to comprise the largest portion of costs associated with SRWC production. While studies have identified key factors affecting system performance (e.g. machine type, horsepower, standing biomass, operator experience), few studies have sought collected and complied this information from commercial scale operations. This initiative is focused collecting field data from commercial scale willow harvesting operations, providing it to modeling teams who then provide direct feedback on the operation of the system, ways to optimize it and reduce overall costs. Harvests were conducted over 20 days on ten, three-year-old shrub willow fields in Northern New York State using two New Holland FR-Series forage harvesters equipped with FB130 coppice headers. These single pass cut and chip systems harvest the willow, chip it and blown the chips into collection wagons pulled by tractors. The harvesters and wagons were equipped with GPS units and wagon loads of chips were weighed. Data is processed to compare harvester performance (effective material capacity, field capacity, fuel consumption, engine loading) and collection system performance related to harvester power, operator experience, standing biomass, and field conditions. The amount of harvested biomass ranged from 15 to 80 Mg ha^{-1}. Mean material capacity of the harvester filling wagons was approximately 30 Mg hr^{-1}, but exceeded 60 Mg hr^{-1} for fresh biomass. Mean material capacity of the collection system delivering biomass to short-term storage was approximately 15 Mg hr^{-1}, but could exceed 30 Mg hr^{-1}, which is an improvement over earlier trials. Bulk density of willow chips ranged between approximately 175 and 325 kg m^{-3} with significant differences between collection vehicles. Results were shred with modelers who are using the data to identify ways to optimize the system and reduce the cost of harvesting and logistics.

539584 Pittsburgh: Urban Agriculture and the Food-Energy-Water Nexus

Abstract:

Cities rely on external supply chains for food, energy and water. Urban areas are growing and the dependence on reliable supply chains for these resources grows. The interdependence between the infrastructure and connections of providing these resources has become increasingly significant. Urban agriculture provides an opportunity to address Nexus challenges. Urban agriculture benefits include increasing food resilience and improving stormwater management. Ancillary benefits can also include the potential to reduce the amounts of materials sent to landfills (through composting), the reduction urban heat-island effect, and reducing wastewater treatment loads â€“ all of which come with energy reduction benefits.

Pittsburgh lost a substantial portion of its population during the transition away from its industrial roots as the â€˜Steel City.â€™ A legacy of the population decrease includes empty and abandoned lots (e.g. over 8,000), aging infrastructure, and areas of decline that have created food deserts and challenges associated with the interdependence of food, energy and water. Pittsburgh has implemented multiple initiatives to address this challenge that will realize the benefits from urban agriculture (e.g. a 23 acre urban farm; community driven revitalizations that include energy efficient housing, stormwater infrastructure, rainwater management, and agriculture spaces).

This analysis examines urban agriculture strategic options, the potential benefits, and the challenges. Options include agriculture options (e.g. geoponic, hydroponic, aeroponic systems), water management options (e.g. water reuse), waste management (e.g. biochar as soil enhancer and carbon storage), and energy option (e.g. distributed generation).

538328 Quantifying Virtual Phosphorus Flows in Interstate Food Trade: Implications for Environmental Sustainability

Abstract:

Phosphorus(P) is one of the essential nutrients for agriculture. It is primarily supplied to crops as inorganic phosphorous fertilizer or though manure application. On farm use of phosphorus efficiency is very low, ranging between 10-15%. Rest of the applied P gets bonded in the soil which may subsequently become available for crop uptake. Current agronomic P recommendations for farmers exclude the accumulated P due to unknown timescales of release. Sustainable management of P is an important topic due to the limited availability of inorganic P sources and its downstream potential to cause eutrophication. Additionally, production of phosphorus fertilizers is highly energy intensive with limited potential of waste stream recovery due to techno-economic constraints. Therefore, an accurate understanding of regional P application and efficiency, combined with interregional food trade is necessary to understand the complexity of P cycle. We combine data on current phosphorus application practices with recommended and existing soil P levels to assess sustainability of P management on farm. Additionally, we use P application data with food trade to quantify phosphorus embodied in the inter-regional trade. This work will enable to identify trends and hotspots in US agriculture P application practices for management of food-water nexus.

We develop a food trade and associated virtual phosphorus model for the U.S. by evaluating over 60 crop items covering commodity groups of cereal grains, fruits, nuts, and vegetables. Regional phosphorus mass balances are constructed using datasets from US Department of Agricultureâ€™s Natural Resource Conservation Service, International Plant Nutrition Institute, and Census of Agriculture. The interstate food trade network is formulated from Freight Analysis Framework and translated to virtual phosphorus network by using crop yields and phosphate fertilizer application rates from Census of Agriculture. Preliminary results indicate that phosphorus fertilizer application differ considerably for crop ranging from 2-10 lb P per ton crop for grains to 10-68 lb P per ton of crop for nuts. To estimate whether calculated P application rates are excessive or necessary for specific soil type, we compare it with recommended P levels in each state. In 2012, 28 states had to recommended levels, including states with highly eutrophic waterbodies such as Ohio, Florida, California, Texas, and Maine. With addition of total P fertilizer across crops in the given year, the soil phosphorus for these states increases by a fraction of 2-4 compared to recommended levels. From food trade perspective, we find that 13 billion US tons of food and 5.3 million US tons of embodied P were transferred across the U.S. Specifically, grains dominate food and total embodied P transfers within the U.S., indicating the importance of considering scale of production. The implications of the study findings will be discussed in the context of balancing on-farm conservation policies for simultaneously enriching soil with nutrients and avoiding excess run-offs.

533778 Towards a Two-Level Superstructure Optimization Framework for Land Use Based on Food-Energy-Water Nexus

Abstract:

Land use optimization is the result of competitions between different land types [1]. The main reasons of these competitions are the quantitative constraints of land scales and corresponding Food-Energy-Water (FEW) resources under multiple conflicting objectives [2, 3]. Despite FEW resources play critical important roles for sustaining and improving human life, increasing production demands and sustainable concerns intensify these competitions due to their limitation. How to coordinate these competitions is a complex task for reasonable land allocation based on limited FEW resources while promoting efficiencies and sustainability [4, 5]. A nexus thinking by considering the FEW flow and interactions in multi-objective land use systems has become apparently necessary for efficient use of FEW resources and trade-off decisions [6, 7]. However, challenges arise in representing quantitative FEW-Nexus that supports land-use decisions while encountering conflict objectives, limited data, and multiple scales [8, 9].

To address these challenges, we develop a multi-objective optimization framework by considering total profit, food production, resources use, and environmental penalty as multiple objectives for land use systems. The proposed framework suggests a two-level superstructure optimization method: (1) In the first level, models of all the production units are developed by data-driven modeling and global optimization methods based on limited realistic data. FEW flow among them are quantified and interlinked to construct interval models, which can be represented as interval superstructures. The small-scale MINLP problems can be solved efficiently due to the limited combinations of land units in the interval subsystems; (2) In the second level, multiple interval models with optimal land and FEW allocations are used to construct the extended systematic network and represented as a large-scale superstructure, which can be solved as a MILP problem. A series of FEW indices are provided for decision-makers to analysis nexus in the system, carry out quantitative assessment based on different objectives, and achieve trade-off solutions. The framework is illustrated by a case study on the crop-livestock system, which shows that it can provide multiple land allocation solutions and predicts corresponding yields for land units within the systems under flexible scales. Computational results from interval models indicate that we achieve optimal solutions in the interval subsystems, and valuable production models for yield prediction with high confidence. The performance of these models can be improved by increasing feedback data [10]. The systematic network provides an efficient method for extending the land allocation solutions to multiple land scales. For multiple objectives, the proposed FEW index can be applied to select strategies for optimal land allocation that maximizing food productivity and minimizing water and energy consumptions.

Reference

[1] Memmah, M. M., Lescourret, F., Yao, X., & Lavigne, C. (2015). Metaheuristics for agricultural land use optimization. A review. Agronomy for sustainable development, 35(3), 975-998.

[2] Bergstrom, J. C., Goetz, S. J., & Shortle, J. S. (2013). Land use problems and conflicts: Causes, consequences and solutions. Routledge.

[3] Keairns, D. L., Darton, R. C., & Irabien, A. (2016). The energy-water-food nexus. Annual review of chemical and biomolecular engineering, 7, 239-262.

[4] Miralles-Wilhelm, F. (2016). Development and application of integrative modeling tools in support of food-energy-water nexus planningâ€"a research agenda. Journal of Environmental Studies and Sciences, 6(1), 3-10.

[5] Simpson, G., & Berchner, M. (2017). Water-energy nexus-Measuring integration: towards a water-energy-food nexus index. Water Wheel, 16(1), 22-23.

[6] Garcia, D. J., & You, F. (2016). The water-energy-food nexus and process systems engineering: a new focus. Computers & Chemical Engineering, 91, 49-67.

[7] Ringler, C., Bhaduri, A., & Lawford, R. (2013). The nexus across water, energy, land and food (WELF): potential for improved resource use efficiency?. Current Opinion in Environmental Sustainability, 5(6), 617-624.

[8] McCarl, B. A., Yang, Y., Schwabe, K., Engel, B. A., Mondal, A. H., Ringler, C., & Pistikopoulos, E. N. (2017). Model Use in WEF Nexus Analysis: a Review of Issues. Current Sustainable/Renewable Energy Reports, 4(3), 144-152.

[9] McCarl, B. A., Yang, Y., Srinivasan, R., Pistikopoulos, E. N., & Mohtar, R. H. (2017). Data for WEF Nexus Analysis: a Review of Issues. Current Sustainable/Renewable Energy Reports, 4(3), 137-143.

[10] Nie, Y., Avraamidou, S., Li, J., Xiao, X., & Pistikopoulos, E. N. (2018). Land use modeling and optimization based on food-energy-water nexus: a case study on crop-livestock systems. 13th International Symposium on Process Systems Engineering; Elsevier; Accepted.

537263 Land Availability, Utilization, and Intensification for a Solar Powered Economy

Abstract:

Among all forms of renewable energy, solar energy is the most promising one with sustainable availability and abundance to meet human beingâ€™s energy need. However, solar energy is a diffuse resource and harvesting it requires vast tracts of land. While most studies estimated that land availability is not a major challenge for the feasibility of a total solar economy, a few recent detailed analyses have raised the problem in land availability in some densely populated countries, using actual solar farm production data and considering other factors such as storage requirement and intermittency of solar availability. However, according to their estimation, countries with relatively low population density, such as the United States, have sufficient land for solar farms to meet all energy demand. In this study, we perform a detailed land requirement analysis in each state of the US for a solar powered future and propose land intensification strategies to resolve possible land use challenges.

Land requirement analysis in this study takes into account several issues that are usually ignored. First, efficiencies of major energy conversion steps from current primary energy to end use are different from those when solar energy is deployed. Second, available land is limited to only part of the urban area (rooftops) and miscellaneous land. Farmland, forest and other reserved land such as national and state parks cannot be utilized for solar plant construction. In addition, we prefer local photons for local use for consideration of minimizing transmission loss and energy security.

Our land requirement analysis shows that for many states there is insufficient currently available land and that other land areas will need to be used, which is contrary to previous report. We propose that farmland is the best choice for land use intensification as it minimizes additional environmental impact to existing habitat and propose several photovoltaic systems for this purpose. We then estimate the power output of these systems and estimate the amount of farmland for each state that is needed to meet energy needs. By land intensification, the energy demand of each state in the US can be locally satisfied.

529950 Microbial Community Profile Versus Water Quality in Urban Watersheds

Abstract:

Microbial community profiles can serve as an indicator for unbran water quality. The aim of this study was to gain insight into bacterial diversity and infer water quality based on selected operational taxonomic units (OTUs). Here we report bacterial communities of 12 watersheds sampled at different frequencies from year 2013 to 2016, were characterized by sequencing the V4 to V5 regions of 16S rRNA genes using 454 pyrosequencing and paired-end sequencing platforms. An nMDS plot of all samples revealed that bacterial communities of the same watershed tended to cluster together and a few watersheds shared similar communities. Effects of temporal variation were more pronounced on bacterial communities within watersheds ($P < 0.01$; 2-Factor PERMANOVA) than spatial variation. Cyanobacteria of *Microcystis* and *Cylindrospermopsis* dominated certain watersheds that tended to have algal blooms. Pairwise correlation of gut and sewage-associated OTUs revealed a strong positive correlation among *Enterobacter*, *Klebsiella* and *Escherichia* and *Clostridium sensu stricto 8*, suggesting that presence of these taxa as possible indicators of impairment. A comparison of the abundances of 13 potential antibiotic resistant bacteria, revealed four watersheds with samples that correlated with higher abundances of potential antibiotic resistant bacteria compared to the other watersheds. Thus microbial community determined by high-throughput sequencing in conjunction with qPCR can be used to indicate quality of water in urban watersheds.

528536 Mechanisms Whereby Microbes Promote Intermediate Soil Moisture Content

Abstract:

The function of microbial communities is inextricably with the soil microstructure. At the pore scale, water content regulates hydraulic connectivity and thereby modulates microbesâ€™ access to aqueous and gaseous substrates. Furthermore, soil bacteria near plant roots produce extracellular polymeric substances (EPS), a polymeric substance that helps modulate soil moisture and promote plant growth. EPS regulates soil moisture (i) by serving as hydrogel, swelling during wet conditions and remaining hydrated during dry conditions, (ii) by promoting soil particle aggregates via capillary forces, (iii) by altering soil surface properties through creating water repellent surfaces. EPS-mediated moisture retention behavior in soil is complicated but can be understood by systematically controlling the microscale environment. Here, we show the effect of physical microstructure on EPS-mediated moisture retention. We employ physically complex emulated soil micromodels as well as micromodels with uniform macropore or micropore geometries. EPS was collected from stationary-phase *Sinorhizobium meliloti* cultures and suspended at different concentrations in growth media salts or in artificial groundwater. Experimental results showed that the 0.25× EPS solution dried eight times slower than deionized water, and the 1× EPS solution dried 16 times slower than deionized water in identical soil micromodels. Also, the residual saturation of 0.25× and 1× EPS solutions were 6 ± 1% and 35 ± 21%, compared with a residual saturation of 0 for deionized water. We observed no effect on rate or extent of moisture loss for the micromodels with a uniform macropore geometry. Studies with uniform micropore geometry confirmed that higher EPS concentrations were associated with a reduce rate and increased variance of moisture loss at the pore scale. In this study, we provide a systematic method to evaluate the performance of EPS solutions in modulating soil moisture, and show how EPS concentration decreases the rate of water loss and also amplifies the variability of moisture content at the pore scale. Longer term, we anticipate that the impact of this work will be to aid in the development of sustainable agriculture biotechnology for increased food production using less water.

528379 Nitrogen Efficient Fertilizer Materials

Abstract:

Nitrogen cycle management is a critical challenge in modern engineering. NH_3 synthesis from N_2 consumes several percent of total world energy while only a fraction of the synthesized fertilizer nitrogen, such as urea, is absorbed by the plants. Majority of nitrogen initially confined in urea is decomposed and released in the environment as gaseous NH_3 or soil and water mobile nitrate, NO_3^-, resulting in low sustainability nitrogen fixation process. Furthermore, for a complete plant nutrient all of the nutrients (N,P,K,Ca,Mg,S) need to be supplied in one solid granulated or liquid mixture. Aqueous liquid fertilizers are expensive to transport since their dissolved nutrient content is governed by the solubility limits.

Molecular crystals of urea with inorganic acids (H_3PO_4 and H_2SO_4) as well as solid salts ($CaSO_4$, $MgSO_4$, $Ca(H_2PO_4)$ and $Mg(H_2PO_4)$) present a convenient form of high nutrient content fertilizers that can be inexpensively transported to the remote locations and applied directly or via fertigation. Importantly, these urea inorganic acid adducts have been shown to result in a much more efficient nitrogen uptake by plants combined with significantly lower nitrogen losses. The work will be presented that will explore design of these fertilizers that contribute to the enhanced sustainability of the chemical processing. In particular, mechanochemical synthesis methods that avoid use of liquid solvents that would otherwise necessitate evaporation or result in high aqueous solution transportation costs will be presented. Since dissolution of the solid urea adducts for fertigation requires in depth knowledge of the corresponding phase diagrams, these data will also be presented. Finally, molecular structure effects of the urea adducts on the improved nitrogen cycle management will be discussed as well as the synthesis of industrial and anthropogenic waste derived nitrogen efficient molecular fertilizers.

528475 Degradation of Lignocellulose and Production of Chemicals Via Sequential Hydrothermal Liquefaction

Abstract:

Conventional hydrothermal conversion of lignocellulose produces low-quality bio-crude at high temperature and pressure. Huge energy loss occurs resulting from polysaccharides conversion to fuels due to the low energy density of these sugars. Especially, lignin decomposition and conversion to fuels and chemicals has been extensively investigated, however, the feedstocks are most restricted to isolated lignin and model lignin compounds. Additionally, distinct lignin isolation methods lead to diverse results owing to the changes of lignin structure during isolation process. The lack of practical depolymerization processes that can produce a limited number of lignin monomers at high yields has limited the options for upgrading lignin. Currently, few studies cover technology that directly convert raw lignocellulose to chemicals to with organic solvent aid process in absence of metal catalyst. Our newly proposed sequential two-stage hydrothermal liquefaction (SeqHTL) process benefits from (1) no pretreatment or pre-isolation step; (2) simultaneous production of bioactive compounds, C5-C6 sugars and their derivatives, and lignin monomers; (3) no toxic organic solvent added.

The preliminary tests showed that two-stage HTL achieved biomass degradation rate of 57.8% at 240 and 76.6% at 280 , while one stage HTL was 50.0% and 71.4%, respectively. A systematical study has been carried out for SeqHTL: (1) doubling reaction time (from 30 min to 60 min) for both stages increased the solid degradation rate from 57.8% to 70.2%; (2) increasing the biomass load from 1:40 to 1:10 (biomass:water mass ration) slightly decreased the biomass decomposition rate from 57.8% to 52.8%. Further increase on biomass concentration to 1:5 dramatically reduced the decomposition rate to 42.7%; (3) If only increasing second stage temperature by 40 and keeping first stage unchanged, a 73.4% decomposition rate was achieved. If only increasing second stage reaction time by 30 min and keeping first stage unchanged, a 68.2% decomposition rate was achieved. So, in terms of decomposition rate, the second step is of significance.

Besides pure water system, SeqHTL has also been tested in water-ethanol and acid (H3PO4)-salt (KH2PO4) system. No significant improvement was observed by addition of 50% ethanol at 240 , however, a tremendous increase in decomposition rate was relarized from 76.6% to 93.0% by addition of 50% ethanol at 280 . For acid-salt system, the situation turned inverse. At 240 , acid-aid SeqHTL improved the decomposition rate from 57.8% to 68.2%; while at 280 , it only slight enhanced the rate from 76.6% to 78.8%. As for solvent concentrations, the decomposition rate keeps increase gradually first and drops quickly after 50 vol% ethanol concentration at 280 . A maximum rate was obtained at an intermediate concentration, which infers the opposite roles of ethanol in HTL reaction. For acid-salt system, a reverse trend also observed. At 240 , 100 wt% acid, 50% acid/50% salt, and 100% salt gave rates at 68.2%, 67.6%, and 72.4, respectively. At 280 , 100 wt% acid, 50% acid/50% salt, and 100% salt gave rates at 78.8%, 72.4%, and 74.8%, respectively. For both cases, the combination of acid and salt showed the lowest degradation ability. The difference is pure salt favored degradation at low temperature while pure acid favored at high temperature. In addition, any acid-salt system at low temperature showed positive degradation ability, i.e. higher than pure water system, while at high temperature, pure salt and 50% acid/50% salt gave negative degradation ability.

Primary products composition analysis supported that first stage harvested mainly C5 sugar degradation products, while second stage enabled mainly C6 sugar degradation products and all lignin derivatives in aqueous phase. An in-depth analytic study is being conducted and a comprehensive SeqHTL pathways will be proposed sooner.

540226 Reaction Mechanism of Retro-aldol Condensation under Hydrothermal Condition

Abstract:

The reaction mechanism of retro-aldohol condensation has been a mystery. In the organic chemistry textbooks, retro-aldohol condensation is written as ionic reaction. However, the experiment on temperature effect on the reaction rate of retro-aldohol condensation has shown it is a radical reaction. We employed glyceraldehyde, the simplest sugar to find out the reaction mechanism, and determined it was surely a radical reaction.

530939 Catalytic Upgrading of Algae Bio-Oil from Hydrothermal Liquefaction on Ni-Based Catalyst : The Role of Support

Abstract:

Biofuels or bio-based fuels are one of the most promising alternative fuels. It can be derived from natural resources as biomass that is highly abundant availability, and carbon-neutral renewable energy resource. Among the biomass-based fuel, algae-base biofuel has much significant attention due to its ability to high lipid content, high growth rate, and fix CO_2 in atmosphere rapidly. In addition, cultivation of algae do not require fertile or agricultural land, thus no impact on food supply chains due to competition between food and energy sector. To convert algae to biofuels, hydrothermal liquefaction (HTL) is the promising process to produce biofuel from algae or wet-biomass feedstocks, which uses water or other solvent as reactant and reaction medium at sub- or supercritical temperature and pressure. Bio-oil obtained from HTL has high heating value around 31-39 MJ/kg and low water content of 5-10 wt%. Unfortunately, the crude bio-oil cannot use directly in automobile engine because of its negative properties such as high acidity, high viscosity, and high level of nitrogen and oxygen content. Consequently, the process of bio-oil treatment is essential for removing their oxygen and undesirable compound to obtain the high-grade bio-oil.

Catalytic hydroprocessing (including hydrodeoxygenation, hydrotreating, and hydrocracking) is presently considered as the most promising process for bio-oil upgrading due to high quality biofuel which higher heating value, high H/C ratio and low oxygen, nitrogen and other heteroatom content could be obtained. The catalytic hydroprocessing of crude bio-oil has been conducted under high temperature and high hydrogen pressure using heterogeneous catalysts. The most study of bio-oil upgrading has been reported on the effect of operating conditions, such as temperature, pressure and residence time. Moreover, types of catalysts are also importantly studied factor due to their influentially for heteroatoms (N, O, S and metal) removing.

Supported nickel based catalysts are found to be the promising catalyst for this purpose due to a profusion of nickel, and non-expensive. In addition, Ni-based catalyst on deoxygenation of fatty acid was proposed by Snare and co-worker (2006), and many report showed good activity for hydrodeoxygenation (HDO) of triglycerides-based and biomass-based (e.g. guaiacol, phenol etc.). However, the monometallic Ni catalysts are known to be active in methanation reaction which consume high amount of hydrogen and may cause the catalyst deactivation by carbon deposits. In addition, sintering is another factor that effect to stability of Ni catalyst due to high amount of Ni-metal loading. To solve this problem, the bimetallic Ni-based catalysts are investigated. NiCu catalyst was reported to be success in preventing coke formation, enhance hydrogenation activity, and decreasing the sintering of metal active phase. Moreover, there are other metals that could be used as a promoter to improve stability and resistance of Ni-based catalyst such as Re (enhanced HDO activity by increasing the dispersion of active phase) which are barely reported on algae bio-oil upgrading.

Based on authorsâ€™ experimental which compare the activity of hydrotreating of vegetable oil using mono-, bi-, and tri-metallic Ni supported on Al_2O_3, namely Ni, NiCu, and NiCuRe. The tri-metallic NiCuRe/Al_2O_3 demonstrated the highest activity of hydrotreating and stability which no deactivation was observed after 24 h time on stream. The purpose of this work is to investigate the activity of tri-metallic NiCuRe with difference supports for upgrading of algae bio-oil. The different supports, namely commercial Al_2O_3, high surface area CeO_2, ZrO_2 and mixed-oxide CeO_2-ZrO_2 were selected. The Al_2O_3 was selected, as it is commonly use and high surface area; however it can be deactivated by coke formation due to the presence of water in bio-oil. The CeO_2, and ZrO_2 were selected, as the promising support because their properties: the activation of oxygen compound on surface, water resistance, high thermal stability, and improving the dispersion of metal particle. In addition, the mixed oxide CeO_2-ZrO_2 has been reported to enhance thermal stability, surface area, and improve catalytic performance [1]. The CeO_2, ZrO_2 and mixed-oxide CeO_2-ZrO_2 were prepared by surfactant-assist co-precipitation method. Then, NiCuRe was load on the support by co-wetness impregnation. The textural properties of fresh and spent catalysts were characterized by N_2 physisorption (BET surface area), XRD, TPR, SEM and TGA. The hydrotreating of algae bio-oil produced from hydrothermal liquefaction (HTL) of *Nannochloropsis* sp. are performed in batch reactor at reaction condition of 350°C, hydrogen pressure of 5 MPa and reaction time of 4 h to determine the effect of support of catalyst on product properties and catalyst performance. Moreover, the operating parameters i.e. operating temperature, H_2 pressure, and reaction time will be investigated in order to find the suitable operating condition. To determine the activity of catalyst, the upgrading bio-oil yield and the comparison of chemical properties of feed and product were analyzed by gas-chromatographic method (GC-MS), FT-IR, and element analysis CHNS/O. Additionally, the properties of bio-oil in term of heating value, acidity and viscosity are also investigated. The relative activities of different reaction pathways among each supported catalyst are also discussed.

Reference

[1] M.A. Ebiad, D.R. Abd El-Hafiz, R.A. Elsalamony, L.S. Mohamed, Ni supported high surface area CeO2-ZrO2 catalysts for hydrogen production from ethanol steam reforming, RSC Adv. 2 (2012) 8145â€“8156. doi:10.1039/C2RA20258A.

533175 Microbial Network Analysis Using Co-Occurrence Patterns of Methanogens and Bacteria in Full-Scale Biogas Plants

Abstract:

Anaerobic digestion (AD) is a complex microbial process and different types of microorganisms with the symbiotic association are involved. Several reports showed microbial community structures were dependent with feedstocks and co-substrate types. Methanogens use acetate (acetoclastic), methanol (methylotrophic), CO_2/H_2 (hydrogenotrophic) for methane production as the final step in AD process, which are products from organic materials by bacteria. The co-occurrence patterns between methanogen and bacteria in full-scale biogas plants were investigated using microbial network analysis. Microbiome data were collected from 60 samples of 18 commercial biogas plants located in Korea, fed with different types of organic sources such as food waste, activated sludge, swine manure, etc. The positive/negative correlation was analyzed between archaeal family and bacterial phylum using correlation methods. The key microorganisms contributing for high methane production were also investigated to better understand the interaction of methanogens and bacteria in biogas plant and to suggest an idea on the enhancement of biogas production, and the key microorganism could be used as biomarkers to indicators for the performance check of methane production.

531197 A Model-Based Sustainability Study of Energy Consumption, Environmental Pollution, and Economic Growth in China

Abstract:

While China was experiencing a period of fast growing economy, with the annual GDP rate jumped from 5.4% in 1980 to 10.4% in 2010 [1], it was also suffering from the deteriorating air condition such as haze pollution. According to a research conducted by QUARTZ in 2013 [2], the haze pollution in China was so severe that people could not even see their fingers in front of them. Furthermore, the air quality index (AQI) was once detected to reach 999 at 3pm on Jan 6th, 2017 in Daqing. Normally, a measure of 100 of AQI is considered unhealthy, and 300 is thought dangerous [3]. As a result, the air pollution has caused serious problems to Chinaâ€™s society and its people. Estimated 1.1 million people die of air pollution each year, and countless people suffer from chronic diseases such as Pneumoconiosis [4].

The tremendous consumption of fossil fuels like coal, oil and natural gas is one of the major driving-forces for both the rapid economic growth and the deteriorating environment. In particular, the production of cements, steels, glass, electrical power, and cars that support the high GDP growth in China [5] release waste gases like SO_2, CO, NO_2, O_3 that are essential for the haze formation in China [6]. On the other hand, different fossil fuels produce different amounts of pollutant waste gases. 81% of the electricity in China was produced from the combustion of coal [7], which stands for the major emissions of NOx, CO_2, CO and SO_2 [8]. It is thus necessary to provide a rational allocation of different types of fossil fuels that can sustain the economic growth in China but reduce the environmental pollution.

In order to address the aforementioned issues, we developed the following four models on the basis of the economic and environmental data for Beijing during 2001 to 2015: 1) a model to predict the production of SO_2, CO, NO_2, and O_3 against the consumption of coal, natural gas, and gasoline (i.e., the energy-gas model); 2) a model to predict haze pollution days from the concentration of SO_2, CO, NO_2, and O_3 (i.e., the gas-haze model); 3) a model to quantify the production of cements, steels, glass, electrical power, and cars from the consumption of coal, natural gas, and gasoline (i.e., the energy-industry model); and 4) a model to predict GDP from the production of cements, steels, glass, electrical power, and cars (i.e., the industry-GDP model). These four models were then integrated together to predict the haze days and GDP if the allocation of the fossil fuels was not adjusted in the following few years. The results provide meaningful information for energy-allocation policy makers.

Reference

[1] https://www.theguardian.com/news/datablog/2012/mar/23/china-gdp-since-1980

[2] https://qz.com/137562/chinas-northeast-hit-by-air-pollution-so-bad-you-cant-see-your-own-fingers-in-front-of-you/

[3] https://qz.com/879746/pollution-in-china-daqing-sees-its-air-quality-index-aqi-spike-to-an-astounding-999/

[4] https://news.nationalgeographic.com/2017/05/china-air-pollution-solutions-environment-tangshan/

[5] Philip Andrews-Speed. China's ongoing energy efficiency drive: Origins, progress and prospects. Energy Policy, 37(4): 1331-1344, 2009.

[6] Kuang Xiao, Yuku Wang, Guang Wu, Bin Fu, Yuanyuan Zhu. Spatiotemporal Characteristics of Air Pollutants (PM10, PM2.5, SO2, NO2, O3, and CO) in the Inland Basin City of Chengdu, Southwest China. Atmosphere 9, 74, 2018

[7] Raymond Li, and Guy C.K.Leung. Coal consumption and economic growth in China. Energy Policy, 40: 438-443, 2012

[8] C.F. You, and X.C. Xu. Coal combustion and its pollution control in China. Energy, 35(11): 4467-4472, 2010

533651 Linking Molecular Structure to Functional Group and Chemical Literature Using a Chemical Reaction Database

Abstract:

The evaluation of potential alternatives for chemicals of concern requires understanding the implications of the chemical of concern during the manufacture, use, and disposal phases of its life cycle. Identifying potential reaction pathways and synthesis routes is an important component of assessing the manufacturing phase of a chemical. This presentation provides an overview of a tool that links a chemicalâ€™s structure and functional groups to information about the synthesis route and ultimately the manufacturing process for that chemical. The structure of the chemical of concern is entered into the application using either a SMILES string or a molecule MOL file. This information is indexed against functional groups listed in the chemical reaction database to determine the functional groups that are present in the molecule. Based on these functional groups present in the molecule, synthesis routes are identified using named chemical reactions. This information can then be correlated to input and output materials for the named reaction, along with reaction conditions, solvents, and catalysts. Additionally, the reaction database contains links to internet references and appropriate keywords for the reaction, further expanding its comprehensiveness. The chemical literature will provide further data on solvents, catalysts, reaction conditions and other reaction products that can then be used to compare various reaction pathways for the chemical of concern.

538235 Sustainability Identification for Infinite-Dimensional Systems

Abstract:

Sustainability Analysis of Infinite Dimensional Systems

Masih Jorat

Dept. of Chemical and Biomolecular Engineering, UCLA, Los Angeles, CA, 90095

Vasilios I. Manousiouthakis

Dept. of Chemical and Biomolecular Engineering, UCLA, Los Angeles, CA, 90095

The need for a multifaceted assessment of human impact on the environment has led to the emergence of comprehensive concepts like sustainability and sustainable development. Setting the priorities for sustainable initiatives requires that a value system first be established. This can only be accomplished within the context of a clearly stated definition of sustainability. In other words, the development of sustainability strategies cannot be pursued in an efficient manner, without a comprehensive, rigorous, and unbiased definition of sustainability. The prevalent common factor among sustainability definitions is the endurance of the system's dynamic behavior over time. Through utilizing the concept of positive invariant sets, the novel concept of Sustainability Over Sets (SOS) [1] provides a conceptual framework within which the time evolution of a system can be analyzed. SOS formalizes the incorporation of human input into the sustainability assessment process, first by defining a set in state-space which exactly quantifies human input regarding what is sustainable, and by then modifying the question "is a system sustainable?" to "is a system sustainable over a predefined set?" which is amenable to definitive (yes or no) answers. A system is sustainable over a set (SOS) in the system's state-space, if the system's state trajectories initiated within the set, remain for all time within the set, in other words the system's vector field is directing inwards at the set's boundaries. The previously presented studies of the (SOS) concept were limited to sustainability assessments of the systems, whose behavior is modeled through a set of ODEs. In this work a developed version of the SOS concept will be presented which has the ability to assess the sustainability of infinite dimensional systems which their dynamic behavior has been captured by a system of Partial Differential Equations (PDE's). In this regard we present necessary and sufficient conditions for sustainability of systems, whose behavior is captured by parabolic and hyperbolic PDE systems, commonly arising in chemical engineering applications, such as diffusion and convection-reaction systems over desired sets. furthermore, the practical applicability of the introduced conditions will be demonstrated through multiple case studies.

[1] Manousiouthakis VI, Jorat M. Sustainability over sets. Environmental Progress & Sustainable Energy. 2017 Oct.

520688 Environmental Genome: New Database for Public Health and Chemical/Materials Manufacturing

Abstract:

Environmental sustainability of consumer, industrial, and military products includes understanding the manufacturing processes and tools for improvement of the approximately 100,000 chemicals-in-commerce. These materials are the scaffolding of all global products. All of these chemicals and materials have full scale manufacturing processes, energy requirements, and unique environmental emissions that represent public health implications. There are public health and environmental impacts (local and global) that need data such as the Environmental Genome (EG) to fully diagnose and change. With 70%-90% of all chronic disease attributed to the environment, these emissions in manufacturing are a significant factor. Chemical engineering technology has been used to map chemical manufacturing and make a direct contribution to public health. The Environmental Genome Initiative is focused on mapping the 100,000 industrial chemicals used to build our societal products. With data on about 2% of the full chemicals-in-commerce available, the impact of mapping the EG will grow substantially. It is the first effort to completely describe the flow of the entire set of individual chemicals represented in the global chemical sector. The EG database is a powerful tool for the development of analytics and computer science applications in fields such as national security, maternal and personal health, product design, and chemical manufacturing. This paper describes the origins and progress in mapping the Environmental Genome and some of the first analytics developed.

523928 Toward a Leading Indicator of Catastrophic Shifts in Complex Systems: Assessing Changing Conditions in Nation States

Abstract:

The 20th century was characterized by substantial change on a global scale. There were multiple wars and unrest, social and political transitions, technological innovation and widespread development that impacted every corner of the earth. In order to assess the sustainability implications of these changes, we conducted a study of three advanced nations particularly affected during this time: France, Germany and the United States (USA). All three nation states withstood these changes and yet continued to thrive, which speaks to their resilience. However, we were interested in determining whether any of these countries underwent a regime shift during this period and if they did, whether there was advanced warning that transition was imminent. This study seeks to evaluate systemic trends in each country by exploring key variables that describe its condition over time. We use Fisher Information to assess changing conditions in the nation states based on trends in social, economic and environmental variables and employ Bayes Theorem as an objective means of determining whether declines in Fisher information provide early warning signals of critical transitions. Results indicate that while the United States was relatively stable and France experienced a great deal of change during this period, only Germany appeared to undergo a regime shift. Further, each country exhibited decreasing Fisher information when approaching significant events (e.g., World Wars, Great Depression), and reflected unique mechanisms linked to dynamic changes in each nation state. This study highlights the potential of using trends in Fisher information as a sentinel for evaluating dynamic change and assessing resilience in coupled human and natural systems.[1]

[1] Vance, L., Eason, T., Cabezas, H. and M.E. Gorman, â€œToward a leading indicator of catastrophic shifts in complex systems: Assessing changing conditions in nation states,â€ Heliyon 3 (2017) e00465.

DOI: 10.1016/j.heliyon.2017.e00465.

526363 Experimental and Theoretical Studies of the Electrochemical Reduction of CO_2 on Cu

Abstract:

The electrochemical and photoelectrochemical reduction of CO_2 to chemicals and fuels has been the subject of considerable interest in recent years. We have been pursuing the subject through a series of experimental and theoretical studies with the objective of identifying the factors controlling the distribution of products formed, with particular interest on establishing how to enhance the formation of C_{2+} hydrocarbons and oxygenates. This work has revealed that the activity of Cu for forming such products depends on the surface composition and morphology of Cu, the composition and concentration of electrolyte cation and anions, and the magnitude of the applied potential. We have also found that mass transfer of CO_2 to the catalyst surface and OH^- anions from the catalyst surface can also modify the distribution of products observed. This talk will talk will illustrate how information obtained from experiments and theory can be used to elucidate the fundamental processes governing product distribution.

538675 Highly Efficient CO Electroreduction Catalyst Based on Polycrystalline Cu Particles

Abstract:

Electrochemical conversion CO_2 to value-added chemicals and fuels using renewable energy is a promising technique for energy storage and to potentially diminish problems associated with CO_2 emissions. Many catalysts are able to reduce CO_2 to CO with high selectivity and activity. However, to achieve more energy-dense products requires further reduction of CO. Cu is the only currently known material exhibiting appreciable activities towards CO electroreduction[1,2]. Recent studies show grain boundary terminated Cu surface is able to effectively improve the selectivity and required energy bias for CO electroreduction[2,3]. Despite significant progress, the optimal activity achieved is only less than 1 mA/cm^2 which is attributed to the poor transportation of CO caused by its low solubility in the electrolyte (~1 mM). Herein we report our latest progress in CO electroreduction using polycrystalline Cu particle catalysts to achieve high efficiency. Our efforts in providing mechanistic insights including reaction intermediates and pathways will also be discussed.

[1] Hori Y, Takahashi R, Yoshinami Y, et al. The Journal of Physical Chemistry B, 1997, 101(36): 7075-7081.

[2] Li C W, Ciston J, Kanan M W. Nature, 2014, 508(7497): 504.

[3] Verdaguer-Casadevall A, Li C W, Johansson T P, et al. Journal of the American Chemical Society, 2015, 137(31): 9808-9811.

538294 Operando Spectroscopic Investigations of Oxide Derived Metal Catalysts for CO_2 and CO Reduction

Abstract:

Electrochemical reduction of CO_2 driven by renewable electricity is a promising approach to convert CO_2 into valuable chemicals and fuels. Copper is the only metal catalyst for this reaction with significant selectivity towards hydrocarbons and oxygenates. CO is a known intermediate in the reduction of CO_2, and thus its conversion is also of notable interest. Oxide derived copper (OD-Cu) has been shown to favor the more desirable C2 and C3 products at lower overpotentials than bulk copper. Despite numerous studies and proposed mechanisms, e.g., the presence of grain boundaries, surface pH and surface roughness, the exact nature of the active phase is still a topic of discussion. In this work, we use operando attenuated total reflectance surface enhanced infrared reflection absorption spectroscopy (ATR-SEIRAS) to probe different types of sites available on Cu and OD-Cu surfaces using CO as a probe molecule. We show that multiple types of CO-binding sites are available on both Cu and OD-Cu surfaces. Further, the types of binding sites present are also strongly potential dependent. We identify strong correlations between the presence of several types of surface sites with the ability of the catalyst to produce C2 and C3 products at -0.4 V vs. RHE.

531291 Molecular-Level Insights into Electrocatalytic Carbon Dioxide Reduction at Cobalt Macrocycles

Abstract:

Electrolysis of carbon dioxide in aqueous media to generate carbon monoxide may provide a modular route to making a key chemical feedstock under ambient conditions. As a component of syngas, carbon monoxide can be converted into higher value hydrocarbons, alcohols, acids, and esters. Cobalt macrocycles, including cobalt phthalocyanines and cobalt porphyrins, are known to be selective catalysts for converting carbon dioxide into carbon monoxide. In our work, we have been able to develop immobilized versions of these catalysts with high turnover frequencies for reducing carbon dioxide. These catalysts have helped to elucidate a rich picture of the nature of the rate determining step, how it can be tuned between an electron transfer and proton-coupled electron transfer, and the interplay of kinetics and transport in the observed reactivity of cobalt macrocycles. Furthermore, we have been able to deconvolute both inductive and electrostatic effects on the activity of these catalysts, providing clear design principles for transition metal macrocycles which are immobilized at electrode-electrolyte interfaces.

527408 Mass Transfer Effects in CO_2 Reduction Electrocatalysis

Abstract:

Significant interests have risen in the development of catalytic nanomaterials for the electroreduction of CO_2. While extensive studies have been reported on tuning of surface structures to improve the chemical kinetics, less attention has been paid to the mass transfer effects in the CO_2 reduction reaction on nanoscale electrocatalysts. We report here an systematic investigation of CO_2 electroreduction on highly dense Cu nanowires, with the focus placed on practical high-flux conditions. Mass transfer effects are found to play an important role in this case, giving rise to diffusion-limited CO_2 reduction activity and selectivity. By correlating the observed transport phenomena to the calculated CO_2 consumption rate and simulated surface concentration of chemical species (including local pH) on the nanowires, an upper limit is revealed for the CO_2 conversion rate on the nanostructured electrodes, which also causes the drop in Faradaic efficiency of CO_2 reduction at large current densities. Our work emphasizes the necessity of considering mass transfer effects in the design of advanced electrocatalysts for CO_2 reduction as well as for understanding their structure-performance relationships.

536098 Single Atom Catalysts for Electrochemical Reduction of CO_2

Abstract:

Electrochemical reduction of CO_2 offers a sustainable route to mitigate the problem of rising in CO_2 concentration in the atmosphere. Among the various electrocatalysts studied for the conversion of CO_2 to hydrocarbons, supported metal catalysts have played an important role in enhancing the rate of chemical reactions compared to bulk unsupported metals. Broadening this distribution of metal to make supported isolated atoms is a novel approach to heterogeneous catalysis. The reduction in the size of the catalyst not only enlarges its specific area, but also increases the number of under coordinated sites, and the HOMO-LUMO gap. This can improve the activity and selectivity of the catalyst. In this work, Î·6-functionalized graphene which includes transition metal single-atom (SA) layer is used as the cathode for electrochemical reduction of CO_2. The Î·6 chemistry retains grapheneâ€™s 2D structure and high electronic properties (high carrier mobility and high electric conductivity), which contributes to a high catalytic activity with a large surface area. Moreover, the transition metal is grafted on graphene by coordination bonds, so that the adsorption energy of CO_2 reduction intermediates is increased and further the energy barrier for hydrogenation is reduced. Such single metal atom supported electrocatalysts offer a higher activity to transition metals as compared to their bulk counterparts. The use of SA catalysts also means that lesser quantities of metals being used per unit weight of the electrode resulting in cheaper catalysts with superior performance. Due to only recent advances in the synthesis of stable SA catalysts, the application of single-atom catalysis for electrochemical reduction of CO_2 seems promising. To study the behavior of single-atom catalysts for CO_2 reduction, we present a design of a compact electrochemical cell with carbon supported single-atom transition metal atoms on graphene as the Rotating Disk Electrode (RDE). An RDE is chosen as it minimizes the cathode polarization and enables the measurement of intrinsic kinetic activity. A product distribution of CO_2 reduction products at near-neutral pH, their partial current density, and Faradaic efficiencies at different potentials will also be reported in this work.

524710 Photoelectrochemical CO_2 Reduction at Plasmonic Nanostructured Silver Electrodes

Abstract:

Electrochemical CO_2 reduction could help mitigate carbon emissions while producing valuable fuels or chemical feedstocks. However, inefficiency and low selectivity toward a hydrocarbon product have limited the commercial viability of this process. Plasmonic hot-carriers and the strong local electric fields produced by plasmon excitation may open new pathways for CO_2 reduction, resulting in increased product selectivity and lower overpotential. Our research aims to understand the fundamentals underlying plasmon-generated excited charge transfer from voltage-biased cathodes to CO_2 reduction intermediates to selectively and efficiently form desirable hydrocarbon products.

We have designed a custom gas flow cell to probe the mechanism of plasmon-enhanced CO_2 reduction on nanostructured noble metal electrodes fabricated by nanosphere lithography. This temperature-controlled photoelectrochemical cell is specifically designed to allow quantitative analysis of gaseous and liquid products formed at an illuminated working electrode. The geometry of the cell enables front-illumination of the photoelectrode and maximizes the electrode surface area to electrolyte volume ratio to increase liquid product concentration and hence enhance *ex situ* spectroscopic sensitivity towards them. Gas is bubbled through the electrolyte in the working electrode chamber during operation to maintain a saturated reactant concentration and to continuously mix the electrolyte. Gaseous products are detected by an in-line gas chromatograph, and liquid products are analyzed *ex situ* by NMR. Product distribution and photocurrent are measured while varying electrochemical potential, illumination intensity, light wavelength, temperature, and the plasmonic catalyst. We demonstrate that plasmonic photocurrent on voltage-biased silver (Ag) nanopyramid cathodes is selective for CO_2 reduction over the competing hydrogen evolution reaction. Our results suggest that further plasmonic enhancements in selectivity and activity towards specific CO_2 reduction reactions are possible by tuning the electrode structure and composition.

This material is based upon work performed by the Joint Center for Artificial Photosynthesis, a DOE Energy Innovation Hub, supported through the Office of Science of the U.S. Department of Energy under Award Number DE-SC0004993.

The work is supported by the National Science Foundation Graduate Research Fellowship under Grant No. 1106400.

Work at the Molecular Foundry was supported by the Office of Science, Office of Basic Energy Sciences, of the U.S. Department of Energy under Contract No. DE-AC02-05CH11231.

541778 Food-Energy-Water Nexus Systems Engineering

Abstract:

Due to population growth, economic development, urbanization, climate change, and natural resource degradation, the global demand for food, energy and clean/freshwater is rapidly increasing [1]. As demand grows the stresses and interdependences between these three resource systems, which are commonly referred to as Food-Energy-Water Nexus (FEW-N), are becoming more apparent [2,3]. Decision making can be very challenging due to multiple dimensions of the biophysical water, energy and food systems, and the players and stakeholders connected with them [4]. A holistic systems engineering approach is thus clearly needed to navigate the multi-faceted FEW-N space, identify opportunities for synergistic benefits and systematically explore interactions and trade-offs [5].

In this work, we present the foundations of a systems engineering framework and quantitative decision-making tools for the analysis and trade-off optimization of stressed interconnected resource FEW-N networks. The framework combines data analytics and mixed-integer modeling and optimization methods establishing the interdependencies and potentially competing interests amongst the food, energy and water elements in the system, along with policy, sustainability and feedback from the various stakeholders. A multi-objective optimization strategy is followed for the analysis of the trade-offs empowered by the introduction of composite FEW-N metrics as means to facilitate decision making and compare alternative process and technological options. The versatility, potential and applicability of the proposed framework is demonstrated by three nexus case studies, (i) a complex power generating system in the South-Central Texas region [6], (ii) a dairy production and processing plant [7], and (iii) a land use allocation problem on a regional farming system in China [8].

References

[1] D'Odorico, P., Davis, K. F., Rosa, L., Carr, J. A., Chiarelli, D., Dell'Angelo, J., ... & Rulli, M. C. (2018). The Global Foodâ€Energyâ€Water Nexus. Reviews of Geophysics.

[2] McCarl, B. A., Yang, Y., Schwabe, K., Engel, B. A., Mondal, A. H., Ringler, C., & Pistikopoulos, E. N. (2017). Model Use in WEF Nexus Analysis: a Review of Issues. *Current Sustainable/Renewable Energy Reports*, *4*(3), 144-152.

[3] McCarl, B. A., Yang, Y., Srinivasan, R., Pistikopoulos, E. N., & Mohtar, R. H. (2017). Data for WEF Nexus Analysis: a Review of Issues. *Current Sustainable/Renewable Energy Reports*, *4*(3), 137-143.

[4] Daher, B., Mohtar, R.H., Pistikopoulos, E.N., Portney, K.E., Kaiser, R. and Saad, W. (2018) Developing Socio-Techno-Economic-Political (STEP) Solutions for Addressing Resource Nexus Hotspots, Sustainability 2018, 10, 512; doi:10.3390/su10020512

[5] Garcia, D. J., & You, F. (2016). The water-energy-food nexus and process systems engineering: a new focus. *Computers & Chemical Engineering*, 91, 49-67.

[6] Diangelakis, N. A., & Pistikopoulos, E. N. (2017). A multi-scale energy systems engineering approach to residential combined heat and power systems. *Computers & Chemical Engineering*, 102, 128-138.

[7] Avraamidou, S., Milhorn, A., Sarwar, O., & Pistikopoulos, E. N. (2018). Towards a quantitative Food-Energy-Water Nexus metric to facilitate decision making in process systems: A case study on a dairy production plant. *28th European Symposium on Computer-Aided Process Engineering*: Accepted.

[8] Nie, Y., Avraamidou, S., Li, J., Xiao, X., & Pistikopoulos, E. N. (2018). Land use modeling and optimization based on food-energy-water nexus: a case study on crop-livestock systems. *13th International Symposium on Process Systems Engineering*: Accepted.

Renewable Carbons from Food Waste for Separation and Catalysis Technologies

This study explores the production of renewable and sustainable carbons from waste biomass resources, for adsorption and catalytic applications. Specifically, the research will explore the production of carbons derived from **food waste** and their utilization as sorbents of aromatic contaminants in **water** and catalyst supports for the production of clean **energy and fuels**.

The idea of using renewable and sustainable resources to replace the carbon derived from fossil fuels is not new. Researchers have been working on the utilization of various biomass resources, such as rice husk, coconut shells and others for years.[1–5] However, the utilization of food waste as a carbon resource has not been extensively explored. Food waste could be the ultimate resource for carbon production, since it is produced in excessive volumes and it represents one of the most significant environmental problems, which is urgently looking for a solution. The critical question is whether the activated carbons derived by food waste have the potential to be used for purification and/or catalysis purposes. Food waste is heterogeneous in nature and contains various types of minerals, which may provide advantages or disadvantages, depending on the application.

In our lab, we receive the food waste from the Dining Halls of the University of Connecticut and after pretreatment, that includes washing, grinding and sieving, we perform pyrolysis and activation to turn the waste into carbons. The produced carbons are then characterized and they are tested as sorbents and catalyst supports. Our goal is to understand how the conditions of pyrolysis and activation affect the properties of the carbons and, consequently, their effectiveness as sorbents and catalyst supports. During the pyrolysis reaction, temperature and residence time are the most important parameters that affect the yield and the properties of the produced biochar. Activation of the biochar is performed either using steam, physical activation, or using chemicals (e.g. H_3PO_4), chemical activation. During the former method of activation, temperature and residence time are, again, among the most important parameters that affect the surface area, the carbon content and the porosity of the produced carbons. During the latter method, the impregnation ratio between the chemical and the biochar is the most critical.

During this talk we will discuss the properties of the activated carbons produced by food waste and how these properties are affected by the reaction conditions and the method of activation. Surface area, porosity, carbon content, surface functional groups, morphologies are the properties of interest which will be discussed. Then, we will discuss the potential of using the activated carbons produced by food waste, as sorbents of aromatic hydrocarbons in water and catalyst supports for the hydrodeoxygenation of bio-oil.

References

(1) Tripathi, M.; Sahu, J. N.; Ganesan, P. Effect of Process Parameters on Production of Biochar from Biomass Waste through Pyrolysis: A Review. *Renew. Sustain. Energy Rev.* **2016**, *55*, 467–481.

(2) Kwapinski, W.; Byrne, C. M. P.; Kryachko, E.; Wolfram, P.; Adley, C.; Leahy, J. J.; Novotny, E. H.; Hayes, M. H. B. Biochar from Biomass and Waste. *Waste and Biomass Valorization* **2010**, *1* (2), 177–189.

(3) Ahmad, M.; Rajapaksha, A. U.; Lim, J. E.; Zhang, M.; Bolan, N.; Mohan, D.; Vithanage, M.; Lee, S. S.; Ok, Y. S. Biochar as a Sorbent for Contaminant Management in Soil and Water: A Review. *Chemosphere* **2014**, *99* (November 2013), 19–23.

(4) Tan, Z.; Lin, C. S. K.; Ji, X.; Rainey, T. J. Returning Biochar to Fields: A Review. *Appl. Soil Ecol.* **2017**, *116* (September 2016), 1–11.

(5) Azargohar, R.; Dalai, A. K. Biochar as a Precursor of Activated Carbon. *Appl. Biochem. Biotechnol.* **2006**, *131* (1–3), 762–773.

541779 A Computational Framework for Sustainable Waste Management and Simultaneous Recovery of Nutrients and Energy

Abstract:

To protect human health and the environment, communities should have sustainable infrastructures to manage waste. Critically, livestock waste may cause some air quality degradation from ammonia and methane emissions, soil quality detriment due to in-excess nutrients and acidification, and water pollution issues resulting from nutrient and pathogens runoff to water bodies, which leads to eutrophication, hypoxia, and harmful algal blooms (HABs).

In addition, the recovery of carbon from organic waste offers an alternative for replacing fossil fuels with renewable sources for energy production. Despite the significant environmental benefits by performing pollution management of these organic materials, the deployment of technologies for obtaining value-added products and renewable energy from livestock waste is not a current practice due to the high investment costs and low product market values.

This contribution describes a multi-scale approach for the development of a supply chain framework to identify and evaluate techniques and logistics for the management of solid waste generated from primary sources of nutrient pollution. This effort will inform stakeholdersâ€™ waste management options while assessing potential economic and environmental benefits. Other potential benefits include the reduction of nutrient pollution and HABs, recovery of valuable products, and adoption of nutrient-reducing technologies in different locations (e.g., community, state, regional, etc.). In addition, the proposed framework captures techno-economic and logistical issues and can accommodate diverse types of material management incentives obtained at federal and state levels.

537690 Density Functional Theory Insights into Corrosion and By-Product Formation Caused By Drinking Water Disinfection

Abstract:

Density Functional Theory insights into corrosion and by-product formation caused by drinking water disinfection

Currently, treatment plants most commonly disinfect drinking water with free chlorine to remove pathogens. Free chlorine, however, is prone to forming carcinogenic disinfection by-products from reactions with organic matter. Ill-considered application of disinfectants and complications from cleanup of their by-products, recently resulted in catastrophic corrosion and discharge of lead into the water supplies of Washington D.C. and Flint, Michigan. To avoid a recurrence of such mistakes, the strategy for drinking water disinfection in vulnerable, lead-containing supply networks was reconsidered. Specifically, we evaluated the merits and drawbacks of replacing free chlorine with chlorine dioxide in systems with lead infrastructure.

Free chlorine, the most commonly used disinfectant in drinking water treatment, is a powerful oxidant that readily reacts with viral and bacterial pathogens, but also with other water constituents, including dissolved organic matter and metallic infrastructure surfaces. Reactions with organic matter produce hazardous by-products, while reactions with infrastructure surfaces can increase the concentration of metals, including lead, in drinking water. Some water treatment facilities have substituted free chlorine with chloramines, an alternative disinfectant, to meet disinfection by-product regulations set by the Environmental Protection Agency. However, chloramines produce their own set of toxic by-products and have lower disinfection efficiency than free chlorine. Removal of disinfection by-products and the precursors to said by-products can exacerbate corrosion in the metal pipes of water distribution systems.

Unlike free chlorine, chlorine dioxide has a significantly lower propensity to form organic by-products, and it is as potent, and in some cases, more potent than free chlorine for the inactivation of bacteria and viruses. In a properly managed dose, chlorine dioxide can deliver effective disinfection of drinking water[1]. Chlorine dioxide is more expensive to use for drinking water treatment than free chlorine and it can decay in water to form chlorite and chlorate, two potentially hazardous inorganic by-products. It has been assumed that chlorine dioxide will behave similarly to free chlorine with regards to lead corrosion mechanisms because of their similar oxidation reduction potentials, but there have been few studies to confirm.

Using three different minerals, a series of experiments were performed to measure the kinetics of chlorine dioxide decay and by-product formation, considering chlorine dioxide concentration, and pH. In the presence of lead minerals, chlorine dioxide does not form chlorate as a by-product. These results are curious as copper minerals were shown to produce chlorite and chlorate through a disproportionation reaction. The lead mineral reaction was also significantly faster than the copper reaction and produced only chlorite.

To explain the disparity in by-product formation between copper and lead minerals, computational chemistry was employed. Density Functional Theory was used to calculate activation energy barriers and minimum energy structures for both lead and copper systems. These calculations provided insights into reactant, products, intermediates, and transition structures to explain experimental findings.

For experimental work, spectrophotometry was used to standardized chlorine dioxide concentrations at 359nm. Ion chromatography, by a Dionex 5000 Reagent-Free ion chromatograph, was used to measure by-products (chlorite and chlorate) and specified anions. All experiments were batch reactions, conducted in the dark under constant agitation of a stir plate. Computational studies were preformed using the Vienna Ab Simulation Package and Gaussian. Simulations were done on solvated surfaces and in the gas phase. The influencing factors on the kinetics of chlorine dioxide decay and by-product formation in the presence of lead minerals will be presented and discussed. Our data will be contrasted with the results of published results on other metals including nickel, copper, and iron.

Aging lead infrastructure will continue to be an issue in the United States. The replacement of distribution systems could take decades and at high environmental, economic, and social costs due to our poor understanding of corrosion and by-product formation mechanisms. There is a demand for alternative disinfectants that have the capability to protect society from waterborne diseases without polluting our environment with lead and toxic by-products. This work is intended to provide guidance, on the viability of chlorine dioxide as a disinfectant, to avoid costs and dangers associated with lead corrosion.

References

(1) Gagnon, G. A.; O'Leary, K. C.; Volk, C. J.; Chauret, C.; Stover, L.; Andrews, R. C. *J. Environ. Eng.* **2004**, *130* (11), 1269–1279.

(2) Edwards, M.; Triantafyllidou, S. *J. / Am. Water Work. Assoc.* **2007**, *99* (7).

(3) Dodrill, D.; Edwards, M. *Am. Water Work. Assoc.* **1995**, No. C, 74–85.

(4) Liu, C.; Von Gunten, U.; Croue, J. P. *Environ. Sci. Technol.* **2013**, *47* (15), 8365–8372.

535908 Parameter Estimation of a Model of Advanced Oxidation Processes

Abstract:

In a previous work, a mathematical model of advanced oxidation processes (AOPs) was developed.[1] It was based on the mole balance equations of species and the kinetic information offered by the reaction mechanism. The model described the treatment of ciprofloxacin (CIP) in aqueous solution using the ozonation process. Intermediate compounds of CIP degradation were neglected, and the key model parameters were the volumetric mass transfer coefficient in the liquid phase (k_La) and the liquid hold up (Î²).

The present work focusses on parameter estimation of the model describing the CIP oxidation considering the reactions of intermediate compounds and total organic carbon (TOC) with the ozone molecule and hydroxyl radical. These reactions were inserted in the reaction mechanism according to a possible degradation mechanism of CIP reported in the literature.[2] A sensitivity analysis was performed in order to evaluate the influence of kinetic constants on the model predictions. The kinetic constants were varied one by one and the predicted concentration profiles of CIP and ozone in the gas phase at the column outlet were analyzed. The model parameters k_La, Î² and kinetic constants were estimated from available experimental data minimizing the differences between the experimental and predicted concentrations of CIP and ozone in the gas phase.[3] In addition, reparameterization for the kinetic constants was used. The optimization problem was solved using the fmincon solver of MatlabÂ®. Values for the initial guesses of the model parameters were taken from the literature, and upper and lower bounds for the parameters were defined.[4,5] In all cases, the accuracy of the model predictions was estimated by means of the root mean square error (RMSE).

Once the parameters were estimated, simulations of the process taking into consideration the experimental conditions were performed. The model was capable to fit adequately the experimental CIP and gas phase ozone concentrations. Process simulations with the estimated parameters considering the intermediates (k_La, Î² and kinetic constants) were compared with simulations using estimated parameters (k_La and Î²) neglecting the reactions of intermediates and TOC. The comparison allowed to study the role of intermediate compounds. These compounds consume ozone affecting the ozonation process. When the intermediates were neglected, the model overestimated the ozone concentrations in the gas phase for all the experimental conditions. The model was not capable to fit adequately the ozone concentrations in the gas phase even working with estimated parameters from the available experimental data. It is necessary to take into consideration the reactions of intermediate compounds of CIP degradation and TOC in order to obtain a good agreement between the model predictions and the experimental data. The mathematical model of advanced oxidation processes with the estimated parameters can be used to test different experimental conditions and design experiments that lead to a maximization of degradation.

References

1.Abreu Zamora MA, Teixeira ACSC, Le Roux GAC. Modelling and simulation of advanced oxidation process: application to the treatment of ciprofloxacin in aqueous solution by ozonation process. *Comput. Aided Chem. Eng.* 2017; 40: 553-558. In: 27th European Symposium on Computer Aided Process Engineering.

2.De Witte B, Dewulf J, Demeestere K, Van Langenhove H. Ozonation and advanced oxidation by the peroxone process of ciprofloxacin in water. *J. Hazard. Mater*. 2009; 161(2-3):701-708.

3.Bardauil Baptistucci C. Degradação do antibiótico ciprofloxacina em solução aquosa por meio de processo oxidativo avançado baseado em ozônio. *Master's Thesis*. Chemical Engineering Department, University of São Paulo, 2012.

4.Beltrán FJ, González M, Acedo B, Rivas FJ. Kinetic modelling of aqueous atrazine ozonation processes in a continuous flow bubble contactor. *J. Hazard. Mater*. 2000; 80(1-3):189-206.

5.Dodd, MC, Buffle MO, Von Gunten U. Oxidation of antibacterial molecules by aqueous ozone: Moiety-specific reaction kinetics and application to ozone-based wastewater treatment. *Environ. Sci. Technol*. 2006; 40(6):1969-1977.

ACKNOWLEDGEMENT: Antonio Carlos S. C. Teixeira would like to express his acknowledgement to FAPESP for its financial support through grant number 2013/50218-2

538185 Kinetics and CFD Model Validation for Combustion of Coal Char Using Cu-Based Chemical Looping with Oxygen Uncoupling (CLOU) Carriers

Abstract:

The chemical-looping combustion with oxygen uncoupling (CLOU) process uses gaseous or solid hydrocarbon fuels and is a promising carbon capture and storage (CCS) technology. In CLOU, the uncoupling of gaseous O2 from the oxygen carrier CuO is thermodynamically favored at temperatures above ~ 1000Â° C and pressures less than ~ 0.1 bar O2. The CLOU carrier chosen for this study was the CuO/Cu2O oxygen carrier system, while the choice of fuel was coal char obtained from the pyrolysis of Powder River Basin (PRB) coal. The overall reduction rates of the CuO oxygen carriers were investigated using a drop-tube fluidized bed reactor in the temperature range of 850Â°C â€“ 1100Â°C. The intrinsic activation energy and Arrhenius prefactor for the CLOU reaction were determined from these experiments, along with the apparent activation energy and Arrhenius prefactor for the overall CuO reduction. In addition, a computational fluid dynamics (CFD) model was developed and validated for prediction of the overall CuO reduction kinetics based on the amount of PRB coal char charged to the reactor and the known kinetics of the individual CLOU, combustion, gasification, and CuO reduction reactions occurring within the system. The PRB coal char added to the reactor consumes the O2 uncoupled from the CuO, driving the thermodynamic equilibrium of the CLOU reaction in the direction of Cu2O product formation such that this reaction is now favored at temperatures as low as 900° C.

530065 Dependence of Photocatalytic Degradation Pathway on Surface Planes of a Catalyst: A Case of Diuron Degradation on ZnO

Abstract:

Heterogeneous photocatalytic reaction has been used as a technique for elimination of toxic contaminants in water. The process relies on oxidation of the adsorbed pollutant by highly active radicals, mainly hydroxyl radical, which are produced when a catalyst is activated by light having energy higher than the bandgap of the catalyst. The attack of the hydroxyl radical on an organic compound leads to its degradation toward total mineralization, but it usually involves several intermediates. Interestingly, the intermediates from the degradation of a compound reported in various papers are usually inconsistent. This discrepancy should be addressed so that appropriate catalyst can be designed, especially for the degradation of toxic compounds because some intermediates may be more toxic than their parent compounds.

In this work, the dependence of the photocatalytic degradation pathway on surface planes of the catalyst is demonstrated. Diuron [3-(3, 4-dichlorophenyl)-1, 1-dimethyl urea], which is one of the most highly persistent and highly toxic herbicides, was used as a model compound. Zinc oxide (ZnO) was chosen as the catalyst because it can be synthesized such that the dominating surface on a particle can be easily controlled. A hexagonal crystal of wurtzite ZnO is consisted of two main sets of surfaces; (i) zinc-terminated (0001) and oxygen-terminated (000-1) polar surfaces locating at the top and bottom planes of the crystal, and (ii) mixed-terminated (10-10) nonpolar surface as the side planes. Herein, two types of ZnO, i.e., conventional hexagonal particles with polar surfaces as the dominating planes, and nanorods with the mixed-terminated surfaces, were studied.

The conventional particles were prepared by the sol-gel technique at room temperature, after which the calcination at 500°C was performed. On the other hand, the nanorods were synthesized by the hydrothermal technique at 140°C. Zinc acetate was used as the precursor for both syntheses. Various characterization techniques, including X-ray diffraction (XRD), X-ray photoelectron spectroscopy (XPS), UV-Visible spectroscopy, selected area electron diffraction, and nitrogen adsorption/desorption, were performed to confirm that both catalysts are nonporous single-crystalline ZnO in wurtzite phase with similar surface compositions. Their optical bandgaps are the same at 3.1 eV. The major differences between these catalysts are their morphologies, sizes and specific surface areas. The conventional ZnO powders are micron-sized particles with (0001) or (000-1) plane, while the nanorods exposes mostly the (10-10) surface. Because of the much smaller particle size of the nanorods, their specific surface area is much higher than that of the conventional particles, i.e., 17 m^2/g versus 1.4 m^2/g.

The adsorption experiments were conducted in batch in the dark, using diuron aqueous solution with concentrations in the range of 0-25 mg/l. The Freundlich model was found to represent the experimental data well. The fitted parameter relating to adsorption intensity shows unambiguously that diuron adsorb more strongly on the conventional ZnO than on the ZnO nanorods. The diuron-surface interaction is suggested to be electrostatic. Interestingly, the greater amount of diuron was adsorbed on the conventional ZnO particles than on the nanorods, although the former has one-order of magnitude lower surface area than the latter.

Molecular calculations using density functional theory were applied to support the experimental findings. The adsorption configuration of diuron was found to be surface dependent. On the zinc-terminated surface, diuron adsorbs in planar configuration because both ends of diuron, which are negatively charged, are attracted to the positively charged surface zinc atoms at the same time. In case of the oxygen-terminated surface, diuron tilts its molecule almost side-way so that the electron-rich amide oxygen moves away from the negatively charged surface oxygen.

On the mixed-terminated surface of ZnO, molecule of diuron is bent so that the most negatively charged part of the molecule, i.e., amide oxygen and nitrogen attaching to the aromatic ring, could approach positively charged zinc atoms on the surface. At the same time, the aromatic end of the molecule is lifted from the surface because of the repulsion between chlorine atoms attaching to it and surface oxygen atoms. If zinc atom on the surface is defined as the adsorption site, it can be seen that only half of the surface atoms of the mixed-terminated surface are adsorption sites, unlike the polar surfaces where all atoms are able to be the adsorption sites. Furthermore, the adsorbate molecule that already adsorbed on the surface will present steric effect toward adsorption on nearby sites. This steric effect will become more severe toward the adsorption capacity on the mixed-terminated surface, where not all atoms are the adsorption sites. That explains why the adsorption capacity on the conventional ZnO particles is much higher than that on the nanorods.

The photocatalytic degradation experiments were conducted in a microreactor to eliminate mass transfer resistance for the transport of diuron from bulk liquid to the surface of the catalyst. The outlet stream from the reactor was collected and analyzed by liquid chromatography equipped with tandem mass spectroscopy (LC-MS/MS). For the degradation on the ZnO nanorods, the first set of intermediates detected using the shortest residence time (1 min), which are assumed to be the direct intermediates from the degradation of diuron, are formed by hydroxylation and demethylation at the methyl group. These reactions take place at the aliphatic moiety of diuron, which is the part of the molecule adsorbed on the mixed-terminated surface of the nanorods. On the other hand, on the conventional ZnO particles, the intermediates detected are the results from diuron being attacked by hydroxyl radicals on both aliphatic and aromatic sides simultaneously. As previously discussed, diuron adsorbs in a planar configuration on the zinc-terminated surface, while it turns perpendicularly to the oxygen-terminated surface. Both of these give rise to equal probabilities of attack by the hydroxyl radicals on either parts of diuron. The structure of the subsequent intermediates should therefore be consequently affected. Hence, we propose that the difference in pathways from using different catalysts is caused by the adsorption configurations of diuron onto the different surfaces.

Finally, the toxicity of the generated intermediates was investigated. Both the phytotoxicity and cytogenotoxicity were evaluated in comparison with diuron. The results show that the initial intermediates formed from diuron degradation on the ZnO nanorods are less toxic than those generated on the conventional ZnO particles. Nevertheless, with prolonged residence time, e.g., 15 min, the toxicity of the degradation product was dramatically reduced to a level similar to that of deionized water.

522623 The Formation of Peroxymonocarbonate (HCO_4^-) and Its Impact on the Degradation of Organic Contaminants during Hydrogen Peroxide (H_2O_2) *in Situ* chemical Oxidation (ISCO)

Abstract:

Hydrogen peroxide (H_2O_2) reacts with iron and manganese adsorbed on surfaces and on minerals to produce hydroxyl radical and other reactive oxygen species during *in situ* chemical oxidation of contaminated groundwater. Under the conditions used for remediation, high concentrations of hydrogen peroxide (e.g. up to ~10 M) are encountered. Using ^{13}C NMR we have shown that under these conditions, an alternative pathway for H_2O_2 loss occurs in which H_2O_2 reacts with HCO_3^- to produce HCO_4^-, at a HCO_3^- concentration of 10~1000 mM, HCO_4^- was formed and quickly reacted with phenol to produce di- and tri-hydroxyl phenols. The rate of transformation of phenol was determined by the concentration of HCO_3^-. HCO_4^- also reacted with substituted phenols in an electrophilic behavior. Results of this study indicate that HCO_4^- is a potentially important but previously unrecognized oxidative species generated during H_2O_2 ISCO and that it may be possible to adjust treatment conditions to enhance its formation as a means of accelerating rates of contaminant removal.

538340 Hydrothermal Liquefaction ofÂ MunicipalÂ Solid Waste Using Binary Transition Metal Oxide (BTMO) Nanoparticles

Abstract:

Sol-gel derived iron based binary transition metal oxide (BTMO) nanoparticles obtained at calcination temperatures of 600-800°C were investigated as catalytic materials for the hydrothermal liquefaction (HTL) of the municipal solid waste. Prior to liquefaction reaction, the powdered BTMO nanoparticles were analyzed using x-ray diffraction, scanning electron microscopy and BET surface area analyzer. The hydrothermal liquefaction of a slurry containing 10-20 wt.% solid waste was carried at temperatures of 250-325°C. Following the HTL reaction, the recovered aqueous phase was characterized by HPLC and GC-MS. The analysis indicated the presence of alcohols, ethers, esters, ketones, phenols, and carboxylic acids. From the recovered biochar, bio-oil was extracted using acetone/ethanol. The higher heating value (HHV) of the oil was found to be approx. 35 MJ/kg. The biochar was sonicated to convert into submicron particles and graphitized in an inert atmosphere and analyzed by BET surface area analyzer. The powdered carbon obtained has utility as a soil amendment material, filtration media, and electrode for a supercapacitor. Detailed synthesis method for BTMO catalyst, yields of different products and coproducts after HTL, utilization of biochar as electrode material, and HHV of the oils recovered will be presented.

531012 Economic and Environmental Sustainability of the Production of Chemicals from a Pyrolysis-Based High Density Polyethylene Refinery

Abstract:

Current global plastics production is estimated to be of 332 million metric tons per year, which is expected to double in 2050.[1] Consequently, an increase in plastic waste generation is expected, for which the most common disposal alternative is landfill storage that has the inconvenience of a limited capacity and public health issues related to it. Mechanical recycling is another important option for disposal but it reduces the quality and utility of the material every time it is processed. Thus, chemical recycling emerges as a promising alternative due to the possibility of producing fuels, chemicals, and monomers that can be used to produce new plastics or other industrial products. Pyrolysis of plastic has been widely studied at laboratory and pilot plant scales showing a wide range of products obtained at different temperature and reactor configurations. Although a deeper understanding on the effects of process variables and chemical mechanisms has been achieved, little information is available about the environmental and economic implications of this technology from an industrial scale. Therefore, in this study previously published data[2] of waste High-Density Polyethylene (HDPE) pyrolysis at temperatures of 625, 650, and 675 °C over a range of vapor residence time is being used to design a process for the production monomers, aromatics, and aliphatic hydrocarbons. In the proposed process, waste HDPE is shredded into fine particles and introduced to a fluidized bed reactor using helium as the fluidizing gas and sand as the heating agent. This process is being simulated using Aspen Plus V. 8.8 and incorporates novel separation technologies and efficient integration of energy.

Techno-economic and life cycle assessment (TEA and LCA) are an essential component of this study and the results are going to be compared with production costs and environmental impacts from the current fossil-based pathways. As observed in previous waste refinery case studies, this process could potentially improve the environmental performance of the production of chemicals but with a possible increment in the cost. The results from the study will provide important information to guide future implementation of this technology from an industrial scale.

References

1. Plastics Europe. The plastic Industry. https://committee.iso.org/files/live/sites/tc61/files/The%20Plastic%20Industry%20Berlin%20Aug%202016%20-%20Copy.pdf (accessed July 19, 2017).

2. Gracida-Alvarez, U.R.; Mitchell, M.K.; Sacramento-Rivero, J.C. & Shonnard, D.R. Effect of Temperature and Vapor Residence Time on the Micropyrolysis Products of Waste High Density Polyethylene. *Industrial and Engineering Chemistry Research,* **57**, 1912-1923 (2018).

540813 Insights into the Anaerobic Digestion of Catfish and Shrimp Processing Wastewaters

Abstract:

Anaerobic digestion has been explored for the treatment of industrial wastewaters and for their potential for biogas production. This talk presents some key insights into digestion of wastewaters from catfish and shrimp processing facilities. Research on bench scale and pilot scale systems have tested effects of various pretreatment methods such as nutrient amendment, acid pretreatment, alkali pretreatment, and ozonation on digestion performance. Process designs have been created for the integration of operational digestion systems into the processing systems of the industrial partners. There are various economic, environmental and business aspects dictating the selection of design alternatives. There are numerous challenges and solutions in implementing a digestion system from bench scale to pilot scale and eventually to operational industrial scale.

541022 Synthesis of a Sustainable Multifunctional Biodiesel Additive from Lipid-Enhanced Sludges

Abstract:

Underground mining use biodiesel blends to reduce harmful exhaust emissions from diesel engines. In practice, diesel and biodiesel are stored separately and blended on site. Unfortunately, the storage and handling of neat biodiesel (B100) becomes challenging during the cold weather season due to biodieselâ€™s undesirable cold flow properties and poor oxidative stability. This study investigated phenyl-branched fatty acid methyl esters (PB-FAME) synthesized from lipid-enhanced sludges as a multifunctional additive to improve biodieselâ€™s cloud point (CP), an indicator of cold flow performance, and oxidative stability index (OSI). Different dosages of the PB-FAME were added to canola biodiesel. Results indicated that canola biodiesel with PB-FAME has a lower CP and higher OSI than neat canola biodiesel.

539306 Monitoring and Classification System for Water Recycling

Abstract:

Minimizing water consumption and disposal has huge benefits in terms of safety and cost savings. However, to maximize the amount of water that is recycled at a site generating multiple complex waste streams, an intelligent water monitoring system is needed to ensure that water from various sources is processed appropriately. Mainstream Engineering has developed a chemometric water monitoring system (CWMS) to fully characterize water prior to recycle or disposal. This technology relies on statistical methods that relate data to properties of interest. Specifically, we used a combination of several rapid, reagentless sensors to characterize the water according to several classes.

We demonstrated the CWMS for a variety of wastewater sources including laundry, kitchen, and latrine. We then validated our models with additional samples from other water sources such as wastewater treatment facilities to demonstrate the robustness of the system. After specifying all of the components, we designed a prototype system. Principal component analysis was used to classify water and draw distinctions between surface non-potable water, greywater and blackwater. We were also able to eliminate the need for dedicated COD, BOD_5, turbidity and chlorine sensors by processing spectroscopic data with chemometrics.

522532 Water Recovery and Reuse in Soluble Coffee Production Using a Dynamic Membrane Process

Abstract:

Through a grant from the U. S. Environmental Protection Agency, Pollution Prevention program, we have partnered with NestlÃ© USA to explore the use of novel membrane technology to improve the manufacturing efficiency of soluble coffee manufacturing operations. Our case study at their Freehold, NJ factory, evaluates water recovery methods for a spray dried and freeze dried coffee manufacturing plant. The particular focus of this paper is on a dynamic / shear-enhanced vibratory field membrane method to recover water for plant operations, thus reducing operating costs and environmental footprint. This research helps address a major issue in the food manufacturing sector â€“ the efficient use of water and waste reduction.

The production of spray dried and freeze dried coffee involves many highly complex processes that rely on large amounts of water. These operations include process water for cooling, water to be heated to produce steam, water for process equipment operations, water for intermediate production steps, and water for cleaning and sterilization. When utility generation is included, such as operation of cooling towers, water consumption can be significant.

Many of the problems associated with effective recovery, involve the separation of dissolved, colloidal, and particulate matter, to obtain water of suitable quality for reuse. Suspended and colloidal matter is one of the biggest causes of fouling and performance degradation in operating these water recovery systems. This causes additional expense, and in many cases industry abandons this approach, since it will not produce sufficient quantity and quality of water for extended periods of time. One potential solution that is described in this paper, is a vibratory-field membrane process for food/biomass waste processing.

Our study describes the benefits and limitations of this technology, and how laboratory investigations help guide industrial use and scale-up considerations. Various membrane types were evaluated to help inform decision makers based on the reuse goals. The study evaluated the overall savings and reduction in life cycle emissions for the recovery of water from a waste stream, for use in cooling tower operation. The assessment included savings, both economically and environmentally, in the reduction in fresh water intake, waste pretreatment, and final disposal. These are compared to the increased cost and operational factors of a new recovery system. The case study found that the greatest cost and environmental impact reductions were associated with the reduction in wastewater discharge. Long-term economics were more highly influenced based on equipment size. This case study is applicable to the food and beverage manufacturing sector.

525339 How Active Is Too Active?Â a Catalyst Selection Study

Abstract:

When selecting a catalyst formulation as part of a new reactor build, the process can be designed around the catalyst such that nearly any commercially-viable catalyst can be used. For these systems, higher activity catalysts generally give rise to higher product rates or reduced capital investment. However, most of industryâ€™s active processes are not newly built, but have been in service for many years (if not decades). For these systems, higher activity catalysts can also enable higher product rates, *but typically only to the extent constrained by the original process design*. Indeed, many if not most of todayâ€™s active systems are now multiply constrained, such that it is not always true that increasingly higher activity catalysts will yield better performance.

As one example, many fixed-bed, vapor-solid reactor systems are constrained by heat transfer. For these systems, a circulating coolant absorbs the heat of reaction thereby controlling the temperature rise and making possible higher product rates. In this context, there is generally a maximum achievable product rate corresponding to the maximum heat removal capability of the system. This maximum product rate is itself a function of the catalyst (in particular, the catalyst activity and selectivity).

In choosing from next-generation, high activity catalysts for a system constrained by heat transfer, the impact of varying activity and selectivity must be balanced, taking into account also the free variables for optimization and their bounds. Here, we utilize detailed reactor simulations to investigate the catalyst selection problem for one such industrial system. Our steady-state simulations incorporate mass and energy balances and reaction kinetics while predicting temperatures and concentrations that vary in two spatial dimensions. We begin by demonstrating how such simulations, when informed by laboratory experiments, can predict the heat transfer-constrained operation as observed in the plant. Then, we utilize the tuned reactor model to explore how process optimization accounting for a catalyst change (e.g., changes to the base reaction rates, activation energies, and partial reaction orders) leads to varying and sometimes counterintuitive changes in reactor performance. To finish, we develop simplified analytical models from first principles and compare their predictions to those generated using the full reactor model. These simplified models can be utilized as a coarse guide for catalyst selection in the case that more detailed and highly tuned reactor models are not available.

530554 Glycol Loss Minimization for a Natural Gas Dehydration Plant Under Upset Conditions

Abstract:

Low-temperature separation with Mono Ethylene Glycol (MEG) injection process is a common dehydration technique for natural gas processing, which needs to maintain a hydrocarbon dew point. However, dehydration systems based on glycol frequently result in a significant loss of glycol during plant upset conditions such as rapid flow fluctuation, refrigeration system failure and inadequate control of stripper column [1]. These losses occur due to carryover with gas, hydrocarbon and vaporization from stripper column overhead. For this reason, it is important to develop process operation methodologies for reducing glycol losses and thereby reducing emissions.

In this research, steady-state and dynamic models are developed using the Aspen Plus simulation framework. A parametric analysis of inlet flow fluctuation, temperature fluctuation of the low-temperature separator (LTS), stripper column overhead temperature fluctuation and reboiler temperature fluctuation were conducted under steady-state conditions. Simulation results indicated that that most of the glycol is lost via vaporization through the stripper column overhead and some glycols are carried over with gas and hydrocarbon from LTS. Dynamic simulations were utilized to develop a new control strategy for optimum column operation along with an improved stripper model. It is shown that the simulation results can be used to study plant upset conditions, minimize glycol loss, improve product gas specifications and reduce plant operating costs. In particular, it is shown that the improved design has the potential to reduce glycol losses by over 70%, lower emissions and provide plant stability.

References

1. Haque, M. E.; Ethylene Glycol Regeneration Plan: A Systematic Approach to Troubleshoot the Common Problems. *Journal of Chemical Engineering*, *IEB, 27, No. 1, 21-26, 2012.*

Hydrodeoxygenation of Karanja Oil for the Production of Green Diesel: Process Design with Heat Integration and Economic Analysis

Swarnalatha Mailaram and Sunil Kumar Maity,
IIT Hyderabad, Hyderabad, India

Introduction

Major concern on environmental effects and depletion of fossil fuels led to search for the alternative resources for energy. Biomass is the best possible alternate source in terms of its availability in nature and it also has the potential to transform to all forms of energy including transportation liquid fuel which is the leading energy consuming sector with ~27% share globally [1]. Much research and technological development has been done so far in the production of biofuels ex: bioethanol, biobutanol, biodiesel etc. Biomass derived oxygenated biofuels suffer with low mileage because of the presence of oxygen and these biofuels are not compatible with existing infrastructure. Transesterification is one of the technology producing bio diesel which is limited to blending with petro-diesel due to the presence of oxygen which lowers the fuel mileage. Therefore, there is a need of hydrocarbon fuels similar to petro-diesel, which is compatible with existing combustion engine. Inedible vegetable oil consisting of long hydrocarbon chain with low oxygen content and lesser complexity of chemical structure compared to lignocellulosic biomass is the promising biomass to produce transportation fuels. Karanja oil production potential is more in south east Asia with 78.6wt% C_{18}, 11.3wt% C_{16}, 3.7wt % C_{20}, 6.4wt% C_{22} fatty acid composition [2]. Hence, Karanja oil is chosen as the feedstock.

Hydrodeoxygenation process

Hydrodeoxygenation (HDO) is the route which removes oxygen content in the triglyceride in the form of water to form hydrocarbon diesel range fuel called as green diesel. Green diesel has higher energy density (44 MJ/kg), high cetane number (70-90) and can directly be used in existing combustion engines [3][4]. UOP has recently developed a two-step ecofining process aiming to produce green diesel by HDO [5]. Studies on various oils by HDO process has been performed by researchers. HDO is generally carried out at high hydrogen pressure (25–150 bar) over supported metal catalyst in the temperature range of 523 K to 773 K [6]. Generally, HDO of Karanja oil can be envisaged through two different routes: (i) direct HDO and (ii) two-step HDO (hydrolysis of vegetable oil followed by HDO of fatty acid). In the Direct HDO, hydrogenation of unsaturated triglyceride in Karanja oil gets saturated triglyceride and undergo HDO to form diesel range hydrocarbons (C_{18}-C_{22}) termed as green diesel, water, traces of CO_2 with propane as a co-product. In the second route, Karanja oil is first hydrolyzed with water to form fatty acids and glycerol (co-product). These fatty acids then undergo HDO forming green diesel, water and traces of CO_2 [3][4][7].

Process design with heat integration

In the present work, process design for both the routes were performed using Aspen Plus software for various capacities of plant ranging from 0.01 to 0.15MMT of Karanja oil /annum. Schematic process for direct HDO process is shown in Fig. 1. In direct HDO, karanja oil and hydrogen enter the HDO reactor operating at 623K, 100bar. Excess hydrogen is removed in a separator following reactor. The product stream containing green diesel, water, CO_2, propane is sent to downstream processing. Water, green diesel and gases (propane and CO_2) are separated and gases are further sent to absorber and stripping unit in which monoethanolamine (MEA) is used as a solvent to separate CO_2 from propane and propane is the revenue generating co-product. As shown in Fig. 2., In two step HDO, karanja oil is hydrolysed with water in hydrolysis reactor at 533K and 56bar to form fatty acids and glycerol. Glycerol and water is separated and sent to the glycerol refining section to obtain 99.99wt% glycerol which is the co-product in this process. Fatty acids formed are sent to the HDO reactor to form where green diesel, water and traces of CO_2 are formed. Green diesel is separated from gases and water. Heat integration for both the processes has been done to reduce the utility requirement and optimize the heat transfer within the plant. Pinch analysis has been chosen to obtain the minimum energy consumption in the plant and design heat exchange network for both the processes. Resulting grand composite curve (GCC) by pinch methodology, was useful to obtain the minimum hot and cold utility requirement and heat recovery within the system. Based on the possibility of heat recovery in the process and pinch analysis criteria, heat exchange network has been designed. This helped in redesigning the processes with the heat exchange network system.

Economic analysis

The economic evaluation has been carried out for different plant capacities ranging from 0.01 to 0.15MMT/annum for both direct HDO and two-step HDO process. All possible cost contributing factors were considered to obtain capital and operating expenses for both the processes. Capital investment for both the processes is practically identical. Impact of plant capacity on both processes is huge. The production cost was reduced with increasing plant capacity. The reduction in the production cost was, however, negligibly small with increasing plant capacity beyond 0.12 MMT/annum. Hence, 0.12 MMT/annum has been considered as the optimum plant capacity. The production cost of green diesel was estimated as 0.84 $/kg and 0.798$/kg for direct and two-step HDO respectively for the plant capacity of 0.12 MMT Karanja oil per annum. Also, contribution of various costs to production cost has been studied. Feedstock alone is contributing around 73-77% to the production cost which is very high. Contribution of co-product credits has significantly reduced the production cost of green diesel in two-step process. Utility consumption reduction with the help of heat recovery in the process has shared 1.8-4.18% which was ~35% contribution without heat recovery. Price of co-products has also a significant impact on the production cost. Production cost was decreased by 0.5% and 1.6% for 10% increase in the price of co-products for direct and two-step HDO respectively. The discounted cash flow analysis was further carried out to evaluate the economic performance and determine the minimum selling price of green diesel. The price of green diesel excluding taxes was found to be 1.186 $/kg and 1.21 $/kg for an acceptable rate of return of 8.5% for direct and two-step HDO respectively. The green diesel can be considered as economically competitive with the retail price of the petro-diesel.

Summary

Conceptual processes of direct and twostep HDO is designed using Karanja oil as feedstock. Maximum energy recovery using pinch methodology has been done. Heat exchange network has been designed for both the processes. Heat recovery within the system with the use of pinch analysis has reduced the utility consumption in the process. Thereby, reducing the production cost of green diesel significantly. Economic analysis for modified processes with heat exchange network was evaluated. Capital, operating expenses, production cost of green diesel is estimated. Feedstock cost plays key role in reducing the cost of the diesel. If reduced, drastic decline in the cost can be observed. Discounted cash flow analysis was used to find the minimum selling price of green diesel.

References

1. Annual Energy Outlook. U.S. Energy Information Administration. Available at: http://www.eia.gov/forecasts/aeo/index.cfm: 2015.
2. P. Kumar, S.R. Yenumala, S.K. Maity, D. Shee, Kinetics of hydrodeoxygenation of stearic acid using supported nickel catalysts: Effects of supports, Appl. Catal. A Gen. 471 (2014) 28–38.
3. S.R. Yenumala, S.K. Maity, D. Shee, Reaction mechanism and kinetic modeling for the hydrodeoxygenation of triglycerides over alumina supported nickel catalyst, React. Kinet. Mech. Catal. 120 (2017) 109–128.
4. Kalnes TN, Marker T, Shonnard DR, Koers KP. Green diesel production by hydrorefining renewable feedstocks. Biofuels Technol 2008:7–11.
5. UOP LLC, A Honeywell Company. Honeywell Green Jet Fuel: Sustainable fuels for cleaner skies. 2011. Available at: https://www.uop.com/?document=honeywell-green-jet- fuel-brochure&download=1
6. Glisic SB, Pajnik JM, Orlovic´ AM. Process and techno-economic analysis of green diesel production from waste vegetable oil and the comparison with ester type biodiesel production. Appl Energy 2016;170:176–85.
7. S.R. Yenumala, S.K. Maity, D. Shee, Hydrodeoxygenation of karanja oil over supported nickel catalysts : influence of support and nickel loading, Catal. Sci. Technol. 6 (2016) 3156–3165.

Figures

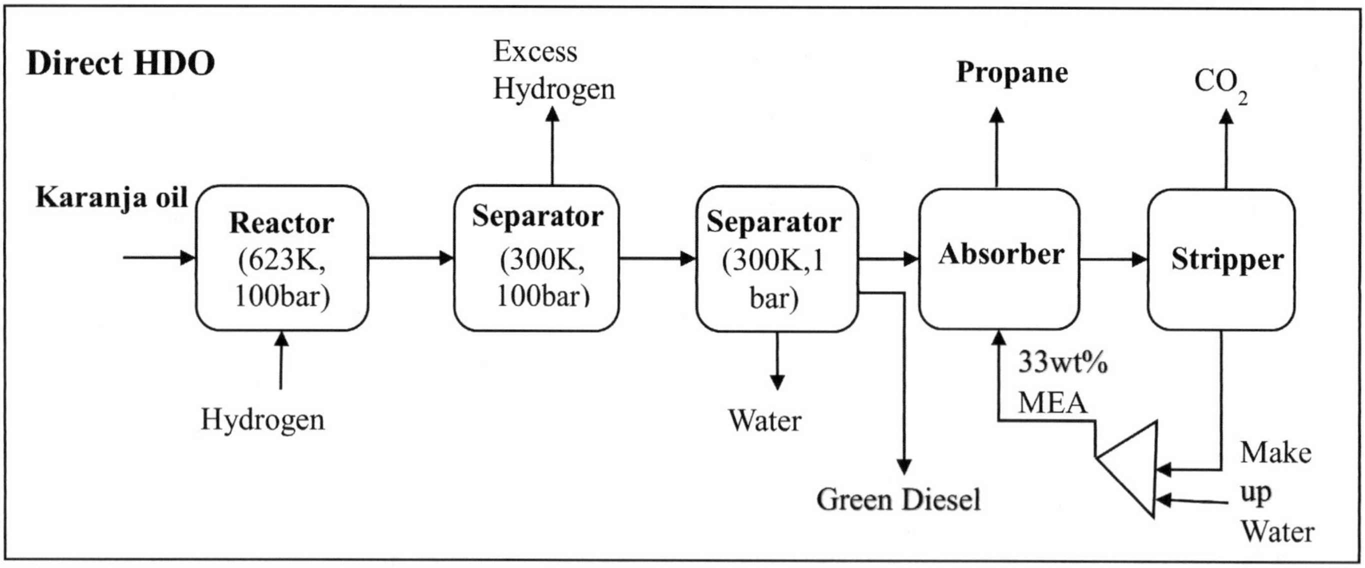

Fig. 1. Schematic process flowsheet of Direct HDO process.

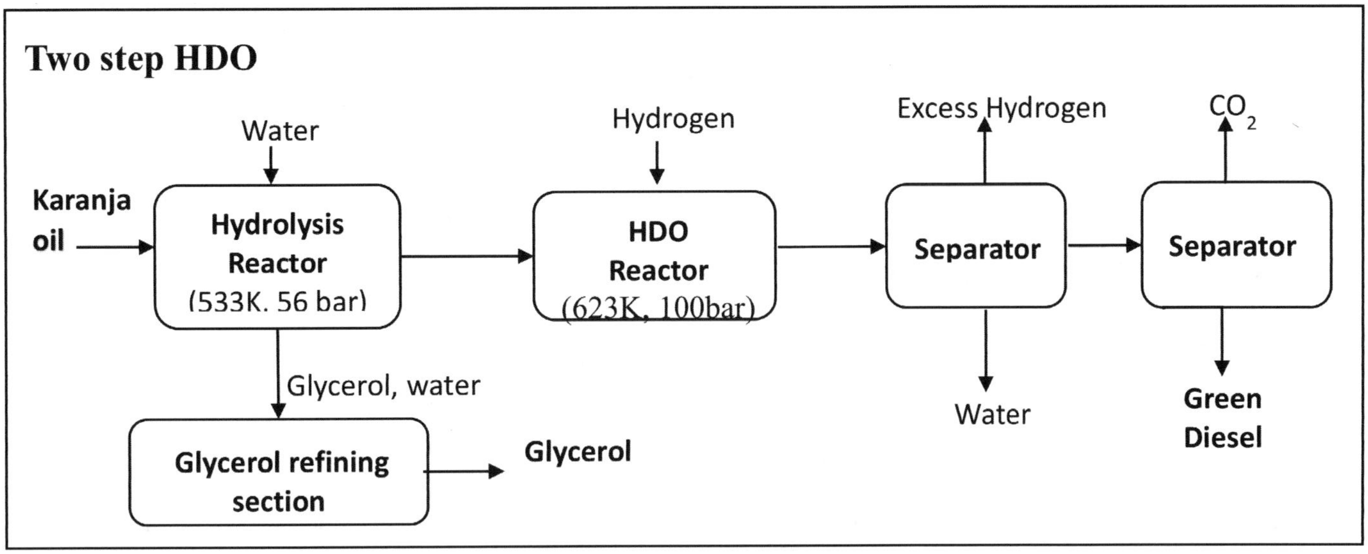

Fig. 2. Schematic process flowsheet of Twostep HDO process.

531870 An Integrated Process for Desulfurization (HDS-Extractive-Oxidative)

Abstract:

An integrated process for Desulfurization (HDS-Extractive-Oxidative)

Mokhtarian, M.R. Dehghani, F. Banisharif

Chemical Engineering Department, Iran University of Science and Technology, Narmak, Tehran, Iran

Abstract:

Available sulfur compounds in the fuels such as thiophens are considered as one of the main source of air pollution. They converted to SOx and released to the atmosphere as a harmful chemical for human health. Hence, removal fuel desulfurization is mandatory from environmental point of view as well as public health issues. Nowadays, in many countries, stringent regulations are being enacted to reduce the sulfur content to lower than 10 ppm for both diesel and gas oil. During recent years many attentions have been paid to extractive-oxidative desulfurization (EDS-ODS) while economic evaluations show that they are not economically feasible compared to the hydro desulfurization (HDS) process. On the other hand, HDS process is considered as an energy insensitive process and needs to be operated at High pressure and temperatures. To present an economic process, in this work an integrated process of HDS and EDS-ODS were modeled. The HDS section was modeled based on operational data from Tehran refinery and extractive-catalytic oxidative desulfurization (ECODS) part were modeled based on our novel pilot experimental data. Gas oil as the feedstock with total sulfur content of 22600 ppmw was fed to the process. Optimal conditions were obtained using the response surface methodology and box-behnken design (BBD), based on experimental design results, hydro desulfurization from the 22600 ppmw to 1930 ppmw level was done under the following conditions: hydrodesulfurization temperature: 311 â„ƒ, hydrodesulfurization pressure: 24 bar and hydrogen purity of 35 wt%. Then gas oil with the total sulfur content of 1930 ppmw was entered into the oxidative-extractive desulfurization unit. The effects of different operating parameters such as temperature, oxidant to sulfur molar ratio and the type of solvent in the extraction stage were investigated. The best performance was obtained at oxidant to sulfur molar ratio of 4:1 and acid to oxidant volumetric ratio of 1:1. It must be mentioned that hydrogen peroxide (30%) and acetic acid (95%) were used as oxidant. Finally, a liquid-liquid extraction process was utilized for deep desulfurization step. Acetonitrile used as solvent at 1:1 ratio towards gas oil. It was found that the proposed process has lower costs and also is inherently safer than that of conventional HDS, ODS and EDS.

Keywords: Optimization, Hydrodesulfurization, Extractive- Catalytic oxidative desulfurization, Ultra-low Sulfur Fuel

540773 2018 Outlook for Energy: A View to 2040

Abstract:

The Outlook for Energy is ExxonMobil's global view of energy demand and supply through 2040. It also highlights the dual challenge of providing affordable energy to support prosperity while reducing environmental impacts – including the risks of climate change. We use the Outlook to help inform our long-term business strategies and investment plans.
The Outlook anticipates significant changes through 2040 across the world to boost living standards, reshape the use of energy, broaden access to abundant energy supplies, and accelerate decarbonization of the world's energy system to address the risks of climate change.

- Global energy needs will rise about 25 percent over the period to 2040
- Electricity demand will nearly double in non-OECD countries
- Worldwide electricity from solar and wind will increase about 400 percent
- Natural gas will expand its role, led by growth for electricity generation
- Growth in oil demand will be driven by commercial transportation and chemical needs
- Efficiency gains and growing use of less-carbon intensive energy sources will contribute to a nearly 45 percent decline in the carbon intensity of global GDP

Energy is fundamental to modern life, and as the world's population approaches 9 billion people in 2040, we are challenged to improve living standards everywhere. Understanding the factors that drive the world's energy needs – and likely choices to meet those needs – is the mission of the Outlook. By sharing the Outlook with the public, we hope to broaden that understanding. Energy matters to everyone, and we all play a role in shaping its future.

541543 Energy Decarbonisation Scenarios

Abstract:

With the energy sector being associated with 2/3 of the annual greenhouse gas emissions, decarbonizing the energy is a key step to reduce their level. The National Determined Contributions (NDC's) announced so far by the participating countries do not meet the objectives of the Paris-climate agreement, as far as reaching a peak in emissions in the near future or decreasing to a level that is compatible with the stated long-term targets.

Using different scenarios for energy demand and supply, options for energy sector decarbonisation are evaluated. Meeting the 2-degree target requires a major transformation in the power sector, efficiency improvements – specially in the buildings and in the industry, and a revolution in transport towards fuel and modal shifts. With a significant increase in renewable energy, system integration and optimization will also be key in reducing the cost of the energy transition. Meeting more ambitious climate goals ("well below 2 degrees") will push further the technological innovation and implementation requirements throughout the entire energy value chain. A scenario aiming at zero net-emissions by 2060 is presented, along with a comparison to the 2-degree case.

542625 Fundamental Research Needs to Advance Energy Technologies

Abstract:

The U. S. Department of Energy (DOE), Office of Science, Office of Basic Energy Sciences (BES) supports fundamental research in chemical and materials sciences to provide the foundations for new energy technologies and to support DOE missions in energy, environment, and national security. This presentation will discuss opportunities for fundamental research to impact DOE's energy mission "to catalyze the timely, material, and efficient transformation of the nation's energy system and secure U.S. leadership in energy technologies" with a focus on the way we generate, store and use energy nationally. I will provide an overview of BES strategic planning over the past decade that identified priority research directions for advancing energy applications, highlight key scientific advances in these areas, and discuss some future opportunities for modern science, particularly chemical sciences, to accelerate the transformation of the U. S. energy portfolio.

533201 A Multi-Objective Approach for the Integrated Design of Flexible Algae Biorefineries

Abstract:

The merits of microalgae as an alternative source of biomass are well established over the years, especially because of their versatility in terms of products. This potential in addition to the utilization of carbon dioxide, which microalgae need for their growth, render the development of a large-scale application extremely promising either as a standalone venture or in integration with other plants. Specifically, the co-production of specialty chemicals from microalgae is gaining ground as the related markets expand and profit margins grow. A plethora of microalgae has been studied and researched, especially for applications in the pharmaceutical, nutraceutical and cosmetic sector.

The work addresses current issues in algo-biorefinery systems following a multi-layer methodology based on optimization, design and knowledge engineering. The synthesis of value chains is generic and demonstrated in the D-Factory project (FP7 collaborative project, which aims at developing a commercial scale biorefinery process) against a background of 15 products and 12 technologies. By employing mathematical modeling, superstructures of potential processing paths and product portfolios are constructed according to customized constraints. The approach offers high-capacity screening with featuring techno-economic, environmental, market and business-related criteria. The work hosts models of different scales, enables bounds on the number of options to screen and delivers the best set of solutions according to each different criterion.

Further integration of synthesis with process design is also important. Scaling up algal biorefineries is challenging especially due to volatile microalgae productivities (location, environmental conditions, seasonality etc.). One is interested in the design of flexible plants that adjust production without significant reconstruction of the plant. Feasible options are investigated in strong interactions with real-life applications that include studies of stand-alone as well as integrated processes. Conceptual and commercial tools are employed in order to fathom the overall biorefinery plant. Efficiency targets are established resulting in energy, water and subsequent cost savings. The economic potential of algae biorefineries is highlighted following the assessment of potential investments which vary in terms of processes and product portfolios.

534153 Biodiversity Enhances Multifunctionality in a Life Cycle Assessment of Microalgal Biofuel Production

Abstract:

Life cycle assessments (LCAs) comparatively analyze and interpret the sustainability of fuel production pathways and are imperative for improving energetic efficiencies and reducing environmental impacts. Accuracy and interoperability within the field of algal biofuel LCAs have proven challenging due to variance in empirical data collection, methodologies, models, and specious ecological assumptions. Here, we describe a contiguous well-to-wheel LCA for the production of algal biofuel that employs an ecologically based growth model to study the impacts of biodiversity on life cycle sustainability metrics. Our approach incorporates data from collaborative publications that considered algal cultivation in indoor mesocosms under two temperature regimes as well as growth in outdoor ponds. The studies integrate empirical input parameters from each stage of algal biofuel production, including elemental composition of algae and biocrude, phosphorous and nitrogen nutrient use efficiencies, as well as biomass and biocrude yields for mono-, bi-, and poly-cultural algal cultivation. Under optimal laboratory growth conditions, we found that *Selenastrum capricornutum*, a dominant species in biomass accumulation and culture stability, yielded an energy return of 0.99 and emissions of 122 kg CO_2eq/MBTU. However, upon translation to outdoor ponds, a biculture of *S. capricornutum* and *Chlorella sorokiniana* outperformed all other culture compositions with an energy return of 0.60 and emissions of 167 kg CO_2eq/MBTU. Despite markedly worse life cycle metrics, the success of the biculture over the best monoculture arose due to an unprecedented and beneficial multifunctionality between input parameters. Our results necessitate the consideration of ecological constraints within LCAs as well as the propensity of select algal polycultures for favorable energetics and reduced emissions during scaled biofuel production.

537587 The Effect of High-Intensity Ultrasound on Cell Disruption and Lipid Extraction from Concentrated and Viscous Slurries of *Nannochloropsis* Sp. Biomass

Abstract:

The effect of ultrasonication on the cell rupture of marine microalgae species Nannochloropsis sp. was studied as a function of solids concentration and treatment time. The concentrated and viscous wet-biomass (11, 20 and 25% solids concentration) was subjected to ultrasonication (20 kHz high intensity ultrasound) at 11.4 for up to 5 minutes. The cell rupture was visualized using multi-ion beam microscopy that showed clear increase in cell rupture with an increase in ultrasonic treatment time. The non-polar lipids were extracted from the ultrasonic treated wet-biomass using hexane extraction at room temperature. The lipid extraction efficiency was measured as a function of solids concentration and treatment time. Compared to hexane extraction without cell rupture, sonication led to a significant increase in the extraction efficiency from ~11 to 62% within 5 minutes of sonication. The lipid extraction efficiency decreased slightly with increase in solids concentration from 11 to 20% but it decreased significantly when the solids concentration is increased to 25%. This is attributed to the attenuation of high intensity ultrasound in viscous biomass. In order to elucidate the effect of viscosity on ultrasound attenuation, we have measured the viscosity of biomass at various solids concentration as a function of shear rate. We found that the Nannochloropsis sp. was shear thinning. We also measured the speed of sound in the biomass as a function of solids concentration to calculate the attenuation of ultrasound. The speed of sound decreased with increase in solids concentration. The increase in viscosity and decrease in speed of sound with increase in solids concentration resulted in the attenuation of ultrasound resulting in low lipid extraction efficiencies at 25% solids concentration. The efficiencies from ultrasound cell rupture at 20% solids concentration were on par with the efficiencies from conventional method of High Pressure Homogeniser (HPH). The lipid oxidation during ultrasonic treatment was characterized by HPLC and found that the lipid profile did not change with ultrasonication. The specific energy requirements were calculated for ultrasonication cell rupture as a function of solids concentration, ultrasound power and treatment time and compared them to HPH and other cell rupture methods.

538836 Borate-Promoted CO_2 Capture for Microalgae Cultivation

Abstract:

A major barrier to economic viability of microalgae biofuels is the high cost of concentrated CO_2 supply cost. Furthermore, global algae biofuel generation would be seriously restricted if micro-algae farms are restricted to the vicinity of CO_2 sources along with the additional limitations of low slope lands and favorable climate. One potential solution to decrease CO_2-related costs is to cultivate microalgae cultures in high pH media (pH >10) that can uptake ambient/atmospheric CO_2 at high rates. In addition, adding high concentrations of carbonate/bicarbonate to the media (i.e. media with high carbonate alkalinity) provides high dissolved inorganic carbon (DIC) for photosynthesis. The presence of rate promotors like boric acid further aids in CO_2 hydration and absorption and can lower the requirements for physical mass transfer and mixing costs in high pH media. In this study, we report results from studies that compare rates of DIC uptake in borate-promoted and promoter-free alkaline solutions exposed to the atmosphere. The experiments were set up by maintaining constant system parameters including the initial pH and total alkalinity in both the systems. Along with the initial and final DIC content, pH was continuously measured until both the system reached equilibrium with the atmosphere. A mathematical model, developed from chemical equilibria first principles, was fit to the measured pH values to determine the rate of DIC increase, mass transfer coefficients and the enhancement factor. Our results showed that the rate of DIC increase in the borate promoted systems was at least 50% higher than the non-promoted carbonate/bicarbonate solutions. Our ongoing studies are investigating the cultivation of the alkaliphilic *Chlorella sorokiniana str SLA-04* in borate promoted carbonate/bicarbonate systems to achieve high biomass productivity without concentrated CO_2 inputs.

543008 Towards Carbon-Negative Bioenergy and Bioproduct Systems using Renewable Electricity

Abstract:

While U.S. production of energy products from the shale gas and oil boom have markedly grown in the last ten years, we remain a net importer of petroleum. Just as technology advances in shale gas and oil recovery processes have decreased our reliance on imports, technology advances in renewable and sustainable approaches are needed to further address national security, rural prosperity, and climate change issues. However, support for their development and commercialization has not fully breached our collective cognitive dissonance. Beyond cellulosic ethanol, for which recent implementation should be commended, approaches that integrate biomass energy with alternative electrical energy to produce domestic liquid fuels offer the potential to be our next technological breakthrough. As DOEâ€™s Billion-ton Report reminds us that domestic renewable carbon is limited, carbon efficient bioenergy strategies are preferable when considering the scale of fossil fuel displacement that is needed. Because hydrocarbon compounds are uniquely capable of forming energy containing covalent bonds, retaining carbon in the fuel product stream is paramount, especially when coupling renewable electrical energy through electrocatalytic hydrogenation. When joined with fast pyrolysis in small rurally deployed depots, this â€œenergy upgradingâ€ method enables agricultural and forested communities to gain access to higher value markets for their feedstocks. Co-product biochar, upgraded using low cost, low energy plasma, will also be described for entering high-end activated carbon markets. Alternatively, biological conversion approaches are also able to make fuels and chemicals using renewable electricity. Toward this aim, an acetone-platform biorefinery will be described as one example of a carbon-negative bioenergy system that uses a microbial process to convert carbohydrates and electrocatalysis to valorize lignin.

543010 Perspective on Carbon Utilization for Fuels, Chemicals, and Materials:Case Studies on Thermal and Catalytic Biomass Processing

Abstract:

Biorefinery process economics are strongly impacted by the overall carbon efficiency from biomass to products; however, market dynamics influence the types of products that should be targeted. Thus, trade-offs must be considered between product value (quality and market demand), carbon efficiency, and the associated costs (capex/opex) of accessing those products. This presentation will discuss biorefinery carbon utilization and these trade-offs from the perspective of two case studies on thermal and catalytic biomass processing: catalytic fast pyrolysis and indirect liquefaction through dimethyl ether. Each of these processes proceeds through an intermediate (stabilized bio-oil and dimethyl ether) that can undergo further upgrading and generates carbon-containing â€˜wasteâ€™ streams, creating both opportunities and challenges for improving overall carbon utilization. Approaches and recent advancements in enhancing biorefinery carbon utilization will be presented and discussed within the context of a market-responsive biorefinery concept.

538379 Electrochemical CO_2 Conversion to Valuable Chemicals

Abstract:

Converting greenhouse gas carbon dioxide (CO_2) to value-added chemicals is an appealing approach to tackle CO_2 emission challenges. The chemical transformation of CO_2 requires suitable catalysts that can lower the activation energy barrier to minimize the energy penalty associated with the CO_2 reduction reaction. Cu is the only monometallic catalyst that can produce an appreciable amount of C2+ products from CO2, while the C2+ selectivity of Cu must be further improved in order to be considered for commercialization. Here, we will present our work in developing an integrated two-stage electrolyzer stack system for the conversion of CO_2 to alcohols. The first stage utilizes a silver catalyst in a flow cell capable of reducing CO_2 to CO with faradaic efficiencies of >95% and partial current densities of >200 mA/cm^2. For the second stage, a CO reduction flow cell using alkaline electrolyte can reduce CO to C2+ products with high selectivity and high current density. This work demonstrates the feasible production of multi-carbon alcohols via CO reduction at appreciable rates, and motivates further development of a two-stage electroreduction process of CO_2 to alcohols.

545759 Development of Devices and Selective Catalysts for the Solar-Driven Reduction of CO_2 to fuels

Abstract:

The electrochemical and photoelectrochemical reduction of CO_2 presents an attractive route towards fuel formation from renewable energy. However, the dearth of selective, active and inexpensive catalyst materials remains a key challenge on the way to renewable production of hydrocarbon fuels. Here, we address these challenges through the design of novel devices and selective earth-abundant catalysts for the electrochemical[1–3] and solar-driven[1,4–6] reduction of CO_2. In addition, we provide insight into the origin of selectivity on electrocatalysts based on copper, paving the way towards more rational design of electrochemical fuel forming devices.

We pioneer the application of Atomic Layer Deposition (ALD) to direct the catalytic selectivity of copper catalysts. Copper-based catalysts derived from the reduction of CuO are known to reduce CO_2 but simultaneously exhibit poor selectivity. ALD coating of CuO with thin layers of SnO_2 allows for formation of CO with high selectivity even at low overpotentials, comparing favorably to conventional gold catalysts, known to selectively convert CO_2 to CO. We use kinetic studies, high resolution transmission electron microscopy and gas chemisorption analysis to explain the observed selectivity changes on the basis of decreased binding strengths for both H and CO intermediates on the Sn-containing catalyst surface.

We subsequently combine our new catalyst with an earth abundant oxide anode to effect overall splitting of CO_2 to CO and O_2. Driven by a 3-junction solar cell, this device achieved long-term solar splitting of CO_2 into CO and O_2 at an efficiency of 13.4 %, constituting the current world record.[1]

Going beyond the production of CO, we investigate the further reduction of CO to hydrocarbons. Insights into the factors controlling the branching between different hydrocarbon species and hydrogen as a side product are scarce, precluding any rational design of selective catalysts. To remedy these shortcomings, we devise a platform allowing for the systematic study of the relevant reaction parameters. Exploiting non-aqueous electrolytes, we isolate activation-controlled kinetic data in electrochemical driving force, proton availability, solvent dielectric and reagent concentration. Our observations provide unique insight into the mechanism of hydrocarbon production on copper electrodes, demonstrating that the competition between H and CO for surface sites plays an important role in governing the product selectivity between methane, ethylene and hydrogen. These results establish a new paradigm in controlling hydrocarbon product selectivity in electrochemical fuel formation.[2]

References

[1] M. Schreier, F. Héroguel, L. Steier, S. Ahmad, J. S. Luterbacher, M. T. Mayer, J. Luo, M. Grätzel, *Nat. Energy* **2017**, *2*, 17087.

[2] M. Schreier, Y. Yoon, M. N. Jackson, Y. Surendranath, *Angew. Chem. Int. Ed.* **2018**, DOI 10.1002/anie.201806051.

[3] G. P. Lau, M. Schreier, D. Vasilyev, R. Scopelliti, M. Grätzel, P. J. Dyson, *J. Am. Chem. Soc.* **2016**, *138*, 7820–7823.

[4] M. Schreier, P. Gao, M. T. Mayer, J. Luo, T. Moehl, M. K. Nazeeruddin, S. D. Tilley, M. Grätzel, *Energy Env. Sci* **2015**, *8*, 855–861.

[5] M. Schreier, L. Curvat, F. Giordano, L. Steier, A. Abate, S. M. Zakeeruddin, J. Luo, M. T. Mayer, M. Grätzel, *Nat. Commun.* **2015**, *6*, 7326.

[6] M. Schreier, J. Luo, P. Gao, T. Moehl, M. T. Mayer, M. Grätzel, *J. Am. Chem. Soc.* **2016**, *138*, 1938–1946.

531445 Bimetallic Nanoporous Pd Alloys as CO Tolerant Electrocatalysts for the Electrohydrogenation of CO_2 to Formate

Abstract:

Electrochemical CO_2 reduction reaction (CO_2RR) has been widely investigated for conversion of water dissolved CO_2 into useful carbon based products e.g CO, formate/formic acid, alcohols etc. Depending upon the binding affinities of the different reaction intermediates, metals available for the CO_2RR show variability in product selectivity, overpotential requirements and sustained activity. Thus development of a suitable electrocatalyst with a high activity/faradaic efficiency (FE) for a single product is a challenge[1]. Among the simple 2 electron reduced products, formate/formic acid is gaining focus because of its high energy density and utility as a chemical precursor. Traditional formate producing metals e.g. Sn, In, Bi, Co etc. require high overpotentials (>700 mV), limiting their effectiveness. Recently, Pd, originally known to produce CO at overpotentials above 0.6 V RHE[2], has recently demonstrated formate FE's near unity at overpotentials below 0.2 V[3]. The gradual poisoning of Pd-based CO_2RR electrocatalysts through the slow evolution of CO, even at near unity formate FEs[4], results in catalyst deactivation and limits the viability of Pd in commercial applications.

Here, we report fabrication of free standing, core-shell, nanoporous bimetallic Pd (np-PdX, where X = Ag, Cu, Ni, Co) alloys that display a compositional dependent CO deactivation rate. The np-PdX electrocatalysts are made through electrochemical dealloying[5] of 4 binary $Pd_{15}X_{85}$ (X = Co, Ni, Cu, Ag) alloys. Electrochemical dealloying drives the formation of Pd skinned nanoporous architectures where the presence of less noble metal under the Pd shell alters the electronic structure of the Pd skin and consequently affects the CO adsorption strengths thereby changing the rate of catalyst deactivation. The np-PdX electrodes have a residual composition of ~ 30 atomic % of the less noble metal and pore sizes ranging from 5 – 10 nm. The tortuous pores give rise to roughness factors above 500 which generates high geometric formate partial current densities (> 40 mA/cm^2). Furthermore, the np-Pd electrodes are free standing and obviate the use of any binder and/or support which otherwise causes extra overpotential losses and morphological instability.

Upon performing extended electrolysis, the CO_2RR activity, formate FE and degree of deactivation of the np-PdX elecrodes are compared with a traditional Pd/C electrocatalyst having particle sizes of 5-10 nm. While all the np-PdX electrodes and Pd/C show above 90% formate FE within 200 mV overpotential, there is a significant variation in tolerance to CO poisoning as evidenced by the compositionally dependent time of deactivation. While np-PdAg and np-PdCu show faster deactivation compared to Pd/C, np-PdCo and np-PdNi avoid poisoning for much longer times. The results are explained on the basis of weakening of CO binding strengths due to electronic interactions of the alloying component that shifts the d-band center of Pd[6]. This is also shown by the trend of weighted average CO stripping peak potentials on Pd-skinned polycrystalline PdX (X = Co, Ni, Cu and Ag) alloys. Thus np-PdX electrocatalysts based on a suitable choice of alloying component, offer substantial suppression of CO poisoning, much higher formate geometric current densities and enhanced operational stability compared to traditional Pd based CO_2RR electrocatalysts.

References:

1. Kuhl, K., Cave, E., Abram, D., Jaramillo, T. New insights into the electrochemical reduction of carbon dioxide on metallic copper surfaces. *Energy Environ. Sci.* **5,** 7050–7059 (2012).
2. Gao, D. *et al.* Size-Dependent Electrocatalytic Reduction of CO_2 over Pd Nanoparticles. *J. Am. Chem. Soc.* **137,** 4288-4291 (2015).
3. Klinkova, A. *et al.* Rational Design of Efficient Palladium Catalysts for Electroreduction of Carbon Dioxide to Formate. *ACS Catal.* **6,** 8115–8120 (2016).
4. Min, X. & Kanan, M. W. Pd-Catalyzed Electrohydrogenation of Carbon Dioxide to Formate: High Mass Activity at Low Overpotential and Identification of the Deactivation Pathway. *J. Am. Chem. Soc.* **137,** 4701–4708 (2015).
5. Li, X. *et al.* Dealloying of Noble-Metal Alloy Nanoparticles. *Nano Lett.* **14,** 2569–2577 (2014).
6. Ou, L. Design of Pd-Based Bimetallic Catalysts for ORR: A DFT Calculation Study. *J. Chem.* **2015,** 1–11 (2015).

534365 Electrochemical Promotion of Catalysis: Non-Faradaic Effects of Applied Potential on CO_2 Hydrogenation and Ethylene Oxidation Reactions

Abstract:

Electrochemical promotion of catalysis is achieved in gas phase reactions over solid catalysts via the coordination of a dual-function working electrode/catalyst and catalytically inert counter and reference electrodes connected via a purely ionically conducting solid electrolyte (H^+ or O^{2-} conductor). Catalytic reaction rates of CO_2 hydrogenation and C_2H_4 oxidation can be reversibly modified by nearly two orders of magnitude via the application of a potential between the catalyst and reference electrodes. Faradaic efficiencies, defined by the ratio of the change in catalytic rate (dr) to the galvanostatically-defined rate of ionic species supplied to the catalyst surface (I/nF), are far in excess of unity. These non-Faradic processes result in changing selectivity of CO_2 reduction catalysis over Ru at 693 K by increasing CH_4 formation rates and concurrently reducing CO formation rates with the application of a positive potential, and positive potential application modifies C_2H_4 oxidation reaction rates at 693 K over Pt by over an order of magnitude with Faradaic efficiencies in excess of 50. Applied potentials result in the changing reactant equilibrium binding constants as modeled by Langmuir-Hinshelwood-type rate expressions. Potential application results in the back-spillover of ionic species from the solid electrolyte onto the ionic surface to form an electrically neutral effective double layer that can be quantified via analysis of galvanostatic transients in conjunction with C_2H_4 oxidation kinetic modeling. This work investigates and models the fundamental non-Fardaic effects of applied potential upon a catalytically active surface with regards to gas phase reactions via alteration of site densities and energetics.

535146 Nano- to Macro Scale Morphological Impacts on CO_2Â electroreduction Product Selectivity over Cu Catalysts

Abstract:

CO_2 electroreduction has recently attracted a great deal of attention as a method to produce highly valuable chemicals and fuels from the waste CO_2. Among the possible products generated in this process, hydrocarbons such as methane and ethylene are of interest owing to their wide applications. Cu is the only metal catalyst which can produce hydrocarbons in significant quantities. However, it needs a high overpotential and also produces other side products which are not as desirable. Among the factors affecting the product selectivity and efficiency of the Cu catalysts, the surface morphology plays a significant role[1-4]. Although it has been observed that increasing the roughness, which is a morphological parameter, is frequently associated with ethylene formation enhancement, the reason for this relationship is still under investigation[1, 4-5].

In this study, the effect of different nanoscale morphologies on the product selectivity was investigated. The Cu catalysts were synthesis by electropoishing and electrodeposition. We found a trend between electrochemical roughness and the faradaic efficiency (%FE) of methane and ethylene. By increasing the roughness from 1.6 to 7.8, the product selectivity changed from methane formation to ethylene formation. On the Cu/Cu 1.6 catalyst, we obtained 28%FE towards methane and 12%FE towards ethylene at -1.18V vs. RHE. In contrast, on the Cu/Cu 7.8 catalyst, 36%FE towards ethylene and 8%FE towards methane was observed at the same potential. Roughness follows the product selectivity frequently but not all time. We have found that electrochemical roughness can be similar on very different catalyst morphologies (Cu/Cu 1.6 and Cu/Cu 1.5) and result different product selectivities, illustrating more basic underlying effects are in play. We observed high selectivities toward ethylene on Cu rough catalysts that had higher current intensities on the surface of the catalyst. We have also observed that this current effect can be translated across length scale from nano to macro. We have found that by changing the electrode geometry from Cu flag shaped to Cu wire shaped, the product selectivity also can be switched between methane and ethylene. Moreover, it was observed that Cu wire had a higher catalytic activity compared to the Cu flag.

1. Tang, W.; Peterson, A. A.; Varela, A. S.; Jovanov, Z. P.; Bech, L.; Durand, W. J.; Dahl, S.; Norskov, J. K.; Chorkendorff, I., *Phys. Chem. Chem. Phys.* **2012,** *14* (1), 76-81.
2. Karaiskakis, A. N.; Biddinger, E. J., *Energy Technol.* **2017,** *5* (6), 901-910.
3. Reske, R.; Mistry, H.; Behafarid, F.; Roldan Cuenya, B.; Strasser, P., *Journal of the American Chemical Society* **2014,** *136* (19), 6978-6986.
4. Mistry, H.; Varela, A. S.; Bonifacio, C. S.; Zegkinoglou, I.; Sinev, I.; Choi, Y.-W.; Kisslinger, K.; Stach, E. A.; Yang, J. C.; Strasser, P.; Cuenya, B. R., *Nature Communications* **2016,** *7,* 12123.
5. Chen, C. S.; Handoko, A. D.; Wan, J. H.; Ma, L.; Ren, D.; Yeo, B. S., *Catalysis Science & Technology* **2015,** *5* (1), 161-168.

538172 Insights into the Electrocatalytic Conversion of CO_2 into CO, Ethylene, and Ethanol in Alkaline Media

Abstract:

Using CO_2 as a feedstock for the production of intermediates for chemical or fuel production such as formic acid, CO, ethylene, ethanol, and methanol is one of several approaches being explored to help reduce anthropogenic CO_2 emissions, while also reducing society's dependence on diminishing fossil fuel reserves. A range of active electrocatalysts for the selective reduction of CO_2 have been reported. We have explored both Ag-based and Au-based catalyst as well as micro-structured CuAg alloy films, respectively for the selective and energy efficient reduction of CO_2 to CO [1,2] and to ethylene and ethanol [3]. Partial current densities for CO production exceeding 150 mA/cm^2 at cell overpotentials less than 1V now can be achieved routinely at CO energetic efficiencies in the range of 50-60%. Similarly, high surface area CuAg alloy films exhibit Faradaic efficiency for C_2H_4 and C_2H_5OH production of 60% and 25%, respectively, at a cathode potential of just −0.7 V vs RHE and a total current density of 300 mA/cm^2, the highest levels of selectivity at high activity and low applied potential reported to date.

This talk will focus on various mechanistic insights of the CO_2 electroreduction process on these different catalysts, especially in the presence of alkaline electrolyte. For example, the onset cathode potentials, the kinetic isotope effect, and Tafel slopes indicate that the production of CO at low overpotentials in alkaline media is the result of a pH-independent rate-determining step (i.e., electron transfer) in contrast to a pH-dependent overall process. Furthermore, in situ Raman and electroanalysis studies suggest the origin of the high selectivity toward C2 products observed with CuAg alloy films to be a combined effect of the enhanced stabilization of the Cu_2O overlayer and the optimal availability of the CO intermediate due to the Ag incorporated in the alloy.

1. Gold Nanoparticles on Polymer-Wrapped Carbon Nanotubes: An Efficient and Selective Catalyst for the Electroreduction of CO_2, H.R.M. Jhong, C.E. Tornow, C. Kim, S. Verma, J. L. Oberst, P.S. Anderson, A.A. Gewirth, T. Fujigaya, N. Nakashima, P.J.A. Kenis, *ChemPhysChem*, **2017**, 18 (22), 3274-3279.
2. Insights into the Low Overpotential Electroreduction of CO_2 to CO on a Supported Gold Catalyst in an Alkaline Flow Electrolyzer, S. Verma, Y. Hamasaki, C. Kim, W. Huang, S. Lu, H.R.M. Jhong, A.A. Gewirth, T. Fujigaya, N. Nakashima, P.J.A. Kenis, *ACS Energy Lett.*, **2018**, 3, 193-198.
3. Nanoporous Copper–Silver Alloys by Additive-Controlled Electrodeposition for the Selective Electroreduction of CO_2 to Ethylene and Ethanol, T.T.H. Hoang, S. Verma, S. Ma, T.T. Fister, J. Timoshenko, A.I. Frenkel, P.J.A. Kenis, and A.A. Gewirth, *J. Am. Chem. Soc.*, **2018**, accepted.

538456 Insights on the Electrochemical Reduction of Carbon Dioxide Using Solid Oxide Electrolysis Cells

Abstract:

Extensive use of fossil fuels and consequential high levels of carbon dioxide (CO_2) emissions are major contemporary challenges.[1] The most direct way to mitigate this process is to activate reverse chemical pathways in which CO_2 is reduced into high-energy molecules.[2] The electrochemical reduction of CO_2 to CO has attracted increasing attention, since CO represents a valuable intermediate to the production of synthetic fuels using established processes, such as Fischer-Tropsch.[2, 3] One attractive approach to electrochemically reduce CO_2 to CO is the use of solid oxide electrolysis cells (SOECs)[4]. SOECs are solid-state electrochemical systems that, in principle, can facilitate the electrochemical reduction of CO_2 to CO with potentially very high rates due to favorable kinetics at high operating temperatures[2, 4]. While electrochemical reduction of CO_2 using SOECs offers a great deal of promise, these systems are still limited by the high activation overpotential losses induced by the sluggish CO_2 reduction kinetics at the cathode.[1, 4] Moreover, the current knowledge on the electrochemical reduction of CO_2 using SOECs is mainly limited to proof-of-concept experiments, demonstrating the viability of its operation. Lack in understanding of the governing mechanism for electrochemical reduction of CO_2 on SOECs, as well as, the catalytic role played by different cathode components (i.e., metal and the mixed metal oxides) make the optimization of the SOEC cathode activity very challenging. In the present work, we combine experimental and theoretical techniques to understand the chemical/electrochemical steps that govern CO_2 electrolysis on metals under SOEC conditions. Controlled electrochemical studies are used to investigate the role played by the metal and the mixed metal oxide cathode components toward the electrochemical reduction of CO_2. Structure/performance relations are developed in order to identify optimum cathode compositions for this process.

REFERENCES

[1] S.D. Ebbesen, S.H. Jensen, A. Hauch, M.B. Mogensen, High temperature electrolysis in alkaline cells, solid proton conducting cells, and solid oxide cells, Chemical reviews, 114 (2014) 10697-10734.

[2] X.-K. Gu, J.S.A. Carneiro, E. Nikolla, Heterogeneous electrocatalysts for CO2 reduction, in: Catalysis, 2017, pp. 94-121.

[3] K. Xie, Y. Zhang, G. Meng, J.T. Irvine, Direct synthesis of methane from CO 2/H2O in an oxygen-ion conducting solid oxide electrolyser, Energy & Environmental Science, 4 (2011) 2218-2222.

[4] X.-K. Gu, J.S. Carneiro, E. Nikolla, First-Principles Study of High Temperature CO2 Electrolysis on Transition Metal Electrocatalysts, Industrial & Engineering Chemistry Research, 56 (2017) 6155-6163.

542591 Electrochemical Conversion of Ammonia and Nitrogen for Sustainable Food-Energy-Water

Abstract:

Electrochemical Conversion of Ammonia and Nitrogen for Sustainable Food-Energy-Water

Gerardine G. Botte

Center for Electrochemical Engineering Research

Department of Chemical and Biomolecular Engineering

Russ College of Engineering and Technology

Ohio University, Athens

OH 45701

The production of reactive nitrogen fertilizer via the Haber-Bosch (HB) process, combined with industrialization and intensification of agriculture, has created an abundant global food supply to support population growth, but has doubled the amount of reactive nitrogen compared to pre-industrial levels.[1] With those perspectives, ammonia can be seen as an important chemical of use or as an environmental challenge. Ammonia emissions into air (ambient ammonia) and water represent an environmental challenge. Ambient ammonia not only contributes to inorganic PM2.5 (particulate matter with an aerodynamic diameter of less than 2.5 Î¼ m) directly but also plays an important role in secondary organic aerosol formation by interacting with gaseous phase organic acids and forming condensable salts. In addition, ammonia emissions in water are associated with environmental problems such as algae bloom. Various industries and other operations are considered ammonia emitters, but the ones related to sustainable food production such as livestock management and fertilizer application are among the major contributors accounting for 60-75% of the ammonia emissions.[2-3]

On the other hand, ammonia produced commercially via the HB process is usually known as one of the greatest inventions of the 20th Century. The HB process made it possible for the first time to produce synthetic fertilizers and produce sufficient food for the Earthâ€™s growing population. However, the HB process requires high temperatures and pressures and as a result it can only be done in large plants which consume large amounts of power and generate large quantities of CO_2 from the fuels used to power the process, such as natural gas. It is estimated that the HB process accounts for nearly 1% of the entire global power consumption. Due to the sheer scale of the HB process, distribution from the point of production to the point of use becomes an additional carbon and energy burden. Alternative methods of manufacturing ammonia that enable small scale, distributed generation at the point of use using renewable energy and sustainable feedstocks could have a great impact on global CO_2 emissions.

A closed loop approach that includes producing fertilizer at the point of use, applying nitrogen fertilizer as close to the time that crops needs it, and recovering and recycling nitrogen fertilizer can contribute to address the challenges related to ammonia emissions, while decreasing water and energy consumption. Electrochemical conversion of nitrogen and ammonia presents an excellent platform to achieve this closed loop approach. Dr. Botte and members of her research group at the Center for Electrochemical Engineering Research (CEER) are working on the electrochemical conversion[4]/synthesis of ammonia[5] in alkaline media. In this talk, recent advances in the platform process for ammonia and nitrogen removal/recycling and ammonia synthesis will be discussed as well as their application for sustainable food, energy, and water.

References:

1. Grand Challenges - Manage the Nitrogen Cycle [WWW Document], n.d. URL http://www.engineeringchallenges.org/9132.aspx (accessed 5.28.18).

2. G. E. Mansell, "An Improved Ammonia Inventory for the WRAP Domain (Final Report)", ENVIRON International Corporation, Report No. 415.899.0700, March 7 (2005).

3. FAO. Global estimates of gaseous emissions of NH_3, NO, and N_2O from agricultural land, International Fertilizer, Industry Association. Rome. 2001.

4. Botte, G. G., Cooper, M. and Vitse, F., (2009) Electro-catalysts for the oxidation of ammonia in alkaline Media, U.S. Patent No. 7,485,211.

5. B. L. Sheets, G. G. Botte, Electrochemical nitrogen reduction to ammonia under mild conditions enabled by a polymer gel electrolyte, Chemical Communications 54 (2018), 4250-4253.

542589 Encapsulation and Nanoparticle Formation for "Non-Standard" Applications

Abstract:

Encapsulation and nanoparticle formation for "non-standard" applications

Prof. Robert K. Prud'homme[1], Prof. Rodney D. Priestley[1], Leslie Shor[2] ,Douglas Scott[1], Jie Feng[1]

[1] Princeton University, Department of Chemical and Biological Engineering, Princeton University, Princeton NJ 08544

[2] University of Connecticut, Department of Chemical and Biomolecular Engineering, Storrs, CT 06269-3222

Presenting author: Robert K. Prud'homme, prudhomm@princeton.edu

Traditionally, encapsulation and nanoparticles have been associated with high value applications such as drug delivery in medicine. The two main reasons have been the high cost of nanoparticle assembly processes for complex nanoparticles that encapsulate and release actives in a controlled way, and the fact that encapsulation technologies for larger scale macro-particles do not translate to nano-scale particles. Recent advances in low-cost nanoparticle assembly present the opportunity to open up new applications. First, we present an overview of encapsulation and release technologies. Then we provide a review of the rapid-precipitation technologies we have developed that rely on the control of the fluid dynamics, diffusion-limited aggregation, and steric stabilization of nanoparticles. The processes are scaleable, low cost, and produce nanoparticles from 50-400 nm. Several novel applications will be presented: (1) nanoparticle delivery of agricultural actives s to plant roots by protist active transport, (2) triggered indicator inks for sterilization processes, (3) "green chemistry" encapsulation of personal care compounds using purely aqueous processing.

529088 Techno-Economic Uncertainty Quantification of Algal-Derived Biocrude Via Hydrothermal Liquefaction Process

Abstract:

Using microalgae for fuel production via hydrothermal liquefaction (HTL) has received increasing attention in recent years because of its high biocrude productivity and quality, cultivation flexibility, and potential for large scale operation. Most of the current knowledge on algae HTL is based on laboratory benchtop findings and/or simulation models. Additionally the conversion mechanism has not been well studied, while the simplified yield correlations or kinetics models were usually developed based on small batch equipment. The lack of commercial experience and knowledge gaps include, but not limited to, feedstock variability, reaction mechanism, overall reaction heat, optimal operating condition, reactor design, thermal-hydraulic properties of algae slurry, and nutrient recycling. These gaps can be addressed through techno-economic uncertainty quantification as a promising method to understand the process economics and potential risks. However, to the best of authors' knowledge, vary limited stochastic simulations in the open literature considered both rigorous process model with parameter uncertainty and uncertainty regarding to economic assumptions, especially for algae HTL processes.

With the above motivations, techno-economic uncertainty quantification is conducted for algae HTL processes based on abundant operating data of PNNL's continuous HTL reactor systems. Particularly, a "component additivity" model is developed to predict the yields and qualities of algae HTL process of a variety of algal biomass with model uncertainty quantified. Monte Carlo simulations are conducted using a rigorous process and economic model developed in Aspen Plus and Excel. Oracle Crystal Ball is used to generate semi-random Monte Carlo samples based on pre-determined probability distributions of model inputs, and perform statistical analysis of the results. Aspen Simulation Workbook and Excel VBA Macro are used to build dynamic links between those software. This presentation will focus on the framework of techno-economic uncertainty quantification, "component additivity" model of algae HTL reactor, plant-wide process and economic models, and the techno-economic uncertainties of algae HTL processes.

529405 Using Produced Water to Grow Microalgae

Abstract:

Produced water â€“ also known as hydraulic fracking fluid, is a byproduct of the oil and gas production process and is water that is naturally contained in sedimentary formations. This water has a salinity ranging from as low as 1,000 ppm to greater than 100,000 ppm of total dissolved solids (TDS) and has bicarbonate, other essential micronutrients.

The goal of our research is to evaluate produced water as a potential growth medium for microalgae. Our preliminary lab scale studies with the microalgae species *Chlamydomonas reinhardtii, Dunaliela sp. and Nannochloropsis salina*, which are known to grow in saline conditions have shown promise. Pre-treatment of produced water was performed by means of centrifugation to remove solids along with vacuum filtration to remove suspended particles (>5 microns). Experiments were then conducted by growing algae cultures in varying proportions of the diluted, pre-treated PW and a standard growth medium specific to each species. The growth performance in each of the cultures was monitored by the measurement of absorbance at 750 nm each day. Initial growth of algae with low concentrations of produced water was similar to the standard growth medium, however, higher produced water concentrations became growth limiting. These studies also indicate that the algae can thrive in produced water even without any necessity of further treatment. The algae-based biomass grown in the produced water could be used as a feedstock to produce biofuels.

532708 Effects of Impurities in Two-Step Vs. One-Step Hydroprocessing of Algae Oils

Abstract:

Microbial lipids are a promising precursor to renewable diesel fuels, but catalytic hydroprocessing of the crude lipid extracts is an under-studied operation. In particular, impurities in the crude extracts can lead to catalyst deactivation. Hydroprocessing of these lipids to green diesel fuel typically comprises deoxygenation (DO) and hydroisomerization (HI) chemistry, and can be conducted in a two-step or one-step configuration. We have shown that in a two-step configuration using Pd/C for DO and Pt/SAPO-11 for HI, a fuel with acceptable cold weather and boiling range properties can be produced without lipid cleanup steps but under severe conditions. In contrast, a Pt/SAPO-11 catalyst is deactivated in one-step DO/HI. However, multiple oil bleaching approaches can remove nitrogen-containing impurities that may be causing the deactivation.

522805 Catalytic Upgrading of Pyrolysis Vapor at Bench Scale with Platinum on Titania

Abstract:

Vapor phase upgrading can produce a liquid of lower oxygen and higher energy density than can fast pyrolysis alone. Therefore, it is anticipated that catalytic pyrolysis followed by a single stage of hydrotreating will be a more cost-effective route to hydrocarbon fuels than fast pyrolysis with multiple-stage or more-severe hydrotreating.

To this end, a bench-scale bubbling fluidized bed pyrolyzer was operated at 150 g/h feed rate of pine flour with a 100 g fixed-bed catalytic upgrader. A 0.5% Pt/TiO_2 catalyst produced a lower oxygen (< 20%) liquid at 40% carbon efficiency at 400°C and biomass-to-catalyst ratio (B:C) of 3 for numerous cycles. The catalyst was regenerated and reused by oxidation with air and reduction with hydrogen. The oxidation and reduction times were reduced to 2 h and 1 h respectively without compromising performance or catalyst temperature control. Increasing the pyrolysis temperature from 500°C to 550°C allowed operation at B:C greater than 3 at comparable performance. Performance was generally similar to 2% Pt/TiO_2 but at reduced catalyst cost. Projected costs are < $4/gallon of gasoline blendstock. Responses to other parameter changes and prospects for further improvements will be discussed.

531654 Catalytic Pyrolysis of Chitin over H-ZSM-5

Abstract:

Approximately 6â€“8 million tons of waste crustacean shells (i.e., crab, lobster, and shrimp shells) are produced annually from seafood processing facilities around the world. Among them, more than 66,000 tons of waste lobster shells are produced in New England alone. More than 40% of crustacean waste by mass consists of chitin, a long chain polymer of N-acetylglucosamine. It is also the second most abundant natural biopolymer in the world behind cellulose. In this work, catalytic pyrolysis of chitin over H-ZSM-5 solid acid catalyst was conducted using a custom-made batch reactor system. The effects of catalyst loading, reaction temperature, and reaction time were systematically studied. Our experiments show that three distinct phases of pyrolysis products exist: solid char, solid condensate, and volatile compounds. Analysis of molecular composition of non-soluble solid condensate suggests that the condensate likely consists of nonvolatile, insoluble molecules with amine, acid, or hydroxyl functional groups. Increase in catalyst loading and a reaction temperature above 450 °C both favor deoxygenation, resulting in increase in solid condensate yields and its oxygen content. On the contrary, reduced char yields was observed at higher catalyst loading and higher reaction temperature, with the elemental composition of char being more carbon and nitrogen rich and oxygen deficient. No significant effect was found for reaction time longer than 15 min, suggesting that chitin pyrolysis is a rapid process and our results represent equilibrium compositions. This work demonstrates that catalytic pyrolysis of chitin could be an energy efficient pathway to promote the production of nitrogen containing char and value added nitrogen-containing products from crustacean seafood waste.

531330 Assessment of Hydrothermal Liquefaction Oil with Catalytic Upgrading for Renewable Fuel and Chemical Production

Abstract:

Lignin is a readily available byproduct of the Kraft pulping process, and may be processed via hydrothermal liquefaction (HTL) to produce a bio-oil suitable for co-feeding into a petroleum refinery hydrotreatment unit. HTL of lignin is performed in near-critical water and, in addition to the bio-oil, produces an aqueous organic and solid char phase. The aqueous organics are primarily phenolics, which may then be converted into valuable co-products via liquid-liquid extraction and hydrotreatment to BTEX compounds. Three technological scenarios were developed: a current technology case, a state of the art research case, and an optimal case based on product targets provided by refiners. For a large Kraft pulp mill (400 metric tons/day of dry lignin), a renewable fuel production of 65-70 million liters per year, with capital costs of $114-125 million and a final per liter cost of $0.41-0.44 were estimated. The BTEX co-product yield ranged from 16.8-18.0 million liters per year. An economic analysis of the process revealed that the hydrotreatment steps have the highest installed capital costs, while the liquid-liquid extraction process is the largest operating cost. The economic analysis also yielded the minimum selling price of the biofuel as $3.52-3.86 per gallon, and the MSP of BTEX as $1.65-2.00 per liter. At the current level of technology, the selling price of the biofuel and co-production of BTEX do not offset the cost of production. Improved technology to further lower the oxygen content of bio-oil and decrease both capital and operating costs are needed to make HTL-based fuels competitive with fossil fuel-based options.

523665 Fatty Acid Methyl Ester Production Via Ferric Sulfate Catalyzed Interesterification

Abstract:

While biodiesel production is on the rise, so too is the production of the low-value product, glycerol. Interesterification is an alternative reaction that produces fatty acid methyl esters (FAME) and avoids the production of glycerol. In this work, ferric sulfate is used as a potential heterogeneous catalyst for interesterification of model triglyceride, triolein, using methyl acetate. Reaction conditions were optimized by varying reaction temperature, methyl acetate to oil mole ratio (MAOMR), catalyst loading, as well as FAME as co-solvents. Additionally, this work also evaluated the effect of typical triglyceride feedstock contaminants, water and free fatty acids (FFA). The optimal reaction conditions were found to be 120 °C, 20:1 MAOMR and a mass loading of 7.5% ferric sulfate with yields up to 83%. The addition of FAME and water in the system is shown to have potential benefits on reaction rate and yield while FFA addition had mildly negative impacts on initial reaction rate. The results in this work show relatively high yields at mild temperatures and reagents loadings and could be a promising means of alternative FAME production.

537809 Impact of Organic and Inorganic Impurities on Catalyst Performance and Finished Product in the Production of Acrylonitrile from Biomass

Abstract:

Impact of organic and inorganic impurities on catalyst performance and finished product in the production of acrylonitrile from biomass.

Swanand Tupsakhare, Zora Govedarica, Jadid E. Samad, and Amit Goyal

Energy and Environment Division

Southern Research, 5201 International Drive, Durham, NC 27712

Ninety percent of the worldâ€™s carbon fiber production utilizes acrylonitrile as a raw material, growing at 11 to 18 percent per year. Due to the high growth rate of carbon fiber production, any reduction in GHG originating from Acrylonitrile production will be highly impactful. Conventionally, acrylonitrile is produced from propylene ammoxidation and highly dependent on volatility in propylene markets. The current method makes use of an alternative approach using sugars obtained from biomass feedstock. Southern Research in partnership (DOE/EERE program) has developed a method to provide renewable acrylonitrile produced from biomass-derived second generation sugars.

The process produces a direct, drop-in replacement for petroleum acrylonitrile that is both economically competitive and sustainable, lowering greenhouse gas (GHG) emissions by up to 40 percent. The proposed process for producing acrylonitrile from biomass-derived sugars has the promise of changing the economics and environmental footprint of this important specialty chemical. An innovative, thermo-catalytic process has been developed that converts second generation sugars obtained from biomass to acrylonitrile. The objective of this work is to expand the range of biomass feedstock that can be used and to understand how the process is affected by impurities that change when different types of biomass and different biomass-to-sugar (hydrolysis) processes are used. Experimental data will allow better prediction and improvement in the overall cost and application areas for carbon fibers.

This paper will present results for - top biomass candidates ideally suited for acrylonitrile production are screened based on the impurity levels and imposed effects on the process. Organic and inorganic impurities found in wheat straw, corn stover biomass, sugar cane bagasse, sorghum straw and hybrid poplar were identified and tested in lab (using commercial grade sugars) for their impact on catalyst performance, catalyst selectivity to desired products and the concentration in finished products. The impurity levels tested were chosen to be the highest values reported in the literature to allow the applicability of the results to a wide range of biomass sugars.

539749 Integrating Non-Precious Metal H_2 Evolution Catalysts into Water Electrolyzers and Photoelectrochemical Water-Splitting Devices

Abstract:

The past decade has produced significant advances in the development o f non-precious metal catalysts for electrochemical H2 evolution. This paper will describe recent efforts in this area, and importantly, translation to water electrolyzers and devices for photoelectrochemical water-splitting.

535226 When Electrocatalysis Matters and When It Does Not: Unexpected Observations in Water Electrolysis and Flow Battery Energy Storage

Abstract:

This presentation will summarize ongoing projects in our lab on applications of electrocatalysis in hydrogen production and flow battery energy storage. We are working to advance these technologies through a balanced effort in scientific inquiry (e.g., synthetic methods, structure-function relationships) and engineering design (e.g., device specification, performance metrics). A persistent theme in our work thus far has been ambitious experimental targets underpinned by naïve assumptions, where the process of stepping back to re-evaluate our assumptions has nevertheless led to intriguing new conclusions.

In the area of water electrolysis, we set out to improve on the performance and stability of Ni–Mo alloy hydrogen evolving electrocatalysts with an eye toward their use in alkaline electrolyzers. Our work focused initially on understanding composition and reaction mechanisms in nanostructured Ni–Mo alloys while we also worked to develop catalyst ink formulations that would be compatible with fabrication of membrane electrode assemblies. Ironically, the latter studies led to unexpected fundamental insights, where observed reaction rates were not limited by kinetics or mass transfer, but rather by interfacial electronic conductivity even in this nominally metallic catalyst material. Incorporating carbon black mitigated the conductivity limitation, resulting in markedly improved mass-specific activity.

In the area of flow battery energy storage, our initial efforts to pursue engineering design specifications for integrated solar-driven battery systems were quickly hampered by the remarkable lack of consensus on intrinsic rates of interfacial electron transfer for known flow battery redox couples. Thus, we have now developed analytical protocols (adapted from the electrolysis and fuel cell literature) for benchmarking the kinetics of flow battery electrode-electrolyte combinations. Results to date suggest that electrocatalysis plays a vital role in dictating the overpotential performance of essentially all known flow battery active materials—even those based on apparently straightforward 1-electron transfer reactions. We are working to modify these electroanalytical protocols for direct deployment in operational flow battery devices.

530509 Understanding the Role of Adsorbed Hydroxide in Reversible Hydrogen Reactions

Abstract:

The kinetics of HER/HOR on Pt and other catalysts is facile at low pH, however, reaction rates are found to be several orders of magnitude slower in base. The explanation for this unusual pH dependence has been a matter of debate within the scientific community. Traditionally, hydrogen binding energy (HBE) has been used as the lone activity descriptor, independent of electrolyte pH [1]. This alone, however, is not sufficient to describe the pH dependence of the HER/HOR activity on metals other than Pt. Markovic et al. have proposed that hydroxide adsorption is necessary to facilitate water recombination/dissociation in base [2].

In this work, our aim is to determine if adsorbed hydroxide plays an active or spectator role in HER/HOR kinetics combining electroanalytical techniques with microkinetic modeling. We use expressions for time-dependent current and voltage as a function of hydrogen and hydroxide binding strength for the indirect (hydroxide-mediated) and direct (hydroxide-as-spectator) Volmer steps in base to simulate cyclic voltammograms. The simulated results are compared to experimental cyclic voltammograms of H-UPD on Pt(110). The free energy of hydroxide adsorption is controlled experimentally by the electrolyte cation via changes in cation solvation strength [3]. Varying the sweep rate results in greater peak separation that can be used to extract rate constants for the adsorption reaction via traditional electroanalysis [4].

Our computations show that at both slow and fast scan rates, either the direct or indirect mechanism can describe the peak potential splitting and peak current. However, hydroxide binding strength affects the mechanisms differently. In the indirect mechanism, stronger hydroxide binding reduces overpotential and, therefore, peak splitting. In the direct mechanism, stronger hydroxide binding decreases available surface sites, leading to lower exchange current densities and greater overpotential. Comparison with experiments show that from 0.1M KOH to 0.1M LiOH the peak potential location decreases while peak potential splitting increases. This trend shows that stronger hydroxide binding has detrimental effects on H-UPD kinetics, and that the direct mechanism is therefore more likely to dominate observed behavior.

Our results strongly suggest that weakening OH binding strength should increase availability of sites for H adsorption. And therefore, oxophilicity is not an accurate descriptor for alkaline hydrogen electrocatalysts and other parameters, such as water orientation and non-covalent interactions, might be playing a greater role in the overall activity.

References:

(1) Durst et al. Energy Environ. Sci. 2014, 2255.

(2) Strmcnik et al, Nat. Chem. 2013, 1.

(3) Danilovic, N. et al. Electrocatalysis 3, 2012, 221.

(4) Angerstein-Kozlowska et al. J. Electroanal. Chem. Interfacial Electrochem. 1979, 1.

532712 Electrocatalysis at Buried Interfaces

Abstract:

Electrocatalysis underlies many emerging clean energy technologies, such as fuel cells and electrolyzers. However, continued improvements in electrocatalyst activity, durability, and selectivity must be achieved in order for these technologies to make meaningful contributions to a sustainable energy future. In this presentation, I will describe unconventional electrocatalyst architectures for which metallic electrocatalysts are encapsulated by ultra-thin overlayers made of permeable oxides. These oxide layers are synthesized with a room temperature process and deposited as uniform and continuous layers onto thin films of Pt, one of the most commonly used electrocatalyst materials. Through various electroanalytical measurements, we show that these oxide overlayers can be selectively permeable to certain electroactive species and thereby enable efficient and selective electrocatalysis at the buried interface between the oxide overlayer and active catalyst. For this reason, these electrocatalysts may be referred to as membrane-coated electrocatalysts (MCECs). By systematically varying the thickness of silicon oxide overlayers in MCECs, we analyze their transport properties and the influence of mass transfer on the performance of thin film hydrogen evolution electrocatalysts. This talk will also describe characterization of the electroactive buried interface and discuss challenges and opportunities for controlling reaction pathways through modification of the physical, chemical, and/or structural properties at the interface. Importantly, this work highlights the potential of oxide-encapsulated electrocatalysts to serve as a tunable platform for efficient and stable electrocatalysis and/or photoelectrocatalysis of a variety of electrochemical reactions.

References

[1] N. Y. Labrador, X. Li, Y. Liu, J. T. Koberstein, R. Wang, H. Tan, T. P. Moffat, and D. V. Esposito, Nano Letters, vol. 16, 6452-6459, 2016.

[2] D.V. Esposito, *ACS Catalysis,* 2018, vol. 8, pp 457â€“465.

[3] N. Y. Labrador, E. L. Songcuan, C. De Silva, Han Chen, Sophia Kurdziel, Ranjith K. Ramachandran, Christophe Detavernier, D.V. Esposito, *ACS Catalysis, vol. 8, 1767â€“1778,* 2018.

528647 Density-Functional-Theory Studies of Face-Centered-Cubic Tungsten Carbide and Pt Core-Shell Nanoparticles Catalysts for the Hydrogen Evolution Reaction

Abstract:

Platinum is among the best electrocatalysts for the hydrogen evolution reaction (HER)—a potentially fossil-free route for hydrogen production—but the scarcity and high cost of this noble metal catalyst makes it economically unviable at commercial scales. Reduction in Pt loading and overall catalyst cost, without sacrificing catalytic activity, can be achieved by the synthesis of highly stable core-shell nanoparticles of transition-metal carbides and Pt. A large body of prior theoretical work has focused on the hexagonal phase of WC (α-WC) as a support material for Pt monolayers (ML), while the face-centered-cubic phase of WC (β-WC) has attracted lesser attention. Motivated by the recent preparation of stable β-$Ti_xW_{1-x}C$@Pt core-shell nanoparticles, we employ density-functional theory (DFT) calculations to study and contrast the suitability of the α and β phases of WC as support (core) materials for Pt ML (shells). We examine the thermodynamic stability of Pt MLs on α- and β-WC surfaces, carefully accounting for the delicate balance between epitaxial mismatch strains and chemical bonding between cores and shells. Furthermore, we study the effects of alloying β-WC with Ti to modulate the stability and electronic structure of the core-shell structures. Based on our thermodynamic studies, we find that β-WC is a better support material for Pt MLs than α-WC and that alloying with Ti improves the stability of Pt/β-WC interfaces without adversely affecting the activity of the Pt shell. We compare the electronic structures of Pt MLs supported on α-WC and β-$Ti_xW_{1-x}C$ surfaces and compare their activities for HER using the hydrogen binding energy (HBE) as a descriptor of catalytic activity. Of the various cases studied, we find that 2ML Pt supported on β-WC (111) and β-$Ti_{12.5}W_{87.5}C$ (111) show higher HER activities than Pt (111) making these core-shell structures very promising candidates for HER electrocatalysts; considering the improved thermodynamic stability of β-$Ti_{12.5}W_{87.5}C$ over β-WC, the former is predicted to be a particularly suitable core material for Pt shells. Overall, our results provide detailed insight into experimental observations of the high catalytic activity of β-$Ti_xW_{1-x}C$@Pt core-shell nanoparticles.

535328 Kinetic Investigation of Nickel-Iron Layered Double Hydroxide for Hydrogen Evolution in an Alkaline Electrolyte

Abstract:

Nickel-iron layered double hydroxide (NiFe LDH) and other transition metal layered materials have recently gained attention for their simple preparation methods, earth abundance, low cost, and most importantly, high activity toward hydrogen evolution (HER) and oxygen evolution (OER) reactions [1],[2]. In this work, we investigate the kinetics of an active NiFe LDH toward HER in an alkaline electrolyte. The mechanism of HER starts with electrochemical adsorption of a water molecule onto an active site on the cathode to produce the adsorbed species H (a step that is typically referred to as the Volmer step). The hydrogen gas can then be produced from two competing reaction paths: a chemical recombination of the adsorbed species H (Tafel step), and an electrochemical desorption step (Heyrovsky step).

Our NiFe LDH films were deposited on Ni foam substrates via a simple one-step hydrothermal synthesis route in an autoclave reactor maintained at a temperature of 155 C for 12 hr [1]. The films were then used as both OER and HER electrodes in a three-electrode electrochemical cell set up with 1 M KOH solution as an electrolyte.

The deposited film demonstrated an overpotential of 247 mV at a current density of 10 mA/cm^2 toward the HER. Linear sweep voltammetry (LSV) data were obtained at a scan rate of 1 mV/s between the cathodic potential range of -0.9 to -1.7 V vs Hg/HgO reference electrode. The calculated cathodic overpotential measurements were corrected for the voltage drop corresponding to the uncompensated solution resistance. A Tafel plot constructed from the data shows a Tafel slope of -38.4 mV/dec, which indicates that the HER mechanism is controlled by either the Herovsky or Tafel steps [3]. In addition, electrochemical impedance spectroscopy (EIS) measurements were taken at different potentials within the LSV potential range. The EIS data were then fitted to an Armstrong and Henderson equivalent electric circuit using both Z-view software and EIS Spectrum Analyser to obtain the equivalent circuit components.

Using a least squares method, a theoretical model relating current density to the electrochemical reaction rates [4] was fitted with the LSV experimental results (Figure 1 attached). The chemical reaction rate constants for the individual HER mechanism steps - Tafel, Volmer, and Herovsky - were obtained as a result of the fitting procedure. The resulting kinetic model shows a good agreement between theoretically calculated Faradaic resistance and experimental EIS results (Figure 2 attached).

The kinetics data revealed the Herovsky step as the rate determining step, which is in agreement of the calculated Tafel slope. Furthermore, the kinetics data show a dependence of the HER mechanism on the applied potential. For overpotential values between 0 and -0.4 V (corresponding to an applied potential between -0.9 to -1.6 V vs Hg/HgO), the HER mechanism is a sequence starting with the Volmer step, followed by the Tafel step. At higher cathodic overpotential values, the mechanism starts with a Volmer step followed by a rate controlling Herovsky step, with negligible contribution of the Tafel step. It was also observed that in higher cathodic overpotential region, the Tafel step reaction rate is independent of applied potential. This can be attributed to the surface coverage of the adsorbed intermediate reaching a constant high coverage value in this potential region.

Figure 1: Experimental and simulated results of current vs overpotential

Figure 2: Experimental and simulated results of Faradaic resistance vs overpotential

References:

[1] Lu, Z. et al. (2014). Three-dimensional NiFe layered double hydroxide film for high-efficiency oxygen evolution reaction. *Chemical Communications (Cambridge, England), 50*(49), 6479-82. doi:10.1039/c4cc01625d

[2] Luo, J. et al. (2014). Water photolysis at 12.3% efficiency via perovskite photovoltaics and Earth-abundant catalysts. *Science (New York, N.Y.), 345*(6204), 1593-6. doi:10.1126/science.1258307

[3] Zhu, Y., Liu, T., Li, L., Song, S., & Ding, R. (2018). Nickel-based electrodes as catalysts for hydrogen evolution reaction in alkaline media. *Ionics : International Journal of Ionicsthe Science and Technology of Ionic Motion, 24*(4), 1121-1127. doi:10.1007/s11581-017-2270-z

[4] Krstaji, N., Popovi, M., Grgur, B., Vojnovi, M., & Sepa, D. (2001). On the kinetics of the hydrogen evolution reaction on nickel in alkaline solution Part I. The mechanism. *J. of Electroanalytical Chemistry, 512*(1-2), 16-26.

537637 $BiVO_4/WO_3$ Photoanodes for Chloride Oxidation with Simultaneous H_2 Production

Abstract:

The most actively pursued application of photoelectrocatalytic fuel production is the decomposition of water to make hydrogen and oxygen (a.k.a. solar water splitting). Although the physical phenomena are reasonably well understood, the hurdle with solar water splitting is the development of a commercially a cost-effective system that can produce H_2 fuel with an overall energy efficiency of at least 10% or higher for a sustained period of time. One reason water splitting is expensive, is one half of the reaction is the 4 electron oxidation of water which is kinetically sluggish and results in the production of O_2, which has limited economic value. Electrooxidation of chloride ions which are increasingly abundant could provide an attractive, cost-effective option to water oxidation.

The focus of the current work is to carry out chloride oxidation in acidic pH (pH 1) using nanoporous bismuth vanadate ($BiVO_4$) as the light absorber, and use electrons provided from the oxidation step to generate hydrogen at the dark cathode. $BiVO_4$ has been widely regarded as a promising low-cost photoanode for solar to chemical energy conversion. However, most of the demonstrated work with $BiVO_4$ so-far has been limited to neutral or slightly basic pH, where $BiVO_4$ has been shown to be chemically and photoelectrochemically stable. For chloride oxidation, the pH of the electrolyte should be less than 3 to obtain high chlorine production yield. To address the above challenge, we report on a method to electrodeposit tens of nanometer thick amorphous tungsten oxide (WO_3) films on top of nanoporous crystalline $BiVO_4$ electrodes, and benchmark its activity for Cl^- oxidation.

525389 On the Combustion Synthesis and Characterization of $Ga_XZn_{1-X}O_yN_{1-Y}$ for Water Splitting Applications

Abstract:

Utilization of solar energy is a necessary path forward to deal with the looming energy crisis and worldwide effects of fossil fuel usage. A key component of this utilization is how to store excess solar energy for use on demand. One such method is splitting water into hydrogen and oxygen, both of which can be stored and either burned as fuel or used as feedstock components in industrial processes to synthesize high-value materials. Metal oxynitrides have shown great potential as photocatalysts for overall water splitting applications. Incorporation of nitrogen into metal oxides narrows the band gap to enable absorption of light in the visible spectrum. The typical synthesis of a metal oxynitride, however, requires annealing temperatures at or above 900 ^{o}C and nitridation by flowing pure ammonia gas for anywhere from 1 to 110 hours. Such harsh synthesis conditions can limit the discovery of new candidate materials.

We have prepared an array of gallium zinc oxynitrides ($Ga_xZn_{1-x}O_yN_{1-y}$) as a model system using a combustion synthesis method. Gallium zinc oxynitrides can be synthesized in 30 minutes in an air atmosphere at temperatures ranging from 400-750 ^{o}C. XPS analysis reveals that nearly 50% nitrogen substitution in the metal oxide lattice can be achieved, dependent on the ratio of gallium and zinc nitrates used as precursors. UV-visible spectroscopy also demonstrates that the band gap of the $Ga_xZn_{1-x}O_yN_{1-y}$ materials can be controlled simply by altering the precursor ratio.

The as-synthesized products are photoactive under both UV-Visible and visible-only irradiation with and without hole scavengers present and without any co-catalysts. More importantly, photocurrent is measured in a 2-electrode setup with only pure water present. This presentation will discuss the physical and photoelectrochemical characterization of these materials to confirm successful synthesis, the current issues we are facing, paths forward to address these issues, and the ultimate goal of a high-throughput synthetic methodology.

532499 Circular Economy Methods of Preparing Fragrance Compounds

Abstract:

International Flavors & Fragrances Inc. (IFF) is a leading global creator of flavors and fragrances for a wide variety of consumer products. Our research and development processes are driven by the application of circular principles in the very early stages of design. This approach eliminates the traditional "take-make-dispose" model in favor of one that is restorative and regenerative by design – it enables ingredients to come from nature and return to nature or be repurposed in a circular and continually sustainable way.

One way in which we apply circular design thinking is through our use of green chemistry, which has long been an embedded part of molecule discovery and manufacturing processes at IFF. Green chemistry aims to increase "naturalness" in products and processes by mimicking chemistry in nature, avoiding corrosive or hazardous materials, minimizing waste and improving process energy efficiency. Example processes illustrate how we utilize circular design principles and green engineering technologies to enhance the sustainability of our new and existing fragrance compounds.

534550 Biosynthesis of highly stable silver nanoparticles by two distinct strains ofÂ *Chlamydomonas reinhardtii*

Abstract:

Optimizing process parameters for stable inorganic nanoparticle production in a Multi-Vessel Reactor system using microalgae

Ashiqur Rahman[1], Tushar Nemade[1], Shishir Kumar[1], Adarsh Bafana[1], Si Amar Dahoumane[2], Clayton Jeffryes[1,3*]

[1]Nanobiomaterials and Bioprocessing Laboratory (NABLAB), Dan F. Smith Department of Chemical Engineering, Lamar University, Beaumont, TX 77710.

[2]School of Life Science and Biotechnology, Yachay Tech University, Hacienda San Jose, San Miguel de Urcuquí, Ecuador.

[3]Center for Advances in Water & Air Quality, Lamar University, 211 Redbird Ln. Box 10888, Beaumont, TX 77710-0088, USA

[*]Corresponding author: cjeffryes@lamar.edu

Inorganic nanoparticles have many applications, including in catalysis, drug delivery and in biomedical imaging, diagnosis and therapy. Microalgal biosynthesis has been regarded as a potential process for producing unique types of inorganic nanoparticles. However, the continuous, self-sustainable mass production of algal-synthesized NPs remains elusive due to a lack of knowledge about process parameters. This work uses a Multi-Vessel Reactor (MVR) system to optimize temperatures and precursor concentrations for such biosynthesis processes. This reactor can hold up to 34 mini-photobioreactors (PBR), each provided with air flow to increase mixing and gas exchange (mainly for the exchange of oxygen and carbon dioxide) and to keep the algae suspended. In this work silver nanoparticles (AgNPs) were produced using whole, living cell cultures of *Chlamydomonas reinhardtii*. Surface Plasmon Resonance (SPR) excitation in the range of 400–430 nm confirmed the synthesis of AgNPs. Experiments were done at an approximate photon flux of 93 ± 7 μmol m^{-2} s^{-1} with temperatures from 17 °C to 38 °C and $AgNO_3$ concentration ranging from 0.0125 mM to 1.250 mM. The results showed that the maximum synthesis of AgNPs took place at 22 °C. The reaction kinetics, at all concentrations, showed that AgNPs formed in proportion to Ag^+ concentration indicating that the availability of biological reducing equivalents were not limiting. The morphological study revealed that the AgNPs were mostly spherical in shape with mean particle size of 17 ± 13 nm (n = ~250). The results from this study can be used to further evaluate the parameters that will lead to a scalable, fully automated, and permanent photobioreactor for the continuous production of different biogenic inorganic nanoparticles.

Keywords: Multi-Vessel Reactor, temperature, kinetics, concentration.

537388 Catalytic Gasification for Waste Management: Selectivity of Oxidation Reactions for Model Polymers

Abstract:

This research seeks to provide a characterization of catalytic gasification reaction mechanisms for different polymers and catalyst types. While catalytic gasification of waste polymers has significance in a variety of engineering applications, it is of particular relevance to in-situ resource utilization (ISRU) and waste management in space exploration beyond low earth orbit (LEO).

NASA Glenn Research Center seeks solutions to the logistical costs of transporting waste back to Earth as well as the need for propellant generation technologies that can be operated on the lunar surface, an asteroid, or on the International Space Station. Low-to-mid temperature catalytic gasification presents a potential solution to both of these challenges. The carbon content of the plastics, paper, and nylon often disposed of on the ISS can serve as the raw material for the generation of fuel (methane). Long missions beyond LEO will not be possible without technologies that can reduce the dependence on logistic resupply as well as eliminate hundreds of kilograms of trash.

Catalytic gasification permits the low-temperature conversion of polymers to a synthetic gas (syngas) which can then be used to generate energy. The catalyst lowers the activation energy of reaction and makes the process possible at lower temperatures. There are four reactions that take place in the entire process: the polymer oxidation reactions (one producing carbon *dioxide* and another producing carbon *monoxide*), the Water-Gas Shift Reaction, and the Sabatier Reaction.

The studied substrates, Polyethylene and Cellulose, are both long chain organic polymers, and make up a substantial portion of both space and municipal waste composition. Although similar in nature, these substrates exhibit marked differences as it pertains to gasification, and were therefore selected as model substrates.

Experiments were performed on model polymers in the form of batch reactions over different catalysts supported by alumina. Analysis of gaseous products using a gas chromatograph with thermal conductivity detection provided data reflecting the conjunctive performance of all reactions. Application of reaction engineering parameter definitions and experimental data enabled the development of a model for the selectivities of reaction products. Kinetic parameters for the oxidation reactions of polyethylene were retrieved to be used in cooperation with a kinetic model for the gas-phase reactions composing a methodological model for the gasification of solid waste.

535307 Sustainable Synthesis of Glassy Liquid Crystals As Advanced Optical Materials

Abstract:

Glass liquid crystals, GLCs, have found numerous potential optical device applications, including waveplates, non-absorbing polarizers, polarized lasers, notch filters, mirrors, and reflectors. From the molecular structure standpoint, cholesteric GLCs entail chemically bonding independent chiral and nematic pendant groups at an optimized stoichiometric ratio to volume-excluding cores. The more elegant approach incorporates both chiral and nematic characters in a single hybrid pendant, which substantially reduces the synthesis and purification efforts and hence facilitates process scale-up. To further promote chemical sustainability, *p*-hydroxybenzoic acid carrying a chemically activated phenolic hydroxy group toward esterification and etherification is adopted as a building block to obviate the protection-deprotection routines, thereby serving a synthesis scheme in fewer steps than previously possible with much improved atom economy, reaction mass efficiency, and other green chemistry metrics. The new strategies are generally applicable to the synthesis of nematic, cholesteric, and smectic GLCs comprising diverse combinations of volume-excluding core and mesogenic pendant.

538822 Informatics for Green and Sustainable Nanomaterials

Abstract:

Nanotechnology research momentum suggests that informatics approaches are tractable in the era of big data. However, there is still much to convey to researchers interested in harnessing the power of big data and informatics for promoting sustainable nanotechnology. This review focuses on current efforts to automate metric calculations, validate nanomaterial synthesis data, and incorporate informatics approaches to the assessment of nano materials. Perspective on current interlinked informatics and decision making systems is provided. This review links green nanosynthesis to holistic predictive nanomaterial life cycle assessment. Outcomes of applying informatics approaches to nanochemistry include more timely evaluations of green chemistry and sustainability in the context of nanomaterial route optimization and scale up.

530359 Material Property Targets for Nanostructured Adsorptive Membranes in Residential Water Purification Applications

Abstract:

Access to clean water is an ever-growing concern for modern society as it is critical to ensure human health, to protect threatened ecosystems, and to promote economic growth and prosperity. Yet the significant advances promised by nanotechnology for water purification have been slow to manifest because the connections between fundamental scientific research and technology development are often overlooked. There is a ***critical need*** to systematically identify the most promising applications for novel membrane materials and identify the gaps in scientific knowledge that most inhibit their translation to scalable water treatment technologies.

Self-assembled block polymers are one such promising class of material for novel water purification systems. Specifically, recent work has explored using A-B-C triblock polymers composed of polyisoprene-*b*-polystyrene-*b*-poly(*N*,*N*-dimethylacrylamide) (PI-PS-PDMA) to generate membranes with molecularly-engineered performance profiles.[1-4] In these systems, the chemically-distinct nature of the A, B, and C moieties in the triblock polymer together with the self-assembly and non-solvent induced phase separation (SNIPS) casting technique used in membrane fabrication allow for well-ordered, uniform nanoscale pores to form on the surface of the membranes. Moreover, these small pores quickly taper into larger pores which allows for the simultaneous attainment of both ***high size-selectivity*** and ***high throughput*** separations. In addition, because the PDMA moiety that lines the pore walls is covalently bound to the robust matrix of the nanoporous thin film, a remarkably high number of binding sites are generated within the membrane structure.[5] Importantly, the chemistry of the C-block, which lines the pore walls, can be tuned in a remarkably facile manner ***after the creation*** of the nanoporous thin film. Hence, high capacity adsorptive membranes, which contain ligands spanning a broad range of specific binding affinities, can be generated in a straightforward manner. Despite the excellent control over material properties provided by the block polymer platform, the nanostructures and chemistries that optimize performance at a module or system level have not been identified in a systematic manner.

In this presentation, we explore the feasibility of triblock polymers for residential water purification systems to remove heavy metal contaminants. We propose a device-scale model for batch and semi-batch adsorptive separation systems governed by a Langmuir isotherm. Using the model, we explore the ***necessary material properties*** (e.g., isotherm parameters, fluxes, etc.) to achieve system energy and size goals. Preliminary analysis reveals simultaneous 2 to 3 order magnitude improvement in both biding affinity and capacity (material properties) is required for modular residential systems relative to existing membrane materials[3]. Guided by this finding, we discuss molecular and nanoscale design strategies to meet these materials performance targets. Finally, we generalize the systems-scale modeling to arbitrary adsorptive separation systems.

References:

[1] Mulvenna, R. A.; Weidman, J. L.; Jing, B.; Pople, J. A.; Zhu, Y.; Boudouris, B. W.; Phillip, W. A. Tunable Nanoporous Membranes with Chemically-Tailored Pore Walls from Triblock Polymer Templates. *Journal of Membrane Science* **2014**, *470*, 246-256.

[2] Weidman, J. L.; Mulvenna, R. A.; Boudouris, B. W.; Phillip, W. A. Unusually Stable Hysteresis in the pH-Response of Poly(Acrylic Acid) Brushes Confined within Nanoporous Block Polymer Thin Films. *Journal of the American Chemical Society* **2016**, *138*, 7030-7039.

[3] Weidman, J. L.; Mulvenna, R. A.; Boudouris, B. W.; Phillip, W. A. Nanoporous Block Polymer Thin Films Functionalized with Bio-Inspired Ligands for the Efficient Capture of Heavy Metal Ions from Water. *ACS Applied Materials and Interfaces* **2017**, *9*, 19152-19160

[4] Zhang, Y. Z.; Sargent, J. L.; Boudouris, B. W.; Phillip, W. A. Nanoporous Membranes Generated from Self-assembled Block Polymer Precursors: *Quo Vadis*? *Journal of Applied Polymer Science* **2015**, *132*, 41683.

[5] Weidman, J. L.; Mulvenna, R. A.; Boudouris, B. W.; Phillip, W. A. Nanostructured Membranes from Triblock Polymer Precursors as High Capacity Copper Adsorbents. *Langmuir* **2015**, *31*, 11113-11123

533667 Nanomagnetic Illuminators for In Vivo Optical Imaging of Osteoarthritic Knee Joints

Abstract:

Nanoparticle formulations of various sizes and charges are being developed to diagnose and treat osteoarthritis (OA), a maladaptive disease of the joint. However, the transport and *in vivo* bio-distribution of nanocarriers in the joint space remain poorly understood. Developing a biocompatible platform for visualizing and quantifying the distribution of nanoparticles in the joint space could help engineer better nanoparticles for drug delivery and disease diagnosis. Here, we focus on block copolymer stabilized nanomagnetic illuminators obtained by encapsulating near-IR lipophilic dyes. We apply both in-vivo imaging system (IVIS) and electron paramagnetic resonance spectroscopy (EPR) to visualize and quantify the distribution of these multifunctional magnetic carriers in the healthy and osteoarthritic joints of Lewis rats.

Magnetic nanoparticles synthesized by a high temperature synthesis method under aerobic conditions, block copolymer PEG_{5kDa}_$PDLLA_{6kDa}$, and lipophilic tracer 1,1'-Dioctadecyl-3,3,3',3'-tetramethylindotricarbocyanine iodide (DiR) were dissolved in the water miscible solvent THF. Rapid mixing of the solutes against water (anti-solvent) using a confined impingement jet mixer under high local supersaturation condition resulted in kinetically frozen structures of about 200 nm. The magnetic composite nanocarriers were purified via magnetic separation and dye encapsulation was confirmed using absorbance measurements and an *in vivo* imaging system.

Toxicity of the DiR loaded magnetic nanocarrier on fibroblast-like synoviocytes was tested using an MTS assay, a colorimetric assay used to analyse cell metabolism. Greater than 80% of the synoviocytes were found viable after 24 hours of particle exposure in the concentration range of 0.1 to 0.8 mg iron oxide/mL, suggesting negligible toxic effects of the nanocarrier under these conditions.

Biodistribution and particle retention studies were performed on 15-month old Lewis rats with healthy and OA induced knees. 60 ug of DiR loaded magnetic composite nanocarrier was injected and distribution of the nanomagnetic illuminator was tracked for 14 days. Further, the retention of the particles in organs of euthanized rats was visualized by IVIS imaging and quantified via EPR. IVIS images of the organs analyzed for radiant efficiency suggested that particles were predominantly retained in the knee joint and cleared faster in a healthy joint as compared to an osteoarthrtic knee. Based on the EPR measurements on the lyophilized organs, we found that ~60 % of the injected iron oxide nanoparticles remained in the joint space. Our studies involving both visualization and quantification of DiR loaded magnetic particle indicate that this carrier can be explored as a safe option to better understand transport of nanoparticles in tissues of joint space and also to deliver hydrophobic drugs for treatment.

539039 pH Responsive Nanoparticle Films for Biofilm Microenvironment Evaluation

Abstract:

Bacterial infections from biofilms formed on medical devices (e.g. catheters or implants), if not detected sufficiently early, may result in fatal sepsis (Shirtliff and Leid, 2009). However, due to the local nature of such infections, detection using non-invasive sampling techniques can be difficult and the problem is often only recognized when the medical device malfunctions or a more systemic infection develops.

Therefore, an attractive solution is to functionalise the implant surface in such a way that an antimicrobial is self-trigger released in the presence of biofilm on the surface. Due to the dense nature of biofilms provided by their characteristic extracellular matrix, they can often have heterogeneous microenvironments, which differ drastically in their composition from their surroundings. One such difference can be the pH of the biofilm; regions of low pH may form even when biofilm is grown in buffered pH 7.4 medium (Schlafer et al. 2015). Drug delivery platforms using calcium carbonate and calcium phosphate nanoparticles, targeted to both tumours and biofilms, have been developed by exploiting their pH dependent dissolution (Dong et al. 2016). Such nanoparticles are stable at pHs around 7 found in healthy tissue but quickly dissolve at lower pHs.

An attractive class of nanoparticles for spectral monitoring are phosphorescent nanoparticles (nanophosphors). These are multicomponent materials consisting of a host matrix and rare-earth dopant which is responsible for the emitted radiation. By choosing the dopant appropriately the emission wavelength may be tuned as show by Sotiriou et al. (2011). Their optical and physical stability offer some attractive advantages over fluorescent dyes and semiconducting nanoparticles which often suffer from either photobleaching, blinking effects and/or toxicity.

Such nanophosphors can be made using flame spray pyrolysis (FSP) in a single-step and directly deposited in-situ onto substrates. This allows for the facile deposition of thin nanoparticle films which can subsequently be stabilised by in-situ flame annealing. Here, phosphorescent nanoparticles of calcium carbonate and calcium phosphate doped with europium were synthesised using FSP. Upon monitoring their luminescence in solution a clear decrease in luminescence intensity was observed at pH 4, which was attributed to the dissolution of the nanoparticles. The rate at which this luminescent signal decreases was observed to depend on the pH of the solution.

Moreover, in order to yield a ratiometric readout, yttrium oxide (Y_2O_3) which is stable in the pH ranges of interest (pH 4-7) was chosen as the host matrix for a reference nanophosphor system. The most intense peak of interest in the europium doped calcium phosphate and calcium carbonate systems occurs at approximately 620 nm, thus by doping Y_2O_3 with terbium which has an intense emission at 545 nm a reference emission independent of pH can be provided. Y_2O_3:Tb was first deposited onto a glass substrate followed by deposition of calcium phosphate nanophosphors. The substrate was then placed in a custom flow cell and the effect of various buffer solutions assessed. Thereafter, biofilms of *S. aureus* and *E. coli* were grown directly on the substrates and the luminescence monitored to demonstrate the use of this system in detecting the formation of acidic microenvironments at the substrate-biofilm interface.

REFERENCES

Shirtliff, M.; Leid, J. (2009) *The Role of Biofilms in Device-Related Infections*, Springer.

Schlafer, S., et al. (2015) Appl. Environ. Microbiol. 81, 1267â€"1273.

Dong, Z. et al. (2016) Biomaterials, 110, 60â€"70.

Sotiriou, G. A. et al. (2011) J. Phys. Chem. C 115, 1084â€"1089.

537969 Development and Use of Modeling Techniques and Continuous Biomolecule Detection Towards Diabetes Treatment

Abstract:

Maintaining glycemic control is of critical for diabetic patients. A number of technologies are currently on the market to improve glycemic control such as using continuous glucose monitoring (CGM) with an insulin pump. However, CGM alone cannot provide perfect glycemic control. One way of maintaining blood glucose is through the development of glucose-responsive insulins (GRIs), which are synthetic insulin formulations that dynamically change their potency in response to their environmental conditions. GRIs provide great promise; however, their development is time consuming and requires iteration of *in vivo* experiments. We use well developed pharmacokinetic models of glucose-insulin metabolism to map out the kinetic parameter space of GRIs, allowing the time consuming experimental iteration to be conducted mathematically. Another strategy of improving glycemic control, would be to add continuous insulin monitoring (CIM) to the insulin pump, allowing for tighter control loop algorithms to be developed. We show the development of continuous biomolecule sensors based on the fluorescent single walled carbon nanotubes. These nanosensors show sensitivity to a variety of biologically relevant analytes and show promise for the development of a platform that can continuously detect biomolecules *in vivo*.

530987 Optical Nanosensors for Monitoring 3D Oxygen Gradients in *Pseudomonas Aeruginosa* Biofilms

Abstract:

Bacterial biofilms are the cause of chronic infections of wounds, medical implants, and immunocompromised patients such as those with cystic fibrosis. These biofilms often exhibit antibiotic resistance that makes them medically difficult to treat. Being able to monitor target analytes within the structure of the biofilm is critical for understanding biofilm disease biology and its response to treatment. Oxygen consumption can be used as a measure of metabolic activity in facultative aerobes, but current methods such as microelectrodes limit monitoring to one dimension and cannot adequately capture dynamics. However, biofilms are complex three dimensional structures that require higher resolution approaches to gain a more complete understanding of activity. Traditional measurements require physically moving the microelectrode, which is inherently disruptive to biofilm structure and obscures rapid dynamics within the system, which can potentially alter results. Here, we demonstrate optical oxygen-sensitive nanosensors that were used to measure 3D oxygen gradients in *Pseudomonas aeruginosa* biofilms with minimal disruption to the biofilm structure. We created and characterized optical nanosensors for detection of molecular oxygen which can be incorporated into the biofilm structure during growth. Using these, we obtained confocal microscopy data from which quantitative 3 dimensional oxygen gradients could be determined, as shown in Figure 1.

Figure 1. a) Raw confocal microscopy data; b) ratiometric fluorescence signal; and c) rendered 3D oxygen gradient.

Using this approach, we improved on traditional electrode-based 1D methods of measuring oxygen profiles by investigating both the spatial and temporal variation in oxygen concentration during biofilm growth and under antibiotic attack. We observed spatial gradation in oxygen concentration during biofilm growth, attributed to nutrient consumption at the edges of the biofilm. We also studied 3D oxygen gradients during antibiotic attack and found that oxygen was present at greater depths compared to untreated controls, consistent with cell death or a transition to anaerobic respiration. This new approach to biological interrogation provides higher resolution data, with improved temporal resolution, while minimizing the impact of the measurement on the biofilm itself. Increased resolution allows for better *in vitro* study and screening of antibiotic treatment relevant to treating *P. aeruginosa* biofilm infections.

535132 A Colorimetric Plasmonic Nanosensor Hydrogel for the Detection of Spatial Dose Deposition of Ionizing Radiation for Clinical Radiotherapy

Abstract:

Complicated radiotherapy techniques are used to deliver radiation dose to the tumor site without causing heavy damage to the surrounding healthy tissue. This delivered dose is recorded and monitored to improve the wellbeing and quality of the patient. Contemporary sensors require highly trained professionals to operate and are very cumbersome to use. In this talk, we discuss a novel colorimetric sensor that can detect therapeutic levels of ionizing radiation. This is accomplished by the conversion of Au^{+} ions to a characteristic pink color of gold nanoparticles in a hydrogel matrix with the help of cationic micelles and ionizing radiation. A significant increase in the intensity of the pink color was observed with increasing the radiation dose. This color difference in the intensity was quantified using Ultraviolet-Visible spectroscopy and was used as a qualitative and quantitative indicator of the delivered dose. Also, a curing agent was allowed to diffuse into the hydrogel matrix which helped render a spatial dose monitoring capability to the nanosensor. The gel dosimeter was able to detect doses in the therapeutic range for conventional radiotherapy, making it clinically viable. This clinical viability of the nanosensor was successfully studied on canine patients undergoing radiotherapy. The hydrogel nanosensor was also successfully able to detect low doses of radiation when employed for the detection of radioactive isotopes in clinical settings.

532466 Electrospray-Mediated Flash Nanoprecipitation for Synthesizing Drug Delivery Nanocarriers

Abstract:

The objective of this work is to develop a robust, scalable platform for synthesizing nano-scale polymeric drug carriers via Flash NanoPrecipitation (FNP). Many cancer therapeutics used currently are hydrophobic; hence, *in vivo* and clinical studies often encounter limited biodistribution, as well as a rapid clearance. One approach to overcome these challenges is encapsulating the hydrophobic drug molecules inside polymeric micro/nanoparticles. For example, micelles are polymeric nanoparticles capable of encapsulating hydrophobic molecules in the hydrophobic core. Generally, micelles are formed via self-assembly of amphiphilic block copolymers (BCPs), and hydrophobic molecules can be encapsulated in the micelle core during self-assembly. Conventional techniques for producing micellar drug carriers include water addition (WA), thin film hydration, and interfacial instability (IS). However, these methods are primarily limited to the bench scale; and, therefore, are not commercially viable. FNP was developed by Prudâ€™homme to synthesize polymeric nanoparticles encapsulating hydrophobic species at scale. In this approach, the nucleation and growth of hydrophobic compounds in water-miscible organic solvent are induced by rapidly mixing the organic solvent with an aqueous solvent. FNP has shown significant advantages for drug delivery applications, such as high product throughput and encapsulation efficiency. However, a typical FNP process requires large flow rates, ranging between 3 to 120 ml min^{-1}, for achieving sufficient mixing intensity.

Inspired by previous FNP studies, we developed an FNP process that utilizes electrospray atomization to enable rapid mixing, as opposed to impinging jet flow. Liquid-Liquid Electrospray (LLE) is a technique in which, unlike conventional electrospray, a nonconductive solvent is atomized inside a more conductive solvent. Using LLE, water-miscible organic solvent (i.e. tetrahydrofuran) containing BCPs and hydrophobic therapeutics was sprayed into water, inducing atomization and rapid mixing. This resulted in an FNP process yielding highly monodisperse polymeric drug. Unlike traditional impinging jet approachces, the rapid mixing achieved in LLE-FNP is independent of flow rate because fluid atomization is primarily based on the applied voltage. Using this approach, we produced monodisperse polymeric nanocarriers (monodispersity ~ 0.07) encapsulating a variety of cancer therapeutics, including dexamethasone, vorinostat, and lutein; the encapsulation efficiency and drug release characteristics were examined (encapsulation efficiency ~ 30 - 70%).

540732 Machine Learning Approaches for Enzyme Engineering

Abstract:

Enzyme engineering aims to improve a variety of different properties, such as stability, activity, and binding affinity. A variety of experimental and computational approaches have been used to improve such properties, including directed evolution and structure-guided rational design using various tools (e.g., Rosetta). We have recently developed a machine learning approach (called MLProScape) to model the protein fitness landscape and to guide enzyme engineering efforts. MLProScape numerically captures amino acid based physio-chemical properties as features, builds a machine learning model based on these features, and then identifies synthetic designs with improved enzyme properties. We have applied MLProScape to successfully design novel variants of glycosyl hydrolases with improved catalytic activity for hydrolyzing both cellulose and xylan, two major components found in lignocellulosic biomass. In this case, MLProScape made highly accurate predictions for glycosyl hydrolase activity under 5 different conditions, with pearson correlation coefficients varying between ~0.72 and ~0.90 and root mean squared errors varying between ~0.4 and ~16.2 mmol/min/mg. Examples of other MLProScape protein engineering applications will also be presented. MLProScape is capable of identifying residue changes outside of the active site and can handle complex design criteria that aim to improve multiple enzyme properties, simultaneously. As such, this machine learning based approach will facilitate enzyme engineering efforts aimed at improving various properties, including activity and thermostability.

528003 Simulating the Mechanistic Pathway of Transglycosylation Via a Mutant Glycoside Hydrolase

Abstract:

There is scientific, industrial, and medical value in synthesis of bespoke oligosaccharides, but significant limitations exist to existing laboratory methods. A more desirable approach is to employ highly specific enzymes, but the glycosyltransferases nature employs to do this job are poorly suited to industrial use. An enzymatic alternative to glycosyltransferases is transglycosylation via mutated glycoside hydrolases. The *Thermotoga maritima* Î±-L-fucosidase (*Tm*Afc) mutant D224G has been demonstrated to have good non-specific transglycosylation activity, but the molecular mechanism is not yet understood, hampering engineering efforts. In this talk, we describe transition path sampling simulations to elucidate the transglycosylation pathway of *Tm*Afc D224G, and discuss our proposed mechanismâ€™s implications for engineering improved activity and specificity in this class of enzymes in the future.

534845 Engineering a Multifunctional Family 5 Glycosyl Hydrolase into a Transglycosidase

Abstract:

Glycans are amongst the most abundant biomacromolecules on the planet but we are far from elucidating their role in the design and regulation of biological systems. There is a dire need of tools to systematically synthesize well-defined glycan polymers as is currently possible with other biopolymers like DNA, RNA, and proteins. Glycans synthesis is naturally performed by membrane associated glycosyl transferases (GTs), but GTs can be challenging to express sometimes for large-scale synthesis. Alternatively, transglycosidases can be selected from a large repertoire of available glycosyl hydrolases (GH) and rationally engineered to enable chemoenzymatic synthesis of designer glycans. Various factors play important roles in engineering a glycosyl hydrolase to function efficiently as a transglycosidases; such as, modulating substrate and product binding site affinities, regulating water accessibility, and enzymatic reaction conditions. Here, we focus on engineering a multifunctional family 5 GH from *Clostridium thermocellum* (CelE) into a transglycosidase to synthesize diverse glucan-based polymers. Active site mutations and presence of carbohydrate binding modules (CBM) were all found to have a significant impact on the catalytic rate, specificity, and overall mechanism of transglycosylation. All transglycosylation products were characterized using TLC, HPLC, and mass spectrometry to decode the synthesized glycan structures and reaction specificity. We further use molecular dynamics (MD) simulations and small angle X-ray scattering (SAXS) analysis to study the effect of CBM on the mechanism of transglycosylation to rationally design a highly efficient GH 5 based transglycosidase. In summary, we are developing a broadly applicable tool-kit and workflow to rationally convert glycosyl hydrolases into highly efficient and specific transglycosidases.

535619 Computational Insights into the Catalytic Function of Processive Cellulases

Abstract:

Cellulase enzymes naturally produced by fungi and bacteria are utilized industrially for many purposes, including the production of clean sugar streams from biomass that can be subsequently upgraded to fuels, chemicals, and products. The development of structure-function relationships in these enzymes is a collaborative effort between biochemical experimentation, structural characterization, and computational simulation. Here, we detail recent success stories from our group wherein molecular simulations are able to provide mechanistic insight with atomistic and dynamic resolution not generally attainable via experiment. In the present contribution, we probe different mechanisms for how a processive cellulase may dissociate from a cellulose chain on the surface of cellulose. Under certain conditions, dissociation from a cellulose strand has been demonstrated to be the overall rate-limiting step of catalytic action for cellobiohydrolases. Despite the growing evidence to support this conclusion, the literature is largely silent on the question of *how* dissociation occurs. In the published literature, "dissociation" is defined purely by the beginning (complexed upon a cellulose strand with the complete binding tunnel occupied) and the end (enzyme binding tunnel empty, and perhaps also "lifted off" of the cellulose surface into solution) states of the process. However, several possibilities exist for how the enzyme may proceed from the former to the latter state. Given the importance of this process within the context of cellulase performance (and thus potential cellulase engineering), we formulate several paths for this process (utilizing *Trichoderma reesei* Cel7A) and then compute free energy profiles for each. In so doing, we identify engineering targets for the production of processive cellulases of enhanced efficiency.

THE UNIVERSITY OF KANSAS

Department of Chemical & Petroleum Engineering

Enhanced Lipase-Catalyzed Hydrolysis and Modification of Fats and Oils

2018 AIChE Annual Meeting

Akash Anand

Committee members

Chair: Dr. Laurence R. Weatherley

Dr. Kevin C. Leonard

Dr. Arghya Paul

Dr. Prajnaparamita Dhar

Dr. Edward F. Peltier

1. Introduction

The oils and fats industry is one of the largest industries where chemical engineering techniques have been applied to the food sector. These are obtained from both animal and vegetable sources providing the raw materials for the manufacture of both edible and non-edible products. The products of hydrolysis oils and fats reactions include fatty acids and glycerol, which form the basis for the manufacture of a wide range of modified oils and fats, polyols, polyethers with application for the manufacture of personal care products and other specialty chemicals [1]. However in today's modern world there is an increasing demand for high quality and cheap products. This demand has led to the large scale production of fatty acids and glycerol. Hydrolysis of oils and fats is hindered by solubility issues since oil and water are immiscible. Hence, conducting the reaction at high temperatures increases the solubility of water in oil phase [2]. Currently, superheated steam is used to obtain fatty acids from vegetable oils or/and animal fats and the reaction is carried out at 100 – 260°C and 100 –7000 kPa using a 0.4-1.5 wt% water to oil ratio. This process is called fat splitting [3] and different variations of this technology are being used by various industries to produce fatty acids (e.g., the Twitchell process, the Colgate-Emery synthesis, and the Eisenlohr process). All these processes involves high temperature and pressure equipment [4-7]. With increasing public demand for companies to adopt a cleaner and more efficient attitude towards their processes many scientists and researchers are seeking to develop newer technologies. Of particular interest in this area is the use of enzymes to produce an accelerated and controlled release of fatty acids [8].

2. Lipase – Catalyzed Hydrolysis

Enzymes are basically proteins. Compared with inorganic catalysts such as acids, bases, metals and metal oxides, enzymes are very specific, they limit their action to specific bonds in the compounds with which they react [9]. Lipases are a group of enzymes whose primary function is to catalyze the hydrolysis of fats and oils (Lipolysis) to yield free fatty acids, diglycerides, monoglycerides and glycerol as shown in Figure 1 [10, 11].

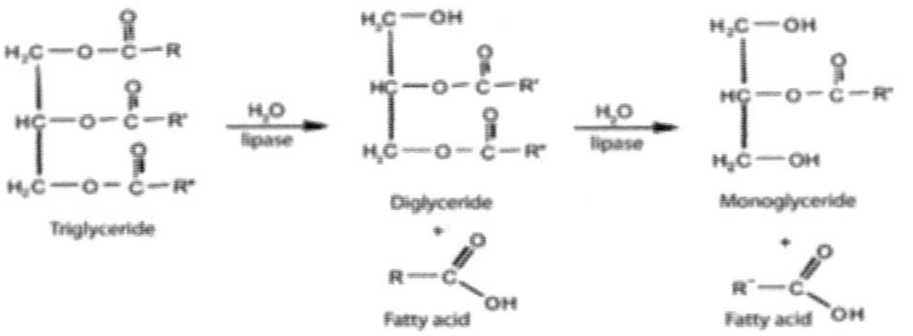

Figure 1. Hydrolysis of triglyceride (oils) catalyzed by lipase

As early as 1958, Sarda and Desnuelle described a sharp increase in lipase activity at substrate concentration exceeding their solubility threshold [12]. Since then the formation of an interface between aggregated substrates and the aqueous solution has been recognized as necessary for the activation of lipases. This behavior known as "interfacial activation" found a structural rationale some years later when an active lipase obtained from the yeast *Candida rugosa* was purified by Yamada and Machida in 1965 and has shown to have a non-specific mode of action [13-16].

3. Hypothesis

The hypothesis of this research is that the enzymatic hydrolysis reaction can be enhanced by immobilization of the enzyme onto a suitable solid support and by the optimization of temperature and pH. It is also proposed that application of an external electric field can enhance both reaction kinetics and reaction selectivity. Furthermore, a mass transfer model in a recirculated continuous reactor with immobilized enzymes could be established with the experimental results and can be used in design of industrial bioreactor and scale up.

4. Enzyme Immobilization

Enzymes, when used in soluble form requires separation from the reaction mixture once the reaction is complete. On account of the complexity of the mixture, recovery and recycling of the enzymes can result in significant loss of total enzyme and specific activity [18]. To overcome these drawbacks enzymes are immobilized onto organic or inorganic support particles by physical adsorption, encapsulation, covalent attachment, or cross-linking. Immobilization can increase the stability of the enzymes and may allow successful performance over a wider range of temperature and pH values [19]. Many supports and techniques for immobilization of lipase have been reported for oil hydrolysis over the years [20-28]. Separation of the immobilized enzyme from the reaction mixture is based on physical separation, allowing continuous use of the enzyme.

5. Glutaraldehyde Cross-linking

Glutaraldehyde cross-linkage is a technique for enhancing the immobilization of the enzymes onto polymeric supports, and in addition shows potential for enhancing the activity [29]. Another potentially significant consequence of applying glutaraldehyde cross-linkage is to enhance retention of the enzyme on the support thereby reducing loss of catalyst activity due to desorption or leakage of the enzyme from the support [30]. Glutaraldehyde has been used extensively as a mild cross-linking agent for the immobilization of enzymes [31, 32]. Another possible use of glutaraldehyde was identified by Haider et al. noting the possibility of it adding a protective layer around the active center of the enzyme [33].

6. Effect of Low Voltage Electric field

Enzymatic splitting of triglyceride esters is limited often by slow kinetics due to mass transfer limitations and by challenges of economic enzyme recycling. Prior work has shown that mass transfer rates and reaction rates can be significantly intensified by creating fine, electrostatic sprays of electrically charged droplets of enzyme solution which are injected into the oil substrate [34, 35]. The principle and sole mechanism for the intensification was assumed to be solely on account of increases in interfacial area and some hydrodynamic turbulence. The possible effect of electrical charge on the lipase activity has hitherto not yet been considered.

7. Experimental Procedures and Results to date

7.1. Materials

The lipase chosen for this study was from Candida rugosa (Type VII, >700 unit/mg solid). Aqueous enzyme solutions used were made up at concentration of 0.5 (%w/v) unless stated otherwise, using 0.1 M phosphate buffer. Vegetable oil and sunflower oil were the main substrate oils. Table 1 and 2 shows the list of support materials which has been used for lipase immobilization. The blended resins create a good balance of hydrophobicity that makes the polymer optimal for immobilization under mild operating conditions. Main advantages of these polymers include high mechanical stability and high enzyme activity at low protein loading. The immobilization criteria were based on finding a low cost method of loading enzyme onto the support. The chosen method of immobilization was adsorption, whereby the enzyme adheres to the surface of the support particles by Van der Waals forces of attraction [35].

Table 1. Properties of supports supplied by Purolite® Corporation

Product Name	ECR8285	ECR1030M	ECR8806M	ECR1090M	ECR1508	ECR1640
Composition	Epoxy/butyl/ methacrylate	Meth acrylic polymer	Octadecyl methacrylate	Macro porous styrene	Macro porous polystyrenic weak base anion	Macro porous polystyrenic strong base anion
Mean pore dia (Å)	400-600	220-340	500-700	900-1100	640	731
Particle size range (μm)	250-1000	300-710	300-710	300-710	-	-

Table 2. Characteristics of Accurel supports supplied by AkzoNobel® Chemicals

Product Name	EP400	EG100	EP100
Composition	Low Density Polyethylene (LDPE)	High Density Polypropylene (HDPP)	High Density Polypropylene (HDPP)
Pore size (μm)	0.1 - 0.5	0.1 - 1.0	0.1 - 1.0
Particle size range (μm)	400 - 1000	3000	200 - 2000

7.2. Immobilization Protocol

The surface of the supports are highly hydrophobic, which indicates that supports surface should be "wetted" before immobilization [36, 37]. The wetting process involved soaking 1 g support particles in 6 ml ethanol. This procedure modified the polymer surface and facilitated adsorption of the aqueous enzyme solution [38]. 10 mL of aqueous enzyme solution were added to 1 g of support particles along with 74 mL distilled, deionized water in a shake flask. The contents of the shake flask were agitated for 20 hours at 25°C in an orbital shaker. Separation of the immobilized support particles and aqueous solution was carried out using vacuum filtration. Immobilized support particles were then washed with 500 mL distilled, deionized water to remove any weakly bound enzymes on the surface of support particles. After washing, the immobilized support particles were dried overnight at 30°C and then assayed for enzyme activity [25, 35, 39]. A graphical summary of this technique is shown in Figure 2.

7.3. Activity Assay (Hydrolysis)

Enzyme activity exhibited by the immobilized enzyme on the polymer support particles was quantified by measuring the rate of hydrolysis of vegetable oil or sunflower oil [34]. All the hydrolysis reactions were carried out by mixing oil, water and immobilized enzyme in 30:10:1 ratio, on a weight basis unless stated otherwise, in a stirred tank reactor operated at 300 RPM [25]. At each and every time interval, 0.5 g of the emulsion was solubilized with 20 ml of 50:50v/v mixture of ethanol and acetone in order to arrest the reaction. The mixture was then titrated against 0.05 M NaOH in the presence of phenolphthalein indicator. Enzyme activity is expressed as the number of moles of free fatty acid produced from the hydrolysis of vegetable oil per minute (molFFA/min) or as moles of free fatty acid (molFFA). A Mettler – Toledo, T5 auto-titrator was used as pH stat.

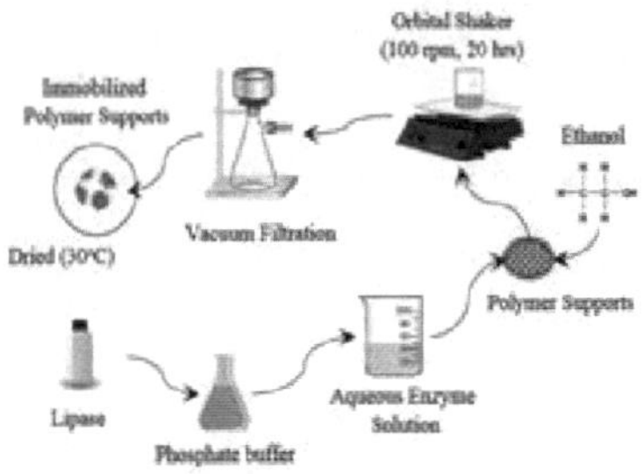

Figure 2. Lipase Immobilization Protocol

7.4. Glutaraldehyde cross-linking

Series of experiments were set up for enzyme immobilized onto polymer supports in the presence of cross-linking agent. The reaction performance in each case was evaluated in order to determine the stability of the cross-linked enzyme. Following a 20-hour contact period of immobilization as previously described, 1 mL of Mc Ilvaine containing 80 μL of 25% glutaraldehyde was added to the immobilization mix and the whole system further contacted for 10 minutes with agitation. 0.2 mL of 20% Sodium hydrogen sulfite ($NaHS_2$) were added and the system was shaken for another 10 minutes. Sodium hydrogen sulfite solution was added to control the activity of the glutaraldehyde. [18, 25, 39].

8. Results for Immobilized Supports

Earlier working comparing the relative stability of the free and immobilized enzyme systems with respect to shelf life, pH and temperature for Accurel supports for the hydrolysis of high oleate sunflower oil, Table 3 and Figure 3 (a and b) showed that: (1) Half-life of the immobilized enzyme was seen to increase up to 7 times that of aqueous enzyme solutions. (2) Lipase immobilized onto EP400 exhibited an increase in enzyme activity over the range of pH 7-9. (3) The temperature stability of lipase was not observed to change on enzyme immobilization and the optimum working temperature of the enzyme was noted at 45°C.

Table 3. Enzyme stability - variation of enzyme activities with time over four week period at room temperature (23°C)

Time (hours)	Enzyme Activity (molFFA)		
	Free Enzyme	EP400 (<400 μm)	EG100 (3 mm)
0	0.10867	0.104	0.00667
168	0.08267	0.1033	0.00533
504	0.00267	0.09333	0.00533
672	0.00133	0.092	0.004

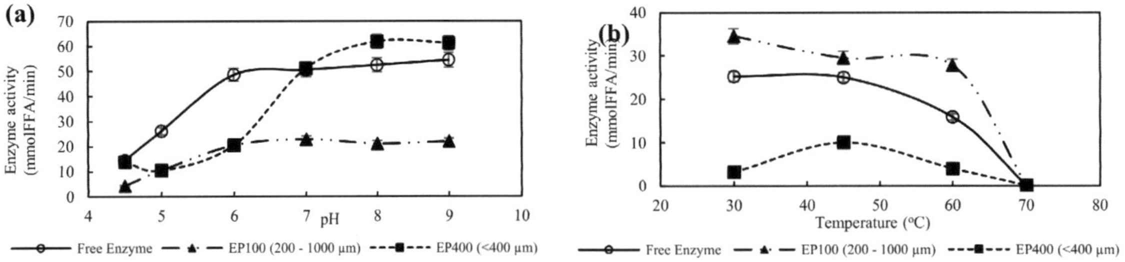

Figure 3. Enzyme activity of free enzyme compared with Accurel EP100 (HDPP) and EP400 (LDPE) immobilized enzyme – (a) varying pH and (b) varying temperature

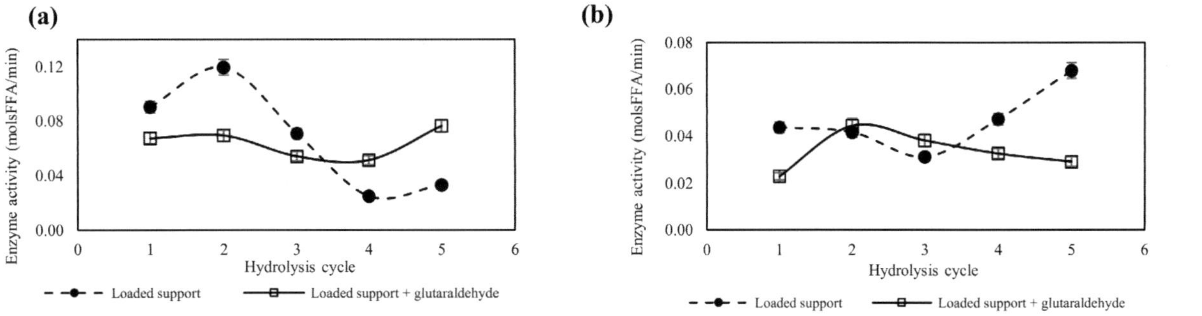

Figure 4. Enzyme activity with respect to different hydrolysis cycle using immobilized support with and without glutaraldehyde cross-linking (a) Accurel EP100 (HDPP) for particle size of 200 – 1000 μm (b) Accurel EP400 (LDPE) for particle size of less than 400 μm

Figures 4 (a and b), shows the enzyme activities measured for enzymes cross-linked onto Accurel EP100 and EP400 supports with glutaraldehyde and comparisons to the loaded supports without glutaraldehyde cross-linking. The activity data are presented for multiple cycles of reaction, with the immobilized enzymes recycled after each activity determination. The results are anomalous, showing significant variation in activity from cycle to cycle. The data for enzyme immobilized onto EP100 with no cross-linking agent, increased in specific activity after one cycle followed by a steep reduction in subsequent cycles. In the case of enzyme immobilized onto EP100 with cross-linking, a much steadier performance from cycle to cycle is observed, suggested improved enzyme stability. Immobilized enzyme, cross-linked onto EP400, showed anomalous behavior from cycle to cycle. In the case of the non-crosslinked catalyst the activity appeared to decrease in the first two cycles, followed by a steep increase in activity after cycle three. Some decrease in activity from cycle to cycle would be expected and the increase in activity after cycle three was unexpected. One possible explanation is based on diffusional effects within the particle structure. It is possible that after each activity experiment, residual reaction was still occurring within the interior surface of the catalysis particle, with fatty acid product for example from cycle three retained in the particle and released slowly during the cycle four activity determination. A further possibility is the continuation of reaction in the inner core of the particle in between cycles. Both of these potential effects, if present, would result in an apparent increase in fatty acid yield during each activity determination beyond cycle three. Figure 4 (a), could be similarly explained by a transient effect. Regardless of the apparent anomalous behavior it is clear for both polypropylene support, and for the polyethylene support that cross-linking imparts a degree of added stability to the immobilized enzyme.

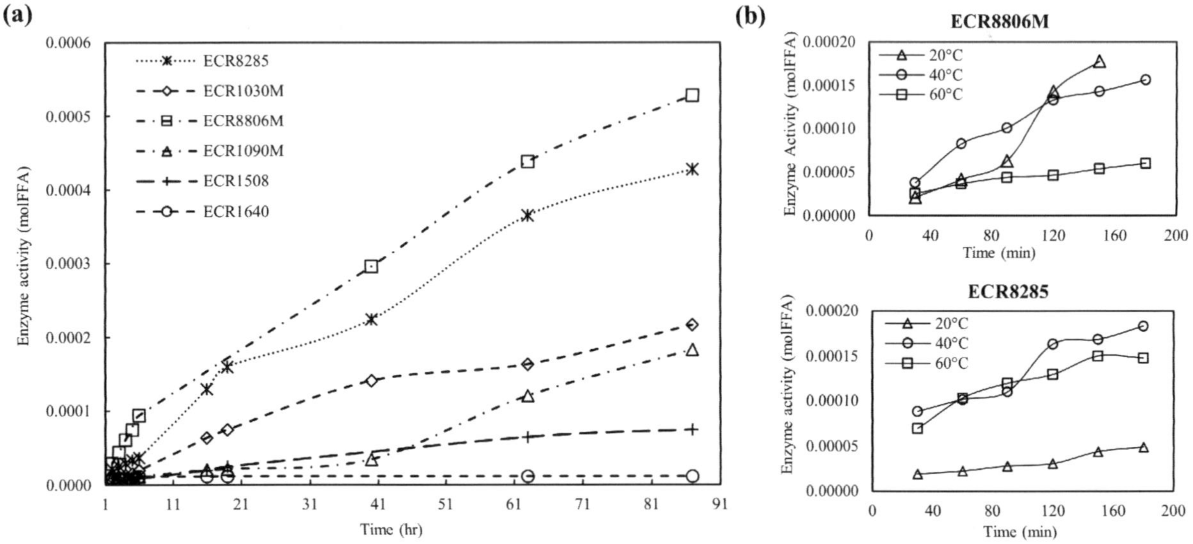

Figure 5. Hydrolytic performance of immobilized lipase on Purolite® supports – (a) Long hydrolysis and (b) Effect of temperature on ECR8806M and ECR8285 resin

Comparison of hydrolytic performance of Purolite supports using vegetable oil hydrolysis shown in Figure 5(a) showed that ECR8806M and ECR8285 performed much better than the all other supports [40]. ECR8806M is an epoxy-supported resin, it shows more stability and activity compared to the other supports because epoxy groups can react with amino or thiol groups of the enzyme and form strong bonds between enzyme and supports which makes it attractive for recycling [36]. Basso et al. showed Octadecyl methacrylate have excellent immobilization capabilities compared to styrene based resin which could explain the poor hydrolytic performance of styrene based resins [37, 41, 42]. From Figure 5(b), ECR8285 and ECR8806M with excellent immobilization capabilities, when examined under three different temperature ranges showed that ECR8285 is more stable at higher temperature range (above 40°C) and ECR8806M at lower temperature range (below 40°C).

10. Electro-Lipase-Activity

Rates of reaction for the enzymatic hydrolysis of vegetable oil in the presence of aqueous enzyme solution were determined in the presence of steady direct current (DC) electrical field. The reactions were conducted in a fixed rectangular geometry batch reactor fitted with planar electrodes as shown in Figure 6(a). The two immiscible liquid phases, oil and aqueous enzyme solution made up at concentration of 1.5 (%w/v), were located between the electrodes with a quiescent rectangular interface of known area. The rates of conversion to the free fatty acid and glycerol were determined for applied DC voltages in the range 0-30 Volts imposed between the two electrodes as shown in Figure 6(b). The results showed clear evidence of significant increase in the rates of reaction in the presence of the electrical field. It was concluded that the electric field has a positive effect on reaction rate, independent of interfacial area [43, 44]. This suggested possible changes in enzyme structure due the electrical field, modification of the nature of the enzyme binding at the liquid-liquid interface, or possibly intensification of local transport rates of substrates and reaction products close to the interface. The latter could involve phenomena such as electro-kinetic transport or electro-osmosis [45-47].

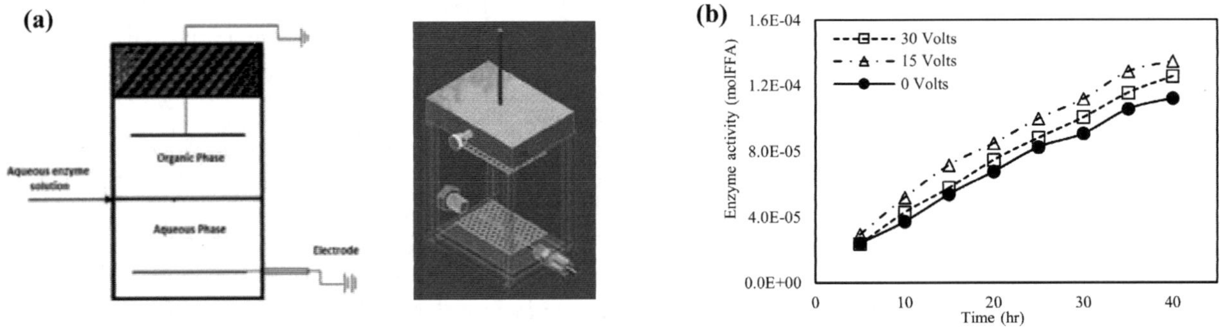

Figure 6. Electrically enhanced enzymatic hydrolysis (EEEH) – (a) Electric field reactor and (b) Comparative performance of free enzymes under electric field

REFERENCES

[1] F.D. Gunstone, R.J. Hamilton, Oleochemical manufacture and applications, CRC Press2001.

[2] K. Ngaosuwan, E. Lotero, K. Suwannakarn, J.G. Goodwin Jr, P. Praserthdam, Hydrolysis of triglycerides using solid acid catalysts, Industrial & Engineering Chemistry Research 48(10) (2009) 4757-4767.

[3] O. Ackelsberg, Fat splitting, Journal of the American Oil Chemists' Society 35(11) (1958) 635-640.

[4] E. Twitchell, Process of decomposing fats or oils into fatty acids and glycerin, Google Patents, 1898.

[5] E.G. W, Process of hydrolyzing oils and fats, Google Patents, 1939.

[6] H. Barnebey, A. Brown, Continuous fat splitting plants using the Colgate-Emery process, Journal of the American Oil Chemists' Society 25(3) (1948) 95-99.

[7] R.L. Holliday, J.W. King, G.R. List, Hydrolysis of vegetable oils in sub-and supercritical water, Industrial & Engineering Chemistry Research 36(3) (1997) 932-935.

[8] W.P. Jencks, Catalysis in chemistry and enzymology, Courier Corporation1987.

[9] A. Zaks, A.M. Klibanov, Enzyme-catalyzed processes in organic solvents, Proceedings of the National Academy of Sciences 82(10) (1985) 3192-3196.

[10] N.N. Gandhi, Applications of lipase, Journal of the American Oil Chemists' Society 74(6) (1997) 621-634.

[11] H. Uhlig, Industrial enzymes and their applications, John Wiley & Sons1998.

[12] L. Sarda, P. Desnuelle, Actions of pancreatic lipase on esters in emulsions, Biochimica et biophysica acta 30(3) (1958) 513.

[13] R. Verger, Enzyme kinetics of lipolysis, Methods in enzymology 64 (1980) 340-392.

[14] Y. Koichi, M. Haruo, High-activity lipase and method of preparation thereof, Google Patents, 1965.

[15] S. Al-Zuhair, K. Ramachandran, M. Hasan, Effect of enzyme molecules covering of oil–water interfacial area on the kinetic of oil hydrolysis, Chemical Engineering Journal 139(3) (2008) 540-548.

[16] M. Hoarau, S. Badieyan, E.N.G. Marsh, Immobilized enzymes: understanding enzyme–surface interactions at the molecular level, Organic & biomolecular chemistry 15(45) (2017) 9539-9551.

[17] R.A. Sheldon, S. van Pelt, Enzyme immobilisation in biocatalysis: why, what and how, Chemical Society Reviews 42(15) (2013) 6223-6235.

[18] C. Mateo, J.M. Palomo, G. Fernandez-Lorente, J.M. Guisan, R. Fernandez-Lafuente, Improvement of enzyme activity, stability and selectivity via immobilization techniques, Enzyme and microbial technology 40(6) (2007) 1451-1463.

[19] C. Laane, S. Boeren, K. Vos, C. Veeger, Rules for optimization of biocatalysis in organic solvents, Biotechnology and Bioengineering 30(1) (1987) 81-87.

[20] H. Noureddini, X. Gao, S. Joshi, P. Wagner, Immobilization of Pseudomonas cepacia lipase by sol-gel entrapment and its application in the hydrolysis of soybean oil, Journal of the American Oil Chemists' Society 79(1) (2002) 33-40.

[21] J.C. Santos, G.F. Nunes, A.B. Moreira, V.H. Perez, H.F. de Castro, Characterization of Candida rugosa Lipase Immobilized on Poly (N-methylolacrylamide) and Its Application in Butyl Butyrate Synthesis, Chemical Engineering & Technology 30(9) (2007) 1255-1261.

[22] Y. Yong, Y.-X. Bai, Y.-F. Li, L. Lin, Y.-J. Cui, C.-G. Xia, Characterization of Candida rugosa lipase immobilized onto magnetic microspheres with hydrophilicity, Process Biochemistry 43(11) (2008) 1179-1185.

[23] M. Monier, A. El-Sokkary, A. Sarhan, Immobilization of Candida rugosa lipase on modified natural wool fibers, Reactive and Functional Polymers 70(2) (2010) 122-128.

[24] M. Basri, W.M. Yunus, Z. Wan, W.S. Yoong, K. Ampon, C.N.A. Razak, A.B. Salleh, Immobilization of lipase from Candida rugosa on synthetic polymer beads for use in the synthesis of fatty esters, Journal of chemical technology and biotechnology 66(2) (1996) 169-173.

[25] L.R. Weatherley, A. Anand, The Performance of Microbial Lipase Immobilized onto Ion Exchange Resins and onto the Natural Zeolite Clinoptilolite, International Journal of Chemical Separation Technology 2(2) (2016).

[26] X.-J. Huang, P.-C. Chen, F. Huang, Y. Ou, M.-R. Chen, Z.-K. Xu, Immobilization of Candida rugosa lipase on electrospun cellulose nanofiber membrane, Journal of Molecular Catalysis B: Enzymatic 70(3) (2011) 95-100.

[27] S.S. Betigeri, S.H. Neau, Immobilization of lipase using hydrophilic polymers in the form of hydrogel beads, Biomaterials 23(17) (2002) 3627-3636.

[28] M.D. Ramos, G.I.G. Gómez, N.S. González, Immobilization of Candida rugosa lipase on bentonite modified with benzyltriethylammonium chloride, Journal of Molecular Catalysis B: Enzymatic 99 (2014) 79-84.

[29] I. Migneault, C. Dartiguenave, M.J. Bertrand, K.C. Waldron, Glutaraldehyde: behavior in aqueous solution, reaction with proteins, and application to enzyme crosslinking, Biotechniques 37(5) (2004) 790-806.

[30] S. Datta, L.R. Christena, Y.R.S. Rajaram, Enzyme immobilization: an overview on techniques and support materials, 3 Biotech 3(1) (2013) 1-9.

[31] F. López-Gallego, L. Betancor, C. Mateo, A. Hidalgo, N. Alonso-Morales, G. Dellamora-Ortiz, J.M. Guisán, R. Fernández-Lafuente, Enzyme stabilization by glutaraldehyde crosslinking of adsorbed proteins on aminated supports, Journal of Biotechnology 119(1) (2005) 70-75.

[32] O. Barbosa, C. Ortiz, Á. Berenguer-Murcia, R. Torres, R.C. Rodrigues, R. Fernandez-Lafuente, Glutaraldehyde in bio-catalysts design: a useful crosslinker and a versatile tool in enzyme immobilization, Rsc Advances 4(4) (2014) 1583-1600.

[33] T. Haider, Q. Husain, Concanavalin A layered calcium alginate–starch beads immobilized β galactosidase as a therapeutic agent for lactose intolerant patients, International journal of pharmaceutics 359(1) (2008) 1-6.

[34] D. Rooney, L. Weatherley, The effect of reaction conditions upon lipase catalysed hydrolysis of high oleate sunflower oil in a stirred liquid–liquid reactor, Process Biochemistry 36(10) (2001) 947-953.

[35] M. Murray, D. Rooney, M. Van Neikerk, A. Montenegro, L. Weatherley, Immobilization of lipase onto lipophilic polymer particles and application to oil hydrolysis, Process Biochemistry 32(6) (1997) 479-486.

[36] M. Mihailović, M. Stojanović, K. Banjanac, M. Carević, N. Prlainović, N. Milosavić, D. Bezbradica, Immobilization of lipase on epoxy-activated Purolite® A109 and its post-immobilization stabilization, Process Biochemistry 49(4) (2014) 637-646.

[37] A. Basso, M. Hesseler, S. Serban, Hydrophobic microenvironment optimization for efficient immobilization of lipases on octadecyl functionalised resins, Tetrahedron 72(46) (2016) 7323-7328.

[38] K. Hernandez, C. Garcia-Galan, R. Fernandez-Lafuente, Simple and efficient immobilization of lipase B from Candida antarctica on porous styrene–divinylbenzene beads, Enzyme and microbial technology 49(1) (2011) 72-78.

[39] A. Anand, L.R. Weatherley, The performance of microbial lipase immobilized onto polyolefin supports for hydrolysis of high oleate sunflower oil, Process Biochemistry (2018).

[40] V.G. Tacias-Pascacio, J.J. Virgen-Ortíz, M. Jiménez-Pérez, M. Yates, B. Torrestiana-Sanchez, A. Rosales-Quintero, R. Fernandez-Lafuente, Evaluation of different lipase biocatalysts in the production of biodiesel from used cooking oil: Critical role of the immobilization support, Fuel 200 (2017) 1-10.

[41] A. Basso, L. Froment, M. Hesseler, S. Serban, New highly robust divinyl benzene/acrylate polymer for immobilization of lipase CALB, European journal of lipid science and technology 115(4) (2013) 468-472.

[42] V.G. Tacias-Pascacio, S. Peirce, B. Torrestiana-Sanchez, M. Yates, A. Rosales-Quintero, J.J. Virgen-Ortíz, R. Fernandez-Lafuente, Evaluation of different commercial hydrophobic supports for the immobilization of lipases: tuning their stability, activity and specificity, RSC Advances 6(102) (2016) 100281-100294.

[43] F. Wu, Y. Jin, D. Li, Y. Zhou, L. Guo, M. Zhang, X. Xu, N. Yang, Electrofluid hydrolysis enhances the production of fermentable sugars from corncob via in/reverse-phase induced voltage, Bioresource Technology 234 (2017) 158-166.

[44] Y. Zhou, Y. Jin, N. Yang, Z. Xie, X. Xu, Electrofluid enhanced hydrolysis of maize starch and its impacts on physical properties, RSC Advances 7(31) (2017) 19145-19152.

[45] C. Albers, L.R. Weatherley, Enzyme reaction kinetics at an electrically charged liquid-liquid interface, (2014).

[46] A. Anand, E. Harmsen, L.R. Weatherley, Free enzyme reaction kinetics at a liquid-liquid interface with an applied electric field, (2017).

[47] A. Anand, J. Hattermer, L.R. Weatherley, Enzyme catalyzed hydrolysis reaction kinetics at a liquid-liquid boundary under effect of an applied electric field, (2017).

[48] F.X. Malcata, H.R. Reyes, H.S. Garcia, C.G. Hill, C.H. Amundson, Immobilized lipase reactors for modification of fats and oils—a review, Journal of the American Oil Chemists' Society 67(12) (1990) 890-910.

[49] Y. Kosugi, H. Tanaka, N. Tomizuka, Continuous hydrolysis of oil by immobilized lipase in a countercurrent reactor, Biotechnology and bioengineering 36(6) (1990) 617-622.

[50] S. Fadiloğlu, Z. Söylemez, Kinetics of lipase-catalyzed hydrolysis of olive oil, Food research international 30(3) (1997) 171-175.

535098 Activity Improvement of D-Psicose-3-Epimerase from *Agrobacterium tumefaciens* CGMCC 1.1488 By Site-Directed Mutagenesis

Abstract:

The enzyme D-psicose-3-epimerase catalyzes the conversion of C3-epimerization from fructose to allulose (D-psicose). Due to the relatively low allulose conversion rate, the improvement of enzymatic activity by site-directed evolution was carried out in this experiment. The DPEase-C58 from *Agrobacterium tumefaciens* CGMCC 1.1488 was supplied to the protein structure comparison and sequence alignment with DTE from *Pseudomonas cichorii*, whose crystal structure and the complex with D-fructose had been determined, to conduct the sites mutagenesis. The two sites of saturation mutation of H228 and T231 were carried out. At first round 5 mutants with 20% relevant higher than wide-type were obtained by high-throughout screening method.

532423 Process Intensification of Algae Oil Extraction to Biodiesel

Abstract:

Dilute algae are often grown in raceways or bioreactors by photosynthesis using carbon dioxide in flue gas. After being harvested, algae oil must be extracted before being transesterified to biodiesel. This process, which generates a carbon-neutral, renewable, liquid fuel, has been shown to be competitive with other liquid fuel sources (Silva et al., 2014). Herein, we introduce the use of carbon dioxide microbubbles to lyse algae cell walls, releasing triglyceride oils, combined with their transesterification using methanol. Process designs are compared with conventional extraction/transesterification processes, showing the profitability advantages (reduced energy utilization and installed equipment costs) of the proposed intensification processes. At lower temperatures and intermediate pressures, carbon dioxide causes methanol to dissolve in the triglyceride phase and triglyceride to dissolve in the methanol phase, increasing the transesterification reaction rate. This paper presents a 3-reversible reaction transesterification model, coupled with multiphase equilibria (VLLE), for a large-scale, continuous biodiesel process. It introduces an algorithm for solving the steady-state reactor model â€“ to be utilized for optimizing the intensified process in future work.

Silva, E. Soliman, F. Cameron, L.A. Fabiano, W.D. Seider, E.H. Dunlop, A.K. Coaldrake, 2014. Commercial-scale biodiesel production from algae. *Ind. Eng. Chem. Res.*, **53**, 5311-5324.

529768 Regional Techno-Economic (TEA) Analysis of the Pyrolysis-Bioenergy-Biochar Pathway for Carbon-Negative Energy

Abstract:

The objective of this project is to evaluate regional economic impacts of an integrated pyrolysis-bioenergy-biochar industry to produce carbon negative energy. Utilization of biochar co-produced from fast pyrolysis process as soil amendment is essential to produce carbon negative energy. Selection of the pyrolysis biorefinery location is important since it involves feedstock availability and prices, biochar demand and credits, soil type and quality, and logistics etc. Therefore, we conducted an economic analysis including regional factors to determine the best biorefinery location to maximize the economic benefits of the pyrolysis-bioenergy-biochar platform.

Three specific land areas of US are investigated depending on the selection of pyrolysis facility locations: Upper Mississippi River Basin (UMRB), U.S. Southeast and California. Three counties including Hamilton (Iowa), Jackson (Florida) and Glenn (California) were selected to represent these three regions. We applied biochar produced from fast pyrolysis of corn stover, switchgrass and forest residue to crop lands for corn, peanuts and rice individually for these three counties. The correlation between biochar application to crop yields change were modeled based on over 100 previous literatures. The mean crop yield increases have been predicted as 2.9%, 10% and 4.5% in Hamilton, Jackson and Glenn. The biochar prices that the farmers are willing to pay were calculated as \$113, \$475 and \$262 per metric ton with a biochar application rate of 5 metric tons per hectare.

Capital cost are also sensitive to the biorefinery location. The location multipliers have been suggested as 0.92, 0.82 and 1.17 for Iowa, Florida and California in Dodge unit cost guide. The feedstock costs vary with the feedstock type in different regions. The capital cost, operating cost and minimum fuel selling price was estimated over three different states to identify the most economic beneficial location to build this pyrolysis-biochar-bioenergy platform.

529654 Modelling Hydrogen Production Via Combined Hydrothermal Liquefaction of Macroalgae *Saccharina japonica* and Hydrothermal Gasification of Aqueous Product Using Aspen Plus Â®

Abstract:

Hydrothermal liquefaction (HTL) of biomass has attracted an immense attention as an economical route for energy production as compared to other conversion pathways. The depolymerization of the feedstock typically takes place at an operating condition of 250-400ËšC with a pressure of 4-20 MPa. HTL can treat wet feedstock, which reduces the cost for its pretreatment and drying, and hence making the conversion more economical in comparison to other processes. This study focuses on the combined hydrothermal liquefaction of *Saccharina japonica* to produce biocrude and hydrothermal gasification of aqueous product to produce hydrogen, which will be used as a feed for biocrude upgrading and other commercial purposes. Macroalgae feed of 140,000 lb/hr was fed to the HTL reactor at operating conditions of 300ËšC and 21 MPa. A balanced reactor was achieved yielding 76.5 wt.% liquid product, 8.59 wt.% of gas phase and 7.95 wt.% of non-soluble products. Hydrothermal liquefaction was simulated using a stoichiometric reactor model prior to which a mass balance between the feed analyses (Proximate, Ultimate, and Sulfanal analyses) and compounds in HTL products was developed. Phase separator separated the product in various phases (gas, aqueous, and organic phase). Gas phase product was processed via steam reforming followed by the water gas shift reaction and finally by pressure swing adsorption (PSA) for hydrogen production. Aqueous phase product was undergone hydrothermal gasification in a Gibbs reactor model, in which all the compounds were considered as products and were later served as a feed for the PSA Unit. PSA produces 4,252 kg/hr of hydrogen and of the off-gas consisting of CH_4, C_2H_6, CO, CO_2, N_2 and a minute fraction of H_2. The techno-economic study was performed to investigate the economic potential for the contribution of this process route towards the energy sector. Sensitivity analyses were also performed which helped in finding the optimal minimum selling price for the hydrogen.

Acknowledgment: This research was respectfully supported by Engineering Development Research Center (EDRC) funded by the Ministry of Trade, Industry & Energy (MOTIE). (No. N0000990)

519829 Process Simulation and Experimental Investigation of Biofuel Production in a High RATE Anaerobic Digestion Process

Abstract:

The non-hazardous waste management hierarchy of the US EPA calls for â€œReduce, Reuse, and Recycleâ€ (or the three Râ€™s) of wastes. The anaerobic digestion process is one of the most important methods that used to treat the wastes and at the same time generate energy out of it. The anaerobic digestion process generates a mixture of methane and carbon dioxide gases which is known as biogas. The biogas composition is about 50-70% methane and 10-30% of carbon dioxide and trace amount of other gases like hydrogen and hydrogen sulfide. This biogas can be used in power generation, heating systems and in combined heating and powering systems. Also, it could be upgraded to improve its quality and make is utilized in all equipment used for natural gas with a minimal adjustment due to the lower BTU contents for methane gas. Three papers were written and submitted regarding the biogas production and the liquid waste water treatment. The first paper focuses on developing a process modeling simulation by aspen plus for the anaerobic digestion process and on conducting a sensitivity analysis to investigate the parameters that could upgrade the biogas quality. The second paper focuses on the effect of hydraulic retention time of the substrate on the biogas production. This investigation conducted in a two-stage high rate expanded granular sludge bed reactor under different variables like substrate concentration, organic loading rate, and operating temperature. The third paper focuses on upgrading the biogas quality and quantity in a two-stage expanded granular sludge bed reactor by investigating the effect of injecting the pre-acidification gas (the first stage), which is mainly produced a gas mixture consist of hydrogen and carbon dioxide) into the second stage (the expanded granular sludge bed reactor).

528976 Stability of Platinum in the Electrochemical Environment: Reconstruction, Roughening, and the Third Peak

Abstract:

Platinum is a widely used catalyst in aqueous and electrochemical environments. It is used in fuel cells and electrolyzers, for example, to catalyze hydrogen oxidation, hydrogen evolution, and the oxygen reduction reactions [1,2]. The activity and selectivity of a catalyst is dictated by its surface structure, which can change in the reaction environment. In our prior work, we have used density functional theory to examine the surface energy (thermodynamic stability) of the low index Pt(111), Pt(100), and Pt(110) surfaces in the presence of adsorbed hydrogen, hydroxide, water, oxygen, surface oxide, and alkali metal cations, spanning the complete electrochemical window of water [3]. We found that the relative stability of these surfaces was altered in the electrochemical environment. While Pt(111) is the most stable surface in vacuum, the strong binding of hydrogen to Pt(100) and of hydroxide and water to Pt(110) make these surfaces more stable at low potentials and high potentials, respectively. We used these results to better understand the reconstruction of platinum electrodes. For example, cycling the potential of a Pt(111) electrode above 1.3 V_{RHE} results in the growth of islands expressing, primarily, (110) step sites [4, 5]; the Pt(110) surface is most stable at potentials where the surface oxide formed at high potentials (> 1.3 V_{RHE}) is reduced [3]. Similarly, the formation of (100) terrace sites upon square wave cycling of spherical nanoparticles to low potentials (below 0 V_{RHE}) [6] may be driven by the greater stability of the Pt(100) surface at low potentials due to its strong binding of hydrogen [3]. In this talk we will show how we have recently extended this work to include the surface energy of the Pt(553) and Pt(533) stepped surfaces. We have additionally considered the thermodynamics of subsurface hydrogen formation and hydrogen induced step-edge roughening. Using density functional theory, we find this step-edge roughening to be both thermodynamically and kinetically favorable. We compare closely to experiment, where it may be responsible for the appearance of the anomalous anodic "third peak" which can be seen in cyclic voltammograms on cycling a stepped or polycrystalline platinum electrode to low potentials (below -0.2 V_{RHE}) [7, 8].

1) H. A. Gasteiger, J. E. Panels, and S. G. Yan, "Dependence of PEM fuel cell performance on catalyst loading," ***Journal of Power Sources,*** vol. 127, pp. 162-171, 2004.

2) H. A. Gasteiger, S. S. Kocha, B. Sompalli, and F. T. Wagner, "Activity benchmarks and requirements for Pt, Pt-alloy, and non-Pt oxygen reduction catalysts for PEMFCs," ***Applied Catalysis B: Environmental,*** vol. 56, pp. 9-35, 2005.

3) I. T. McCrum, M. A. Hickner, and M. J. Janik, "First-principles calculation of Pt surface energies in an electrochemical environment: thermodynamic driving forces for surface faceting and nanoparticle reconstruction," ***Langmuir***, vol 33, pp. 7043-7052, 2017.

4) M. Wakisaka, S. Asizawa, H. Uchida, M. Watanabe. "In Situ STM Observation of Morphological Changes of the Pt(111) Electrode Surface During Potential Cycling in 10 Mm HF Solution," ***Phys. Chem. Chem. Phys,*** vol. 12, pp. 4184– 4190, 2010.

5) L. Jacobse, Y.-F. Huang, M. T.M. Koper, M. J. Rost, "Correlation of surface site formation to nanoisland growth in the electrochemical roughening of Pt(111)", ***Nature Materials***, vol. 17, pp. 277-282, 2018.

6) N. Tian, Z.-Y. Zhou, S.-G. Sun, Y. Ding, Z. L. Wang, "Synthesis of Tetrahexahedral Platinum Nanocrystals with High-Index Facets and High Electro-Oxidation Activity" ***Science***, vol. 316, pp. 732– 735, 2007.

7) K. Kinoshita, J. Lundquist, P. Stonehart, "Hydrogen adsorption on high surface area platinum crystallites", ***Journal of Catalysis***, vol. 31, pp. 325-334, 1973.

8) O. Diaz-Morales, T. J.P. Hersbach, C. Badan, A. Garcia, M. T.M. Koper, "Hydrogen adsorption on nano-structured platinum", ***Faraday Discussions***, Accepted Manuscript, 2018, DOI: 10.1039/C8FD00062J.

537022 Sustainable Synthesis of Electrocatalytic Bismuth-Based Core-Shell and De-Alloyed Nanoparticles

Abstract:

The dissolution of base metal cores made of Ni, Fe, and Co in the hot, acidic environment of proton exchange membrane fuel cells (PEMFCs) has impeded the use of novel core-shell nanoparticles (NPs) where the precious metals Pt and Ru are confined to surface. Since electrocatalysis occurs on surfaces, a substantial reduction in precious metal use could be achieved with core-shell NPs relative to pure systems, as long as the core remains stable during use. Bi is an abundant earth metal with a low environmental and human toxicity and water stability across a wide pH range, particularly under the acidic conditions of a fuel cell environment. These characteristics make Bi an ideal candidate for use as a durable and inexpensive core material to minimize precious metal use in highly dispersed core-shell Bi-Pt nanoparticle (NP) electrocatalyst in PEMFCs. Bi has been found to substantially promote direct alcohol fuel cell (DAFC) efficiency relative to pure Pd by mitigating catalyst poisoning due to irreversibly bound reaction intermediates such as CO and acetaldehyde. Unfortunately, the formation of Bi NPs is difficult and currently based on unsustainable methods employing toxic organic solvents, expensive organic precursors, and dangerous reducing agents such as hydrazine and lithium that frequently yield highly non-uniform structures. In this work, we describe an inorganic ligand metal redox synthesis approach which is used to synthesize a variety of well-defined core-shell as well as Pt-Bi alloy nanoparticle structures under sustainable, all-aqueous conditions. The approach uses SnCl3- as both a reducing agent and inorganic stabilizing ligand to provide unprecedented control over the size and structure of these nanoparticles. Furthermore, the ligand itself exists in a variety of structures that can be used to control electrostatic assembly onto various substrates. The materials to be discussed in this presentation have never been previously reported and their synthesis is unexpected given both that pure Bi has well-known instability in high salt, low pH environment and that Bi-Sn ligand formation has not been previously recognized. The implications of such metal NP synthesis are transformative with respect to the practical realization and wide application of alternative energy technologies given its simplicity and sustainability. We will provide data showing that such structures can double the electrocatalytic oxygen reduction reaction mass activity and increase the specific activity by nearly an order of magnitude relative to pure systems (see graphic below). As important, we observe simultaneous mitigation of carbon monoxide poisoning due to the persistence of Bi and Sn on their surface. These observations are discussed relative to the surface distribution of active sites as characterized utilizing electrochemical adatom stripping.

536531 Atomically Dispersed and Nitrogen Coordinated Metal Site Catalysts for Oxygen Reduction in Acids

Abstract:

One of the bottlenecks hindering the wide implementation of polymer electrolyte fuel cells (PEMFCs) is the high cost and massive use of platinum (Pt) catalysts for boosting oxygen reduction reaction (ORR) kinetics at the cathode in PEMFCs. Fe-N-C catalysts, derived from pyrolysis of nitrogen, iron and carbon precursors together, have been the most promising candidates to replace Pt because of the abundance and low cost of such elements as well as the highly intrinsic ORR activity of Fe-N-C sites forming during the pyrolysis[1]. However, the major challenge for such Fe-N-C catalysts is to obtain sufficient Fe-N-C active sites catalysts to achieve high ORR activity, since undesired metallic species (i.e. Fe/Fe_3C) are easily produced during high-temperature pyrolysis due to the improper design of synthesis and the poor chemistry control in the preparation of catalysts, lowering the actual Fe-N-C active sites density of catalysts. To address this issue, we have developed an facile approach to prepare Fe-N-C catalysts with exclusively atomically dispersed Fe in Fe-N-C single-sites instead of containing metallic Fe particles by using well-defined Fe-doped metal-organic framework (MOF) precursors. The morphology of MOF precursors can be directly transferred into Fe-N-C catalysts with the retained shape after pyrolysis. Such controlled synthesis of precursors and resulting homogenous catalysts allow us to precisely control and tune the composition and morphology of Fe-N-C catalysts to understand how the structure and composition change of catalysts impact their ORR activity.

This presentation will discuss the critical roles of Fe content in the preparation of catalyst and particle size of catalysts on their ORR activity. The Fe content are controlled at the stage of Fe doped MOF precursor synthesis. Trace amounts of Fe was found to significantly boost ORR activity of catalysts, attributed by the highly-dispersed exclusive Fe-N-C active sites. The ORR activity of such Fe-N-C catalysts enhanced when increasing the density of Fe-N-C active sites at low range of Fe content. On the other hand, decreasing in ORR activity was observed when using high Fe content in the synthesis because of the formation of Fe particles in catalysts at high Fe content. Moreover, it is known that the particle size at nanoscale level critically governs the activity of catalysts. Taking advantages of this facile synthesis from MOF precursors, the Fe-N-C catalysts with size from 20 nm to 1000 nm are able to be prepared by adjusting the size of MOF precursors[2]. As a result, reduction of particle size in Fe-N-C catalysts allows us to increase the utilization of Fe-N-C active sites for ORR. Among studied catalysts, the catalyst with particle size around 50 nm shows the best ORR activity with a half-wave potential of 0.85 V *vs.* RHE, only leaving 30 mV gap with Pt/C (60 $\mu g_{Pt}/cm^2$) in 0.5 M H_2SO_4 along with the excellent stability. When the particle size in catalyst was further reduced to 20 nm, massive agglomeration of particles were found in the catalyst with a decrease in ORR activity. These high-performance Fe-N-C catalysts present a promising potential to replace Pt for ORR in future PEMFCs. In addition to extensively studied Fe-N-C catalysts, atomically dispersed and nitrogen coordinated Co and Mn sites catalysts will be briefly introduced in terms of their synthesis, characterization, electrochemical activity and stability in RDE, and fuel cell performance.

526772 Oxygen Reduction Reaction over a Novel 3D Pt-Supported Vertically Aligned Carbon Nanofiber

Abstract:

For both proton exchange membrane and direct methanol fuel cells, oxygen reduction reaction (ORR) is a critical, performance-determining step. While Pt has been established as a standard bearer in the field of ORR catalyst development, practical long-term applications require continuing effort to pursue active, stable, as well as affordable materials as alternatives to precious metals.

In this talk, carbon-based vertically aligned carbon nanofibers (VACNFs), essentially a stack of conical graphitic structures, will be discussed as highly promising ORR catalysts. In alkaline medium, at -0.4 V applied potential, the current density measured on VACNF cathode is twice more than that measured with commercial Pt catalyst (Pt/C). Molecular insights have been gained through the use of Density functional theory (DFT) calculations to: (1) reveal interactions of ORR intermediate species at the catalytically active sites; and (2) elucidate the origin of catalytic activities of VACNF under reaction conditions. Specifically, semi-periodic fishbone-like stacked graphene sheets, bare or terminated with H and OH, have been employed to understand the impact on ORR mechanism and corresponding reactivities. Bindings of O_2, O, and OH at various configurations obtained from DFT are then used as descriptors to assess catalyst performance against simple Pt (111) and Pt (211) surfaces.

In addition, Pt nanocatalysts supported on VACNF (Pt/VACNF), which can push the ORR current density even higher, will be discussed as well. Again, with DFT, a model with Pt anchored at the VACNF edges was proposed to investigate the functionalities of Pt/VACNF for alkaline ORR reactions. In this talk, both associative and dissociative ORR pathways will be analyzed based on the established models. In particular, the overall computed ORR thermodynamics are shown to be favored toward Pt/VACNF than Pt (111), consistent with experimental results. Nevertheless, the unique interactions of ORR intermediates such as O*, OOH*, and OH* with catalytic active VACNF edge sites, revealed by DFT calculations, suggest intriguing but also complex electrochemistries.

528877 PdCu Alloy Nanoparticles As Highly Active Electrocatalysts for Hydrogen Oxidation in Alkaline Electrolyte

Abstract:

Alkaline membrane fuel cells hold promise to reduce cost by employing a non-precious metal cathode catalyst, however, exploring more efficient anode catalyst to improving hydrogen oxidation reaction, which becomes sluggish in alkaline electrolyte, remains a great challenge. Herein, we report that the bcc-phase PdCu alloy nanoparticles synthesized via a wet-chemical route followed by a thermal-treatment exhibit surprisingly higher HOR mass and specific activities than those of Pd/C and Pt/C in a liquid alkaline electrolyte. HRTEM, XPS, HAADF-STEM element mapping and EELS scanning results show that the particle size, surface composition, and chemical state of PdCu nanoparticles do not change during applied thermal treatment. In-situ HE-XRD characterization reveals that low annealing temperature favors formation of a PdCu fcc structure, but at higher annealing temperatures, the bcc structure dominates PdCu NPs. HOR activity of PdCu catalyst can be attributed to bcc phase dominance. Density functional theory (DFT) computations further unravel the enhanced HOR activity on the bcc structure PdCu NPs originates from contributions from both H and OH adsorption. With the incorporation of Cu, the PdCu catalyst exhibits much weaker hydrogen adsorption strength than that on Pd surface. The PdCu fcc and bcc surfaces possess very similar H binding strength, however, the PdCu bcc surface features much stronger OH binding as compared to that on PdCu fcc surface, which is closer to that on the most HOR active Pt model surfaces. Therefore, the PdCu-bcc phase exhibits much higher HOR activity than its fcc counterpart due to synergistic optimization of both H and OH binding strength.

524239 Recent Developments in Electrochemical Synthesis of Hydrogen Peroxide

Abstract:

Recent Developments in Electrochemical Synthesis of Hydrogen Peroxide

Samira Siahrostami

samiras@stanford.edu

[1] SUNCAT Center for Interface Science and Catalysis, Department of Chemical Engineering, Stanford University, 443 Via Ortega, Stanford, California 94305, United States

Hydrogen peroxide (H_2O_2), is an environmentally benign and powerful chemical oxidant being able to eliminate a wide range of harmful bacteria and inorganic pollutants in water. Electrochemical production of hydrogen peroxide (H_2O_2) offers a sustainable way for local synthesis and onsite applications.[1–3] In this talk, I summarize the recent developments and efforts in electrochemical synthesis of hydrogen peroxide guided by both theory and experiment. [1–6]

References

(1) Siahrostami, S.; Verdaguer-Casadevall, A.; Karamad, M.; Deiana, D.; Malacrida, P.; Wickman, B.; Escudero-Escribano, M.; Paoli, E. a; Frydendal, R.; Hansen, T. W.; Chorkendorff, I.; Stephens, I. E. L. S.; Stephens, I. E.; Rossmeisl, J. *Nat. Mater.* **2013**, *12* (12), 1137–1143.

(2) Shi, X.; Siahrostami, S.; Li, G.; Zhang, Y.; Chakthranont, P.; Studt, F.; Jaramillo, T.F.; Zheng, X.; Nørskov, J. K. *Nat. Commun.* **2017**.

(3) Lu, Z.; Chen, G.; Siahrostami, S.; Chen, Z.; Liu, K.; Xie, J.; Liao, L.; Wu, T.; Lin, D.; Liu, Y.; Jaramillo, T. F.; Nørskov, J. K.; Cui, Y. *Nat. Catal.* **2018**.

(4) Chen, S.; Chen, Z.; Siahrostami, S.; Kim, T. R.; Nordlund, D.; Sokaras, D.; Nowak, S.; To, J. W. F.; Higgins, D.; Sinclair, R.; Nørskov, J. K.; Jaramillo, T. F.; Bao, Z. .

(5) Chen, Z.; Chen, S.; Siahrostami, S.; Chakthranont, P.; Hahn, C.; Nordlund, D.; Dimosthenis, S.; Nørskov, J. K.; Bao, Z.; Jaramillo, T. F.; Wilcox, J.; Bao, Z. N.; Rossmeisl, J. *React. Chem. Eng.* **2017**, *1*, 53.

(6) Siahrostami, S.; Li, G.-L.; Viswanathan, V.; Nørskov, J. K. *J. Phys. Chem. Lett.* **2017**, 1157–1160.

538267 Super Critical Transesterification of Fats and Lipids Extracted from Dissolved Air Floatation Sludge

Abstract:

In recent years, biodiesel has gained immense attention as an alternative to diesel due to rise in petroleum prices, environmental concerns and energy security. The high cost of biodiesel is majorly due to the high cost of feedstocks which are generally edible oils and are not economically feasible. These challenges can be solved by utilizing unconventional sources of oils like non- edible oils and waste cooking oils for biodiesel production.

We examine the potential of using dissolved air floatation (DAF) sludge obtained from local rendered meat processing plants and municipal wastes sludge as a feasible feedstock for biodiesel production using non-catalytic methods. Alkali catalysts method of transesterification of oils involve various shortcomings in the form of high water and free fatty acid (FFAs) content which leads to unwanted side products and saponification leading to low biodiesel yield. Using a less efficient acid catalyst removes this problem but reaction times are unfavorable and requires reaction times of up to 48 hours. This problem becomes more pronounced for lipid sources based on rendered meat industry by products which are rich in FFAs and water. Super critical transesterification is a very effective technique to bypass these shortcomings and provides high yield of biodiesel. DAF sludge was characterized and chemical composition and lipid composition was assessed. Subsequently, extracted lipids were transesterified in a lab scale supercritical reactor set up and various factors like methanol to oil ratio, temperature and reaction time were investigated.

536113 Reaction Engineering Routes to Waste Gasification for Sustainable Living Environments

Abstract:

Concern over "green," or environmentally friendly, technologies has risen considerably over the past decade. Thus, over the past 50 years, metropolitan areas have realized a need for waste disposal alternatives as landfills approach capacity and take up valuable land space. Although newer technology has made incinerators more efficient, there is an increasing interest in formulating more 'green' gasification options as alternatives to incinerators.

Gasification converts organic and carbonaceous materials into a combination of gaseous products known as "syngas," or synthetic gas. This research focuses on advancing the current knowledge of a catalytic gasification process as a potential in-situ resource utilization and waste management system. With significance in a variety of engineering applications, this research is of particular relevance towards municipal waste management and as in-situ resource utilization alternative for advancing space exploration.

The catalytic gasification reaction mechanism evolves through a multi-phase reaction mechanism. The process is initiated by a liquid phase oxidation reactions of long chain polymers. The liquid phase oxidation reactions produce carbon monoxide, carbon dioxide, and water. Current long chain polymers being studied in the catalytic gasification mechanism are focused on cellulose and polyethylene as model substrates. The oxidation reactions are complemented by two gas phase reactions: the Water Gas Shift and the Sabatier reaction, the main stages in the pathway of producing hydrogen and methane.

Gasification experiments examine gasification dynamics for different substrates promoted by different catalysts. It utilizes a high pressure batch scale reactor which allows examining controlled multi-phase catalytic isothermal oxidation of solid and melting substrates in the presence high purity air.

The studied substrates, Polyethylene and Cellulose, are both long chain organic polymers, and make up a substantial portion of both space and municipal waste composition. Although similar in nature, these substrates exhibit marked differences as it pertains to gasification, and were therefore selected as model substrates.

Analysis of the liquid and solid reaction residues, along with the gaseous products, enabled the characterization of the gasification kinetic pathway. The results are used in conjunction with a kinetic model for the gas-phase to formulate a methodological model for the gasification of solid waste.

Preliminary results suggest that catalytic gasification provides potential control for the liquid-phase oxidations. Under appropriate conditions, oxidation was found to follow the incomplete oxidation pathway which resulted in significant yields of hydrogen and methane, which are potentially high value products as they can be used for energy recovery.

The results will be presented along with a preliminary extension for a continuous gasification process, which clearly illustrate the potential of catalytic gasification as a sustainable waste management alternative.

537285 Process Analysis of Continuous Catalytic Gasification As a Waste to Energy Alternative

Abstract:

Presently, space exploration waste is either carried onboard for the duration of the mission or discarded into a trash module which is sent back into Earthâ€™s atmosphere for destruction. On the other hand, with an estimated cost of $10,000 per pound, the cost of delivering mass to locations beyond low earth orbit (LEO) locations is prohibitive.

Since approximately 15 pounds of waste are produced daily by a crew of four during a mission, the intrinsic value of this waste can be greatly enhanced by waste to energy technologies. Moreover, conceptualizing waste gasification as an in-situ alternative for processing of trash is also an efficient route to control waste, while maintaining a healthy habitable environment, during long-duration missions (Santiago-Maldonado et al., 2010).

The gasification of long-chain polymers is hypothesized to follow a mechanism of four combined reactionsâ€"two parallel oxidation reactions which are kinetically controlled, accompanied by two series reaction where oxidation products (carbon dioxide and carbon monoxide) are converted into methane and hydrogen. Integrating this process into spaceflight missions would greatly reduce the onboard mass requirements, as these products can be used as propellant fuel. Further, under certain conditions the process could be also be adapted to produce oxygen and water, two essential life support species for space missions.

This paper explores scale-up alternatives to implement low-temperature continuous gasification processes as potential waste management processes. A kinetic model formulated from batch gasification experiments is integrated into a process simulator along with a detailed thermodynamic description.

Emphasis is placed on the preparatory treatment of the waste slurries, the reactor module, and equilibrium stages for recycling reactants and recovering the catalyst. Detailed characterization of low and mid fidelity simulants is presented.

549045 Electrochemical Process for reduction of Nutrient Runoff from Concentrated Animal Feeding Operations Waste Land Application

Abstract:

In order to recover nitrogen and phosphorous from concentrated animal feeding operations (CAFO) waste as a solid slow release fertilizer product, and to remove a large part of these pollutants from the supernatant, an efficient solution consists in forcing the precipitation of struvite. However, the formation of struvite ($MgNH_4PO_4$Â·$6H_2O$) in animal wastewater treatment plants can lead to scaling and thus operational problems reducing the treatment efficiency. Therefore, controlled struvite formation in animal wastewater treatment plants not only presents an opportunity to recover nutrients but also corresponds to the valorization of wastes. The overall objective of the present research is to demonstrate that it is possible to use an electrochemical reactor with rotating cylinder electrode to recover phosphorus and nitrogen from CAFO waste as a solid fertilizer product ($MgNH_4PO_4$âˆ™$6H_2O$). The effect of applied potential to the cathode, turbulent flow, Ca^{2+}, and temperature on the kinetic of precipitation of struvite was evaluated. Struvite electrochemical precipitation appears as an alternative method to recover nitrogen and phosphorous from waste waters. This approach could offer CAFO a modular technology which smoothly integrates into current manure lagoon systems. Key parameters controlling the electrochemical precipitation process of struvite are: solution pH, temperature, rate of stirring, and the presence of foreign ions including calcium.

536269 Integra LCA: An Innovative Tool for Health Risk Assessment of Plastic Waste

Abstract:

Plastic waste management and the associated risks to human and ecosystem health are recognized as key issues in sustainable waste and resource management worldwide. The plastic crisis, has induced a number of jurisdictions to pose bans on use of plastic bags and enhance plastic recycling in the respective municipal waste management systems. Still to date, however, landfilling remains the most common waste management practice in Greece in spite of enforced regulations aimed at increasing recycling, pre-selection of waste and energy and material recovery.

Especially with regard to plastic waste, a lot of studies have exemplified the adverse effects of low-biodegradability plastic material on the health and sustainability of natural ecosystems, disrupting the food web and inducing endocrine disruption and eventually gender alteration (feminization) in sensitive species such as fish. This effect in turn endangers biodiversity and ecosystem sustainability.

As yet, there is a much more limited number of studies focusing on the adverse human health effects of plastic products and waste, the ubiquitous nature of plastic material notwithstanding. Thus, in this study we have developed an innovative tool for integrated health risk assessment of plastic waste. The INTEGRA LCA software couples the integrated external and internal exposure assessment capabilities of the INTEGRA computational platform (Sarigiannis et al, 2014) with life cycle impact assessment (Clift et al, 2000). The integrated software platform allowed us to perform a first-of-its-kind analysis of adverse health outcomes attributable to chronic exposure to persistent organic pollutants associated with plastic material use and disposal.

Our analysis focused on plastic waste generated in the two main metropolitan centers in Greece, Athens and Thessaloniki. A comprehensive review of up-to-date information on plastic products and plasticizers used by the urban population was performed in order to build up the application-specific release/emissions inventory. This review included both plastic products (e.g. PET bottles, PVC material, polycarbonate products) and plasticizers used in food packaging. The latter is of particular importance when considering the total human exposure to persistent toxic compounds found in plastic matrices, such as plasticizers found in the interior of cans, baby bottles, pacifiers and other products coming in contact with the oral route of exposure (Luz, 2012). Compounds of interest in this regard include bisphenol A, phthalates such as DEHP and its metabolites, DINCH, di-(2-ethylhexyl)adipate (Ehlert et al, 2008). The environmental fate analysis performed, included both multi-media contamination of POPs found in plastic waste and contamination of the food web including seafood.

Integration of all human exposure routes and pathways to the toxic compounds contained in plastic was done at the level of systemic internal dose using the intake fraction methodology (Sarigiannis and Karakitsios, 2016). We then parameterized properly the state-of-the-art generic physiology-based biokinetic (PBBK) model in the INTEGRA platform for these compounds and used it to reckon the biologically effective dose (BED) at the target tissues more closely associated with the putative adverse health outcomes considered in our study. Thus, we estimated the level of homeostatic perturbation induced by the BED at the target tissues. The extent of the perturbation was then used as the fundamental metric that was linked with adverse health outcomes reported in the literature to quantitatively assess the related health risk.

Bisphenol-A is characterised by low environmental concentrations, as a result of the limited environmental emissions. In Greece, BPA concentration in urban PM has been estimated equal to 0.06 ng/m^3, which is in a very good accordance with the predictions of the multimedia model incorporated in INTEGRA. Exposure to BPA mainly occurs through exposure to specific consumer products. As expected, children are exposed to higher amounts of BPA, and the mean urinary concentration in Greece have been identified equal to 1.2 and 2.1 μg/g creatinine for adults and children respectively. Using the exposure reconstruction module of the INTEGRA platform, the respective external exposure levels correspond to 0.06 and 0.156 μg/kg_bw/d, which is far below the t-TDI of 4 μg/kg_bw/d set by EFSA. Although DEHP is produced in lower volumes compared to BPA, overall exposure is significantly higher. The measured urinary levels of MEHP, the major metabolite of DEHP have been identified as 7.6 and 2.8 μg/g creatinine for adults and children respectively. Similarly to BPA, contribution of DEHP exposure originates from consumer exposure (mainly use of PVC flooring and use of various plastic toys and equipment, as well as food conduct materials), however, these exposure levels (10 μg/kg_bw/d are in the same order of magnitude with the respective TDI. DEHA is considered one of the major DEHP substitutes. Based on the LCA and considering its broad applications, the estimated daily intake is in the range of 1 μg/kg_bw/d. In Europe, the estimated intake for children based on biomonitoring data, has been estimated within the same range. DINCH is another of the DEHP substitutes. Based on the respective LCA, the predicted exposure estimates of DINCH are close to 0.7 μg/kg_bw/d; this value is within the range the one (0.5 μg/kg_bw/d) estimated from biomonitoring data collected in daycare centers in Germany. Contribution from consumer applications is similar a few orders of magnitude higher that the rest of contributing pathways.

Among the various waste management options, incineration contributes to a higher amount of exposure through air and dust for the local population, as a result of the higher emissions in air from the incinerator. Among the various waste management options, incineration contributes to a higher amount of exposure through air for the local population, due to the higher emissions occurring from the incinerator. Similarly, exposure to dust is also slightly higher. On the contrary, landfilling contributes to slightly higher exposure through food (and drinking water) from the leachates ending up to the ground underwater and eventually to drinking water.

The estimated risks of the plasticizers investigated above seem to be relatively low, with the exception of DEHP, where the estimated intake is in the same order of magnitude with the respective TDI. Beyond that, the cumulative effect of these compounds has to be taken into account, as well as the presence of additional plasticizers (beyond the ones investigated herein) within the plastic matrix. Among the various steps of the respective life cycle, waste remaining in the environment is expected to be particles/fragments abraded from end-use products during their service life and during disposal (e.g. particles abraded from car undercoating, coil coating, shoe soles and fragments of plastic bags etc.). These particles are primarily released to the urban/ind. soil compartment in landfills. However, the smallest fraction may also be distributed to the air compartment or to the surface water environment ending up in the sediment. Overall, the comprehensive analysis of the waste management options indicated that incineration contributes significantly to increased levels of exposure through inhalation for the population living close to the incinerator facility, while on the other hand recycling results in lower exposure from all environmental exposure pathways.

The coupled integrated exposure and life cycle assessment methodology developed in this study and translated into the INTEGRA LCA platform is a significant step towards the direction of comprehensive, precise and transparent estimation potential health risks associated with use, management and disposal of plastics in urban settings. Our first results in the two major Greek metropolitan centers indicate the need to optimize the management of plastic waste in the country and to enhance the awareness of consumers with regard to the long-term adverse health effects of chronic exposure to plastic components. The incorporation of life cycle analysis produces different conclusions than a simple environmental impact assessment based only on estimated or measured emissions. Taking into account the overall life cycle of both the waste streams and of the technological systems and facilities envisaged under the plausible scenarios analyzed herein, alters the relative attractiveness of the solutions considered and enhances the robustness of the health impact assessment.

References

Clift, R., Doig, A., Finnveden, G., 2000. The Application of Life Cycle Assessment to Integrated Solid Waste Management: Part 1-Methodology. Process Safety and Environmental Protection 78, 279-287.

Ehlert, K.A., Beumer, C.W.E., Groot, M.C.E., 2008. Migration of bisphenol A into water from polycarbonate baby bottles during microwave heating. Food Additives and Contaminants - Part A Chemistry, Analysis, Control, Exposure and Risk Assessment 25, 904-910.

Luz C, 2012. Our food packaging and public health, Environmental Health Perspectives, 120(6): A232-A236.

Sarigiannis D.A. and Karakitsios S.P. Complex exposure modeling. In: Mixtures toxicology and risk assessment. J.E. Simmons and C. Rider (eds.), Springer (2016).

Sarigiannis, D.A., Karakitsios, S., Gotti, A., Loizou, G., Cherrie, J., Smolders, R. De Brouwere, K., Galea, K., Jones, K., Handakas, E., Papadaki, K., Sleeuwenhoek, A. 2014. INTEGRA: From global scale contamination to tissue dose, 7th International Congress on Environmental Modelling and Software (iEMSs), San Diego (California), USA, 15-19/6/2014.

522027 Energy-Water Nexus Study for a Mushroom Farming Initiative in Nigeria

Abstract:

A water treatment and recycle system was designed for a mushroom farming initiative in Kaduna, Nigeria. The energy requirements of the proposed system were studied to determine the optimal system to minimize energy use for water treatment. Other considerations were ease of use and capital and operating costs.

The concept behind the cooperative is for 200 women farmers pool their resources to pay for water and energy costs. Electricity will be provided through solar panels. We were consulted to design a water treatment and recycle system for a demonstration of the concept. The concept will likely be replicated in remote areas without an abundance of water, so water recycle is essential. Reliable electricity will also likely be an issue so solar panels will be used.

The first step was to collect samples of water from the potential water source: the Kaduna River. Preliminary tests showed that the water was relatively clean but contained bacteria and some hardness. A treatment and recycle system was designed appropriate for the impurities in the water. Due to limited energy provided by the solar panels, energy requirements were studied for all appropriate treatments to minimize energy use of the final system. The final design and final energy use of the water system will be presented.

523808 Sustainable Optimal Strategic Planning for Shale Water Management

Abstract:

The development of new additives in the shale gas industry, which tolerate the use of high Total Dissolved Solids (TDS) base fluid, has allowed reusing the wastewater in the drilling of subsequent wells (U.S. Environmental Protection Agency, 2016). This practice is currently the most popular and cost-effective option for shale gas water management. Although it minimizes freshwater consumption, producers should take into consideration possible long-term issues and challenges. For instance, the TDS concentration will increase significantly, around 2-3 times, which can represent a cost barrier to reuse the water for fracturing operations. Besides, as the number of drilled wells decrease this practice becomes less attractive. Specifically, the volume of fracturing fluid required to fracture new wells may be less than the volume of water generated by producing wells in the area.

Currently, several works have been reported on the optimization of shale gas water management (Gao and You, 2015; Guerra et al., 2016; Yang et al., 2015). However, most of the works published in the literature consider that the water blending ratio is restricted or the return to pad operations are not allowed. Drouven and Grossmann (2017) assume that the water-blending ratio is unrestricted, they over-estimate the friction reducers expenses. The MILP model that they proposed does not account the salt concentration of impaired water. Moreover, they only distinguish between two types of water: impaired water and freshwater. Hence, the model cannot handle any water management option when drilling operation decrease.

In this work, we propose a mixed-integer non-linear programming (MINLP) model considering the TDS concentration of flowback and impaired water as a function of time and for different water treatment solutions. We estimate the friction reducers expenses as a function of TDS concentration to determine if the level of TDS in impaired water is an impediment to reusing it in fracturing operations. Moreover, the model distinguishes between four types of water: impaired water, flowback water, desalinated water and freshwater. The objective is to maximize the "sustainability profit" in order to obtain a compromise solution between economic, environmental and social aspects. Only a single objective function is necessary since all the indicators are expressed in monetary terms (Zore et al., 2017).

The water management system comprises wellpads, shale gas wells in each wellpad, centralized water treatment technologies (CWT), freshwater sources, fracturing crews, and disposals wells.

After hydraulic fracturing, a portion of water called flowback water returns to the wellhead. The flowback water is stored in fracturing tanks (FT) onsite before basic treatment (pre-treatment) in mobile units or transport to CWT facility or Class II disposal. Pre-treatment includes technologies to remove suspended solids, oil and grease, and bacteria, certain ions that can cause the scale to form on equipment and interfere with fracturing chemical additives. After pre-treatment, the water can be used as a fracturing fluid in the same wellpad or in the neighboring wellpad, or can be desalinated in the onsite TDS removal technologies. The flowback water reused for fracturing operations is called impaired water and is stored in impaired water tanks (IWT). Several desalination technologies can be selected such as multi-stage membrane distillation (MMD), multi-effect evaporation with mechanical vapor recompression (MEE-MVR) or forward-reverse osmosis (FO-RO) hybrid. We consider that the outflow brine salinity in the onsite treatment is close to salt saturation conditions to achieve ZLD operation. Cost and salinity levels restrict the type of desalination unit that can be used for TDS removal. The onsite desalinated water can also be used as a fracturing fluid in the same wellpad, transported to the next wellpad or discharge for other usages. The flowback water can also be transported and treated in CWT. Desalinated water from CWT can select the same routes as the desalinated water in onsite technologies. Freshwater is obtained from an uninterruptible fresh water source. Desalinated water and freshwater are stored in freshwater tanks (FWT) and/or water impoundment. Transportation of freshwater, impaired water, flowback water and desalinated water can only be through trucks. Storage tanks and mobile treatment are assumed to be leased.

The problem is to determine: (1) wellpad fracturing start date (fracturing schedule), (2) number of tanks leased at each time period, (3) number of trucks needed in each time period, (4) flowback destination, reuse (impaired water) , treatment (onsite or offsite) or disposal, and (5) quality of water used to fracture each well.

The optimization problem is formulated as an MINLP model that includes: assignment constraints, material balance in storage tanks, mixer and splitters, logic constraints and an objective function. Salt material balances are modeled using total flows and salt composition as variables (bilinear terms). An advantage of using this representation is that the variables involved in the bilinear terms are well bounded, allowing us to define tight under and over estimators.

Them objective function includes the economic profit, the eco-profit and the social-profit.

- *Economic profit* includes revenues from natural gas minus the sum of the following expenses: wastewater disposal cost, storage tank cost, freshwater cost, friction reducer cost, wastewater and freshwater transport cost and onsite and offsite treatment cost.
- *Eco-profit* distinguishes between eco-benefit (raw material and products that unburden the environment) and eco-cost (raw material al products that burden the environment). Both terms are calculated by using eco-cost coefficients (Delft University of Technology, 2017). In our problem, impaired water and desalinated water used to fracture a neighboring well exhibit unburdening effect on the environment. However, natural gas, freshwater withdrawal, disposal and transportation burden the environment.
- *Social profit* includes social security contributions paid for the employed people to fracture a well, plus the social transfer by hiring people, minus social cost. We only take into account the numbers of jobs on a fracturing crew and the time that they are working to fracture a specific well. Once the well is completed, the number of jobs generated by truck drivers or maintenance team are not considered.

The proposed model is applied to a case study in Marcellus Play with 3 wellpads and 20 wells, three years discretized at one week per time period, four interruptible sources of fresh water, three class II disposal wells, two CWT plants and one fracturing crew. The MINLP problem consists of 2808 binary variables, 14977 continues variables and 12010 constraints.

The solution of the model shows that the capabilities to reuse impaired water to fracture other wells is the best economic and environmental practice for shale gas water management. The model, which can be effectively solved with the proposed decomposition technique, reveals that the level of TDS in impaired water is not an obstacle to reusing it for fracturing purposes. Also, it has been shown that onsite desalination treatment can be cost-effective for operators once no more wells to fracture are available.

References:

Delft University of Technology, 2017. The Model of the Eco-costs/Value Ratio (EVR). URL http://www.ecocostsvalue.com (accessed 12.1.17).

Drouven, M.G., Grossmann, I.E., 2017. Optimization models for impaired water management in active shale gas development areas. J. Pet. Sci. Eng. 156, 983–995.
Gao, J., You, F., 2015. Shale Gas Supply Chain Design and Operations toward Better Economic and Life Cycle Environmental Performance: MINLP Model and Global Optimization Algorithm. ACS Sustain. Chem. Eng. 3, 1282–1291.
Guerra, O.J., Calderón, A.J., Papageorgiou, L.G., Siirola, J.J., Reklaitis, G. V., 2016. An optimization framework for the integration of water management and shale gas supply chain design. Comput. Chem. Eng. 92, 230–255.
U.S. EPA, 2016. Technical Development Document For Effluent Limitations Guidelines and Standars for the Oil and Gas Extraction Point Source Category. Washington, DC.
Yang, L., Grossmann, I.E., Mauter, M.S., Dilmore, R.M., 2015. Investment optimization model for freshwater acquisition and wastewater handling in shale gas production. AIChE J. 61, 1770–1782.
Zore, Ž., Čuček, L., Kravanja, Z., 2017. Syntheses of sustainable supply networks with a new composite criterion – Sustainability profit. Comput. Chem. Eng. 102, 139–155.

534130 A Multi-Objective Energy-Water Nexus Planning Model: A Case Study of the Power Systems in Texas Edwards Aquifer

Abstract:

Increasing power demands and water scarcity as well as environmental concerns give rise for the need of a planning strategy for power generating systems that takes into consideration the nexus between energy and water [1]. Power systems planning problems are complex and comprise of decisions related to: (i) the ability to add additional power plants with different types of cooling and power generating technologies, (ii) the ability to convert the cooling and or power generating technologies of existing power plants, (iii) uncertainties, and (iv) environmental and social impacts [2]. Given that the type of power generating and cooling technology of a power plant directly affects its water usage, the optimization of a power system is a multi-objective problem [3]. Therefore, a systematic methodology based on the Energy-Water Nexus (EW-N) is necessary for modeling and solving the optimization problem on power systems planning [4].

In this work, we develop a novel superstructure for planning strategies of power generating systems based on EW-N connections. The EW-N problem is formulated as a two-stage stochastic multi-objective mixed integer linear program [5]; with "here-and-now" decision variables such as additional power plants with specific power generation and cooling technologies and "wait-and-see" decision variables comprising of and not limited to the capacity and water utilization of power plants. The model determines the optimal conversion and expansion policies for the current infrastructure, based upon uncertain future water availability and power demands. The model is implemented into a case study regarding EW-N in the Edward Aquifer, aiming to provide policy makers with a systematic tool to analyze various scenarios and technology options. Moreover, the model is also flexible in that it can also be integrated into the Edwards Aquifer Groundwater and River System Simulation Model (EDSIMR) [6, 7], which is a regional model for the comprehensive decision-making and resources allocation regarding Food-Energy-Water Nexus in the Edwards Aquifer.

References:

[1] Diangelakis, N. A., & Pistikopoulos, E. N. (2017). A multi-scale energy systems engineering approach to residential combined heat and power systems. Computers & Chemical Engineering, 102, 128-138.

[2] Koltsaklis, N. E., Dagoumas, A. S., Kopanos, G. M., Pistikopoulos, E. N., & Georgiadis, M. C. (2014). A spatial multi-period long-term energy planning model: a case study of the Greek power system. Applied Energy, 115, 456-482.

[3] Mohtar, R. H., & Daher, B. (2016). Water-energy-food nexus framework for facilitating multi-stakeholder dialogue. *Water International*, *41*(5), 655-661.

[4] Garcia, D. J., & You, F. (2016). The water-energy-food nexus and process systems engineering: a new focus. *Computers & Chemical Engineering*, *91*, 49-67.

[5] Birge, J. R., & Louveaux, F. (2011). Introduction to stochastic programming. Springer Science & Business Media.

[6] Gillig, D., McCarl, B. A., & Boadu, F. (2001). An economic, hydrologic, and environmental assessment of water management alternative plans for the south central Texas region. Journal of Agricultural and Applied Economics, 33(1), 59-78.

[7] McCarl, B. A., Dillon, C. R., Keplinger, K. O., & Williams, R. L. (1999). Limiting pumping from the Edwards Aquifer: an economic investigation of proposals, water markets, and spring flow guarantees. Water Resources Research, 35(4), 1257-1268.

534167 Optimal Use of Thermal Membrane Distillation (TMD) for Treatment of Flowback Water

Abstract:

Optimal use of thermal membrane distillation (TMD) for treatment of flowback water

Kaiyu Cao[1], Priscille I. Etoughe[2], Rajib Mukherjee[2,3], Debalina Sengupta[3], Joseph S. Kwon[1,2,3*], Mahmud M. El-Halwagi[1,2,3]

1. *Artie McFerring Department of Chemical Engineering, Texas A&M Univrsity, College Station, TX*
2. *Texas A&M Energy Institute, Texas A&M University, College Station, TX*
3. *Gas and Fuels Research Center, Texas A&M Engineering Experiment Station, College Station, TX*

The flow rate of flowback water from hydraulic fracturing varies with time. Most of the water are produced within the first two weeks after hydraulic fracturing is completed. The concentration of total dissolved solids (TDS) in the flowback water also changes with time. For efficient recovery and reuse of flowback water, treatment with thermal membrane distillation (TMD) for the removal of TDS has been found to be the method of choice [1]. Even though flowback water is a continuous process, in this work, for modeling purpose a multi-period optimization problem has been formulated to account for the time-dependent variation of the flow rate and TDS concentration of flowback water, accounting for the time-; hereeach day is treated as a period. Then, we have integrated a TMD unit, considering the largest capacity of flowrate from a well in any given period. An upper limit on the concentration of TDS in the discharged (reject) water is considered to avoid precipitation. Utilizing the field data of flowback water from a given well for each period, the optimization formulation has been solved to determine the maximum period of time for optimal use of TMD, maximizing annualized profit from recovered water per period. Our result shows that it is not profitable to use water recovery beyond first few days, depending on the formation. The performance of the optimization formulation is measured in terms of percentage of recovery and is compared with cost associated with them.

References:

[1] Elsayed, N. A., Barrufet, M. A., Eljack, F. T., & El-Halwagi, M. M. (2015). Optimal design of thermal membrane distillation systems for the treatment of shale gas flowback water. *Int J Membr Sci Technol*, *2*, 1-9.

536214 The Energy-Water Nexus of Thermoelectric Power Generation and Its Impacts in the Muskingum River Watershed in Ohio

Abstract:

Thermoelectric power plants generate electricity and flue gas scrubbers mitigate emissions from the plants. However, at the same time the power plants withdraw large quantities of water from the watershed, which corresponds to 45% of the total 2010 US water use [1], and emissions from the scrubbers cause deposition of nutrients in the watershed. Therefore, increasing electricity generation could increase water stress and deteriorate water quality. Also, in the watershed where power plants are located, other activities, such as farming, interact with the power plants since these activities also require water as well as energy from the power plants and release nutrients to the watershed. To prevent shifting of environmental impacts across multiple flows [2], the Energy-Water nexus between different activities in the watershed needs to be understood in assessing the impacts of power plants.

Muskingum River Watershed (MRW) in Ohio ranks fourth as the most polluted watershed in the US [3], and two coal-fired power plants and three natural gas-fired power plants are located in the MRW. In this work, a holistic analysis approach is employed to evaluate the water-energy nexus in the MRW. Various technological and ecological systems are considered. Technological systems include mining, thermoelectric power generation, and CO_2 conversion activities. Ecological systems include land use and farming practice. Alternative scenarios for each component are analyzed to address the nexus. An agro-ecological model, which is developed by the Soil and Water Assessment Tool (SWAT), is used to assess water quality and quantity [4], while engineering models are used for the power plants and scrubbing processes.

Fossil fuel power plants not only require a huge amount of water, but also emit 28% of the 2016 US greenhouse gas emissions [5], 67% of the 2014 US SO_2 emissions, and 12% of the 2014 US NO_x emissions [6]. Therefore, ecosystem services, such as water quality regulation, air quality regulation, and climate change regulation, play an important role in mitigating nutrient runoffs and air emissions. To account for the role of ecosystems, Techno-Ecological Synergy framework is applied in analyzing the impacts of activities in the MRW [7,8]. In this analysis, the trade-offs between water quality, water quantity, net electricity generation, carbon footprint, air quality, and net present value are identified. Various scenarios are considered to suggest better solutions that are 'win-win' in terms of multiple objectives.

Considering the Energy-Water nexus in the analysis will avoid shifting of the environmental impacts across multiple flows because interactions between flows are identified and all flows are included as objectives [2]. This work could be applied to any watershed and could be extended to address the Food-Energy-Water nexus by including food productions from farming activities as another objective. Addressing the nexus could help businesses and policy makers in making decisions to improve the sustainability of the watershed.

References

[1] Maupin, M.A. et al. Estimated use of water in the United States in 2010. U.S. Geological Survey, Reston, Virginia (2014).

[2] Bakshi, B. R., Gutowski, T. G. & Sekulic, D. P. Claiming Sustainability: Requirements and Challenges. ACS Sustain. Chem. Eng. 6, 3632–3639 (2018).

[3] Inglis, J. & Dutzik, T. Wasting our waterways: Toxic industrial pollution and restoring the promise of the clean water act. Environment America Research & Policy Center, Boston, MA (2006).

[4] Khanal, S., Lal, R., Kharel, G. & Fulton, J. Identification and classification of critical soil and water conservation areas in the Muskingum River basin in Ohio. J. Soil Water Conserv. 73, 213–226 (2018).

[5] Inventory of U.S. greenhouse gas emissions and sinks: 1990-2016. U.S. Environmental Protection Agency, Washington, D.C. (2018).

[6] 2014 National Emissions Inventory (NEI). US Environmental Protection Agency. https://www.epa.gov/air-emissions-inventories. Accessed on Apr 13 2018.

[7] Bakshi, B. R., Ziv, G. & Lepech, M. D. Techno-ecological synergy: A framework for sustainable engineering. Environ. Sci. Technol. 49, 1752–1760 (2015).

[8] Gopalakrishnan, V., Bakshi, B. R. & Ziv, G. Assessing the capacity of local ecosystems to meet industrial demand for ecosystem services. AIChE J. 62, 3319-3333 (2016).

537140 Systematic Analysis and Optimization of Water-Energy Nexus

Abstract:

Securing scarce energy and water resources from the environment, while at the same time satisfying on-growing water and energy demands for end-use consumption is crucial [1]. The intermediate processing facilities, energy sources (acting also as water sinks) and water sources (acting as energy sinks), receive energy and water natural resources respectively, process, exchange and deliver them as usable electricity, fuels and water for consumption. It is important to quantify the performance and the efficiency of this water-energy nexus, and understand how it is affected by the selection of the energy and water sourcesâ€™ technologies, operating capacities and their connectivity.

To this end, we present a scalable, graph-theoretic representation of a nexus for the design and optimization of complex water-energy systems [2]. We define a nexus as a directed bipartite graph with water and energy flows and show that for specified external grid demands, the optimal nexus configuration with minimum water and energy generation is the one without any redundant subgraphs. We then introduce the water-energy (WEN) diagram that can systematically identify and eliminate those redundant subsystems. This leads to (i) minimum generation/extraction of water and energy resources from the environment, or (ii) maximum yield of water and energy to meet external demands. Our approach is simple to implement and results in optimal nexus configurations that also consider operational constraints, connectivity restrictions and water quality specifications. In the second part, we develop mathematical programming-based algebraic method that provides the same results but for more complicated and large-scale nexus design. This also takes into account the intensity factors of the sources, restrictions on source-sink matching, and quality specifications. The model is further extended to include capital and operating costs of the processing facilities, as well as geographic distances. The approach is demonstrated using case studies on water-energy nexus systems focusing on power generation, seawater desalination, groundwater and surface water at regional and national scales.

References:

[1] D. J. Garcia and F. You, "The water-energy-food nexus and process systems engineering: a new focus," Computers & Chemical Engineering, vol. 91, pp. 49-67, 2016.

[2] S. Tsolas, M. N. Karim, M. M. F. Hasan, " Optimization of Water-Energy Nexus: A Network Representation-based Graphical Approach," Applied Energy, *Submitted.*

538380 Thermo-Economic Optimization Based Comparison of Membrane Distillation Vs Mechanical Vapor Recompression for Shale Gas Produced Water Treatment

Abstract:

With advancements in horizontal drilling and hydraulic fracturing methods, unconventional shale gas extraction is rapidly developing. However, shale gas production is accompanied with a host of environmental challenges. One such critical challenge is the sustainable management of large quantities of high salinity wastewater generated in the process of hydraulic fracturing. While recycling may be a short-term strategy, long term solutions are needed for economical and environmentally conscious management of high salinity wastewaters. Desalination is one possible technological solution to treat the produced water and reduce the brine volume for final disposal while simultaneously producing fresh water for other beneficial uses. Membrane distillation (MD) system is a vapor pressure driven emerging membrane technology capable of handling hypersaline wastewater. Since MD can be operated at low temperature (30-90^{0}C), it has the potential to be integrated with low-grade waste heat sources. Another advantage of MD is the low capital investment of the process. However, low energy efficiency of MD results in high operating costs when waste heat is not available. Mechanical vapor recompression (MVR) is another desalination technology capable of treating hypersaline produced water. Unlike MD, MVR system is a well-established work driven desalination process, in which the heat recovery and energy efficiency is relatively high by utilizing the vapor compression cycle. In contrast with MD, MVR system requires considerable amount of high quality energy sources. Considering the competitive desalination efficiency of these two technologies, there is a need for a systematic way to comparison their energetic and economic performance. To date, very limited work exists on techno-economic optimization and evaluation of MD and MVR technologies for produced water treatment.

We present a mixed-integer non-linear programming (MINLP) formulation of a direct contact membrane (DCMD) process design for produced water treatment while minimizing the total treatment cost. The proposed model simultaneously optimizes the operating conditions (i.e. the temperature, pressures, and flowrate of all streams) and DCMD plant configuration (number of membrane stages, number of modules in each stage, and membranes geometry). Also, the optimization model, takes into account the possibility to improve system performance by including brine recirculation and heat integration. The results of the proposed optimization model are compared with our previous work on MVR optimization for application in produced water treatment. The optimization model is tested for a hypothetical desalination plant with 0.5 million gallons per day (MGD) capacity, in which produced water with 10% salinity is concentrated to 30% salinity of the exiting brine. The results are also validated through ASPEN simulations. Sensitivity analysis is also conducted to analyze the effect of input parameters on the system performance and economics. Tradeoffs between MD and MVR technologies for treatment of shale gas produced water will be discussed.

529522 Application of Adsorbate Solid Solution Theory to Design Novel Adsorbents for Arsenic Removal Using Computer-Aided Molecular Design

Abstract:

Application of Adsorbate Solid Solution Theory to Design Novel Adsorbents for Arsenic Removal Using CAMD

Rajat K. Doshi[1,2], Arti Ayodhyaprasad Rajput[3], Rajib Mukherjee[1], Suresh Gupta[3] , Urmila M. Diwekar[1*],

1. Vishwamitra Research Institute, Crystal Lake, IL, USA
2. Henry B. Plant High School, 2415 S. Himes Ave., Tampa, USA
3. Department of Chemical Engineering, Birla Institute of Technology & Science, Pilani, India

*Corresponding: Urmila@vri-custom.org

Arsenic is a carcinogenic contaminant that pollutes the groundwater, a consequence of poor arsenic disposal. Various techniques are used to remove arsenic, such as oxidation, coagulation-flocculation, ion exchange, and adsorption, among which adsorption is the most efficient method. Current arsenic separating agents have a limited adsorption capacity. The overall objective of this work is to develop a computational tool for the design of novel adsorbents for arsenic remediation using clay materials including beidellite, zeolite, and sepiolite that are cheap and readily available [1]. In the first part of this research, we use the Group Contribution Method (GCM) to predict thermodynamic properties and calculate the UNIFAC interaction parameters between arsenic and other functional groups that are selected from clay materials. In the second part of this research, we utilize a computer-aided molecular design (CAMD) framework that develops new adsorbent candidates with enhanced adsorption capacities based on the group interaction parameters generated in the first part. The efficient ant colony optimization (EACO) algorithm maximizes the adsorption capacity with certain structural possibilities, thermodynamic property correlations, and process conditions. It was found that the newly designed adsorbent has an order of magnitude higher removal capacity than the adsorbents' reported in the literature [2]. The adsorption capacity estimated for the designed adsorbent was simulated in groundwater conditions; however, arsenic contamination is not solely confined to groundwater as it is also highly prevalent in wastewater flows from coal-burning power plants. An expansion of our work will be carried out by conducting experiments on the designed adsorbents. The goal is to test the performance of the designed adsorbent in coal power plant's wastewater flows and understand how the different ionic environment affects the adsorption capacity. The maximum amount of arsenic that the designed adsorbent can adsorbed per unit mass of adsorbent from wastewater found in coal-burning power plants will be presented. Moreover, the novel adsorbent has also been synthetically prototyped to determine the actual adsorption capacity of the novel adsorbent in varying ionic environments and arsenic concentrations. Our initial results from conducting a set of experimentation has demonstrated an enhanced adsorption capacity of the synthetic adsorbent that is higher than any naturally occurring adsorbent.

References:

1. Bektaş, N., Aydın, S., & Öncel, M. S. (2011). The adsorption of arsenic ions using beidellite, zeolite, and sepiolite clays: a study of kinetic, equilibrium and thermodynamics. Separation Science and Technology, 46(6), 1005-1016.
2. Doshi, R. K., Mukherjee, R., & Diwekar, U. M. (2018). Application of Adsorbate Solid Solution Theory To Design Novel Adsorbents for Arsenic Removal Using CAMD. ACS Sustainable Chemistry & Engineering, 6(2), 2603-2611.

542620 The Impact of Shale Gas and Oil on the Chemical Industry

Abstract:

Lower alkanes and aromatics are very useful as both fuels and chemical raw materials. Occasionally supply disruptions constrained availability but also encouraged investigation of alternative feedstocks and chemistries. Recent new technologies for the production of unconventional gas and oil sources from tight shale formations significantly increased availability of both methane and condensate, radically altered process economics, resulted in the construction of much new intermediates capacity, and multiplied the ways traditional and alternative chemistries might be exploited. This presentation will outline historically how natural gas components became useful as chemical feedstocks, how government regulations occasionally led to perturbations in availability and price, how supply crises led to the development of alternative chemistries, catalysts, and ultimately the exploitation of shale resources. More specifically, will be discussed how excess coproducts, especially condensate, from increasing shale gas and oil production for energy are having significant unexpected impacts on the chemical process industries and opportunities for new process chemistry research.

541186 Sustainable Energy and Chemicals: Past, Present, and Future

Abstract:

The last 40 years has seen the U.S. energy situation vacillate from gasoline rationing and concerns over Peak Oil, to projected energy and feedstock prices that were noncompetitive in global markets for chemicals production, to the economic revolution of frack gas and light tight oil. The importance of low cost energy and feedstocks in driving economic opportunity will be discussed relative to this history, along with the importance of sustainability in decision making to gain stakeholder buy-in and acceptance. Strategies for managing career paths and competencies despite changing targets and high uncertainty will be addressed.

542618 Disruptions: What the Future May Hold

Abstract:

Disruptive forces are all around us. They change the way we communicate and work. They change the way we develop goods and services. They also change the types of products that are manufactured. Examples such as the way in which we produce and consume energy or the feedstocks we use to produce product have seen many changes over the course of the last few decades, and these disruptions are expected to continue. The chemical engineering discipline and the expectations for a chemical engineerâ€™s skills has also changed and adapted as a result of these disruptive forces. A discussion of the broad themes associated with these forces and the implications that they may have to chemical engineers in the future will be discussed.

542621 Geopolitical Factors Influencing the Evolution of the Chemical Industry

Abstract:

The basic chemicals and plastics industry is built on a simple principle: find a cheap feedstock and convert it into something more valuable. However, the right feedstock depends on place and time. Certain policies and actions of various governments have influenced the evolution and profitability of the chemical industry during the past 50 years, sometimes in surprising ways. For example, high oil prices and US government funding encouraged the development and commercialization of coal gasification technology during the late 1970s and early 1980s. This enabled China to develop a competitive basic chemical industry using coal as a feedstock during the last 25 years. Similarly, an abrupt increase in oil prices during the first decade of the 21st century encouraged the present shale gas revolution, causing a major shift in the economy of producing basic chemicals and polymers in the US. If history is our guide, then what governments choose to fund and regulate may be as important as market forces in determining the shape of our industry and profession in the future.

542622 Agility & Resilience: How to Maintain Career Competitiveness in the Changing Chemical Industry

Abstract:

The world around is changing. The shift to protectionism, global instability and price volatility of raw materials and feedstocks has the potential to wreak havoc on the chemical industry. With the revolution of shale gas in North America and the sudden drop in feedstock costs, the chemical sector in the US has reasserted its historic dominance. Learn how the role of a chemical engineer can contribute to industry growth and resilience, while adapting to allow for personal agility throughout the career life cycle.

536023 Single-Molecule Characterization of Cazyme Protein Modules Adsorption to Multivalent Glucan Polymers like Cellulose

Abstract:

Protein adsorption to multivalent carbohydrate (or glycan) derived ligands plays a critical, but poorly understood, role in both fundamental cellular (e.g., glycocalyx and viral protein binding) and industrial processes (e.g., cellulosic sugar-based biofuels production). However, we currently have a poor understanding of the molecular mechanisms driving protein adsorption to glucan polymers like cellulose. We also lack suitable experimental tools and predictive adsorption models that can suitably explain the heterogeneous and varying affinity multivalent protein-carbohydrate binding interactions. Here, we study the adsorption-desorption of a family 1 carbohydrate-binding module (CBM), and its conserved aromatic amino acid mutants, to model glucan polymers like cellulose using a combination of bulk ensemble and single-molecule protein-ligand binding methods. Our single-molecule force spectroscopy method allows us to directly probe the structure-function relationships driving CBM adsorption to glucan polymers. Analysis of the CBM-cellulose unbinding rupture forces and total bond lifetimes dataset has provided a molecular basis for distinct binding orientations for CBMs on ordered glucan polymer surfaces like cellulose and the corresponding binding affinity for each orientation. Molecular dynamics simulations provide further insights into the possibility of multiple binding orientations for CBM on cellulose surfaces, which is an often-neglected feature of these interactions. We also report ongoing work with the aromatic amino acid mutants that are critical to stabilizing specific binding orientations and how these binding orientations could vary for various Type-A and Type-B CBMs on cellulose allomorphs. Our findings allow us to generate a comprehensive picture of non-productive binding interaction mechanisms of CBMs with glucan polymers that will ultimately lead to the development of more efficient enzymes for cellulosic biomass deconstruction into fermentable sugars.

535107 Combinatorial Experimental and Computational Approach for the Effective Entrapment of Glucose Oxidase in Hyaluronic Acid Nanogels

Abstract:

Biocatalysis offers important advantages over traditional chemical catalysts when sustainable technology pathways are being considered. The advantages of high specificity, activity, mild reaction conditions, and high product selectivity of the biocatalysts are however outflanked by the limitations that an enzymatic approach will have in commercial-scale processes, namely loss of enzyme stability over time, narrow process operating conditions, and requirement of intensive and costly separation processes from products. Much work is currently done in the field of enzyme immobilization for enzymatic activity retention and reusability, as well as for improvement of product yield, all at low implementation costs. In particular, the immobilization of enzymes onto renewable biopolymers has recently gained attention for both its sustainable approach and its abundant, inexpensive starting materials. Herein we evaluate the suitability of a naturally occurring polysaccharide and main component of the extracellular matrix as both a support as well as an active and protective encapsulator during enzyme immobilization. For this, we combined hyaluronic acid (HA) with model enzyme glucose oxidase (GOx) and evaluated both the enzymeâ€™s functionality at the interface with the polysaccharide as well as its ability to catalyze direct electron transfer controlled reactions. The reaction kinetics at the interface have been elucidated both experimentally using spectroscopic and atomic force microscopy assays and computationally using molecular dynamics simulations to offer both a viable mean for understanding transfer reactions when product formation is being considered as well as for evaluating the effectiveness of immobilization. Our combinatorial approaches provide a feasible strategy to control enzyme-based interface reactions that allow for fast product formation while preserving and extending enzyme functionality and stability. Further, our results provide a viable approach to implement enzyme-based immobilization for different applications ranging from glucose biosensors for the control of diabetes to stabilizers for food preservatives and colors.

523575 Spatiotemporal Dynamics from a Classic Enzyme Cascade with Self-Governing Substrate Competition

Abstract:

The construction of enzymatic reaction networks *in vitro* with programmable and predictable dynamics is a central goal of synthetic and systems biology. Biomacromolecules such as genes and enzymes (proteins) are the basic building blocks in the fabrication of such dynamic systems. Despite the abundance of complex dynamics realized with genetic circuits, and even though enzymes play vital roles in the regulation of metabolic pathways and signal transduction processes, the demonstration of enzyme-based networks *in vitro* has only been explored very recently.[1,2]

In this presentation we will introduce a two-enzyme reaction network consisting of a FAD-dependent oxidoreductase and a peroxidase that can generate tunable complex dynamics.[3] In a classic enzyme cascade of glucose oxidase (GOx) and horseradish peroxidase (HRP), we discovered a new reaction that GOx can take $ABTS^{+\bullet}$ as a competitive substrate with oxygen to oxidize glucose. This reaction becomes dominated when the dissolved oxygen is depleting. Based on this, we can generate multiple pulsed responses including the typical charging/discharging, rectangular and parabolic waveforms in a closed system. The waveforms can be well tuned by changing the concentrations of substrates and enzymes. We also demonstrated a safer version of the green bottle experiment with this network, in which the system toggles between colorless and green when the solution is shaken.

In an open system, this reaction network can spontaneously form and display reaction-driven Rayleigh-Bénard convective patterns. The convection is initiated by the accumulating hydrodynamic instability at the air-solution interface where the enzymatic reactions create a denser layer of product. The patterns manifest themselves in green resulting from the interfacial formation and convective flow of $ABTS^{+\bullet}$. The spatiotemporal patterns are highly dependent on the depth of the solution and will become increasingly unstable with greater depth.

In summary, we demonstrated the smallest enzymatic reaction network with tunable output waveforms and spatiotemporal pattern formation. The complex dynamic behaviors of this network are based on self-governing substrate competition, rather than activation or inhibition. In addition, the ease of direct observation of convective patterns may also encourage further research on reaction-driven Rayleigh-Bénard convection.

References:

1. Semenov, S. N. et al. Rational design of functional and tunable oscillating enzymatic networks. *Nature Chemistry* **2015**, 7, 160–165.
2. Nijemeisland, M., Abdelmohsen, L. K. E. A., Huck, W. T. S., Wilson, D. A. & van Hest, J. C. M. A compartmentalized out-of-equilibrium enzymatic reaction network for sustained autonomous movement. *ACS Central Science* **2016**, 2, 843–849.
3. Zhang, Y., Tsitkov, S. & Hess, H. Complex dynamics in a two-enzyme reaction network with substrate competition. *Nature Catalysis* **2018**, 1, 276–281.

528491 Uncovering the Quantum Mechanical Origins of Enzymatic Catalysis with Systematic QM/MM Methods and Accelerated, Large-Scale Electronic Structure

Abstract:

The large size of most enzymes necessitates hybrid quantum mechanical-molecular mechanical (QM/MM) simulation for mechanism elucidation. Numerous convergence studies of QM/MM methods have revealed over the past several years that key energetic and structural properties approach asymptotic limits with only very large (ca. 500-1000 atom) QM regions. This slow convergence has been observed to be due in part to significant charge transfer between the core active site and surrounding protein environment, which cannot be addressed by improvement of MM force fields or the embedding method employed within QM/MM. I will discuss our recent development of systematic methodologies that enable atom-economical determination of optimal QM regions (typically ca. 200-300 atoms) and provide insight into the crucial interactions in enzyme active sites captured only in large QM regions. The two approaches, based on measuring the change in chemical environment around an active site or the degree of charge transfer between core substrates and the surrounding environment, produce robust, consistent, and readily validated results on test case enzymes (e.g., catechol O-methyltransferase, cytochrome P450cam, and hen eggwhite lysozyme) that asymptotically approach results from very large QM regions. These systematically selected QM regions enable identification of charge-transfer-mediated prolonged electrostatic attraction as a previously overlooked component of enzyme rate enhancement in group transfer across a wide family of methyltransferases. I will present how pairing our efforts with graphics-processing-unit-accelerated QM/MM simulation has enabled our collection of over a nanosecond of enhanced sampling dynamics with range-separated hybrid density functional theory to reveal the magnitude of charge transfer and free energy barriers in QM/MM with QM regions up to 600 atoms in size. These simulations reveal dramatic fluctuations in the charge density mediated through hydrogen bonding interactions in the greater protein environment as a key component of enzyme action that is apparent only with large, systematically selected QM regions.

532386 Towards Scalable Production of Enantiomerically Pure Amines: Enzyme Mechanism and Kinetics

Abstract:

Towards scalable production of enantiomerically pure amines: enzyme mechanism and kinetics

Robert D. Franklin‡, John M. Robbins‡, Joshua A. Whitley‡, Andreas S. Bommarius‡#

‡School of Chemical and Biomolecular Engineering, Georgia Institute of Technology,

Parker H. Petit Institute for Bioengineering and Bioscience,

Engineered Biosystems Building, 950 Atlantic Drive N.W., GA 30322, USA

#School of Chemistry and Biochemistry, Georgia Institute of Technology,

950 Atlantic Drive, Atlanta, GA 30332-2000, USA

Amine dehydrogenases (AmDHs) catalyze the reductive amination of prochiral ketones to produce chiral amine compounds[1-3]. The reaction requires ammonia as the nitrogen source and a hydride transfer from an NADH cofactor, as seen in the reaction scheme below. Chiral amines are frequent intermediates in high value active pharmaceutical ingredients. AmDHs were developed by engineering the substrate binding pocket of amino acid dehydrogenases (AADHs) to eliminate activity towards keto acids and introduce activity toward ketones. Determining kinetics is a necessary step toward to designing a reactor for processing on scale. While significant work has been done to determine the kinetics of AADHs, the same treatment has not been given to AmDHs. In the present work, we demonstrate the first detailed study into the kinetic mechanism of amine dehydrogenases, in both the reductive amination and oxidative deamination directions. Of particular interest are the differences and similarities between the mechanisms and kinetics of AmDHs and their parent AADHs, as well as the inhibition patterns. We also employed isothermal titration calorimetry to determine binding order and relative substrate affinity. Lastly, we investigated temperature- and pH-dependence of the rate. Understanding of enzyme kinetics, and determination of a rate law are key steps toward developing a viable API manufacturing strategy using enzyme catalysts.

References:

1. Bommarius, B. R.; Schurmann, M.; Bommarius, A. S., A novel chimeric amine dehydrogenase shows altered substrate specificity compared to its parent enzymes. *Chem Commun (Camb)* **2014,** *50* (95), 14953-5.

2. Abrahamson, M. J.; Vazquez-Figueroa, E.; Woodall, N. B.; Moore, J. C.; Bommarius, A. S., Development of an amine dehydrogenase for synthesis of chiral amines. *Angew Chem Int Ed Engl* **2012,** *51* (16), 3969-72.

3. Abrahamson, M. J.; Wong, J. W.; Bommarius, A. S., The Evolution of an Amine Dehydrogenase Biocatalyst for the Asymmetric Production of Chiral Amines. *Advanced Synthesis & Catalysis* **2013,** *355* (9), 1780-1786.

522125 Construction of Artificial Metalloenzyme Catalyst and the Study of Size Effect

Abstract:

The incompatibility between enzyme and metal catalyzed reactions is a critical issue in integrating chemoenzymatic cascade processes. In this study, we report the synthesis of temperature-responsive lipase-palladium nanohybrids (Pluronic-CALB-Pd) with the size of Pd NPs ranging from 0.8 nm to 2.5 nm. As the Pd NPs downsizing to sub-nanoclusters, the catalytic performance of Pluronic-CALB-Pd in racemization of (S)-1-phenylethylamine was dramatically improved and even exhibited efficient catalytic activity at 55 ^{o}C which perfectly matched the optimal temperature of CALB. The yield of dynamic kinetic resolution (DKR) of racemic 1-phenylethylamine catalyzed by Pluronic-CALB-$Pd_{1.61}$ was up to 96 % and 99 e.e. at 55 ^{o}C for 12 h. The catalytic performance was greatly enhanced compared to the combination of commercially available Novozym 435 and Pd/C. In addition, the Pluronic-CALB-Pd nanohybrids can be homogeneously dispersed in organic solvents at relatively high temperature (>50 ^{o}C) and be precipitated at ambient temperature for an easy reuse after reaction.

Figure 1. Catalytic performance of Pluronic-CALB- Pd nanohybrids in the racemization of (S)-1-phenylethylamine

539823 Functionalized Magnetic Graphene Oxide Sheets for Efficient Cota Laccase Immobilization

Abstract:

In the recent decades, massive industrial revolution of textile dyes brought a serious environmental pollution, especially for water. Synthetic dyes gave warning not only for the environment but also for human health because of their existence in wool, textiles, papers, and leathers. As environmentally friendly biocatalysts, laccases can degrade a broad range of industrial toxic contaminants. However, the industrial applications are limited due to the low stability and poor reusability of free laccases.

Enzyme immobilization can reduce the restrictions by increasing the stability, durability and realizing continuous operations. Up to now, laccase has been successfully immobilized on various kinds of carriers, such as microspheres, nanoparticles, nanofibers and membrane. Among these carriers, the specific magnetic particles can be produced by immobilization of an affinity ligand on the surface of prefabricated magnetic beads, which can be quickly separated from the reaction medium and controlled by applying a magnetic field; then the catalytic efficiency and stability properties of the enzyme can be greatly improved.

In the current study, functionalized magnetic graphene oxide (MGO) support attached with NÎ±,NÎ±-Bis (carboxymethyl)-L-lysine hydrate (NTA-NH_2) and chelated with Cu^{2+} were synthesized. The Cu^{2+}-chelated MGO (MGO-NTA-Cu^{2+}) exhibited the highest adsorption capacity of 177 mg/g-support for CotA laccase among all synthesized nano-composites. The maximum activity recovery of laccase using MGO-NTA-Cu^{2+} was 114%. The catalytic properties of MGO-NTA-Cu-CotA laccase were significantly improved in comparison with those of free laccase. MGO-NTA-Cu-CotA laccase showed efficient decolorization rate for Congo red (CR) reached 100 % after 5 h reaction at 60 Â°C and pH 8. MGO-NTA-Cu-CotA laccase retained 89.4 % of its initial activity after 10 consecutive cycles. These MGO-NTA-Cu^{2+} supports provide promising potential for large-scale laccase immobilization in practice.

539274 Catalytic Aromatization of Bio-Derived Liquid Under Methane Environment

Abstract:

Hydrodeoxygenation upgrading is the most popular way to produce bio-oil from biocrude. This approach usually involves continuous hydrogen flow and high operation pressure (15~100 atm), resulting in significantly increased capital and operational costs. Compared to hydrogen, which is not naturally available and thus costly, methane can be readily obtained as natural gas. This work studies the technical feasibility of directly using methane as the reducing agent instead of hydrogen to remove oxygen from a series of model compounds including ethanol, acetic acid, acetone, and phenol, representatives of biocrude derived from low cost carbon resources at low pressure (<5 MPa) and moderate temperatures (400 °C) under the facilitation of transition metal modified aluminosilicate catalysts, the produced aromatics abundant liquid will become more cost competitive. Moreover, methane itself is incorporated into the formed aromatics beneficial for extra liquid chemical production along with lower carbon dioxide (CO_2) emissions from the upgrading process. The involved catalytic upgrading mechanism is also intensively studied using isotopic labeling technique coupled with versatile characterizations including NMR, FTIR, XAS, XPS, SIMS, and DSC as well as molecular simulation at DFT level for verifying methaneâ€™s participation and benefiting catalystâ€™s rational design. The outcomes from this research will provide a more efficient and cleaner way to utilize low cost carbon resources and natural gas which are abundant in nature. Furthermore, the successful implementation of the study will be beneficial for the effective reduction of greenhouse gas emission and significant cost reduction of the current bio-oil production process making it more economically feasible.

529825 Wave Liquefaction™ Processing of Carbon Materials for the Production of Value-Added Chemicals and Feedstocks: Spectroscopic Diagnostics and Material Characterization

Abstract:

1.0 Introduction

This paper presents optical spectroscopic diagnostics as used in an engineering pilot implementation of H Quest Vanguard's microwave plasma process applied to coal, and material characterization results from application of the process to H_2 generation from natural gas. Plasma diagnostics for process control will be illustrated. Chemical analyses are presented of coal conversion products. Recent results yielding nanographene as a premium carbon by microwave-plasma processing of methane will be shown.

Microwave-driven plasma-mediated reaction engineering represents process intensification for energy efficient and economical hydrocarbon conversion processes. Examples include coal-to-liquids and natural gas conversion to hydrogen and premium carbon products. Optical emission spectroscopy and analysis provides species identification and determination of temperatures in different energy modes in the non-equilibrium plasma.

Microwave energy, variable between 1 - 4 kW is coupled into a continuous flow reactor. Wave Liquefaction™ process applied to coal implements in-situ product upgrading via hydrogenation and/or methylation reactions enabled by activating hydrogen or methane present in the process gas. Pulverized coal is fed by a particle seeder. Gas ratios are optimized for the particular hydrocarbon feedstock conversion. An alternative application of the microwave plasma processing is methane decomposition to H_2 and premium carbon products. Reactor residence times are sub-second and performed at atmospheric pressure. Filters, traps and cyclone stages downstream of the reactor perform gas/liquid/solid separations.

2.0 Optical Emission Spectroscopy

Optical diagnostics are central to reaction characterization and hold particular value for species identification and temperature determination [1]. Observed intensity ratios or spectra band shapes can yield temperature by Boltzman analysis using spectral constants. Moreover, optical emission serves to identify reactive species and intermediates, albeit indirectly inferred by the observation of their electronically excited counterparts, e.g. CH* and C_2* radicals. For example, the presence of key atomic or diatomic radicals can support postulates of electron impact dissociation, and radical mediated bond insertion or radical capping reactions and provide mechanistic insights from the temperatures associated with the different degrees of freedom – electronic, vibrational, such as from the C_2^* ($d^3{}_g - a^3{}_u$) Swan band emission. In this study spectroscopic data collection was performed using a fiber-optically coupled Ocean Optics HR2000 spectrometer fitted with a UV-NIR grating for a spectral range of 900 nm. All optical emission spectra were captured with equipment spectrally calibrated using a Hg/Ar lamp and corrected for instrument response function using a NIST-traceable tungsten lamp as spectral calibration standard. Temperatures of C_2* are determined using custom spectral band-fitting algorithms, taking into account the electronic baseline and underlying blackbody radiation. Spectral fits are performed at 1 nm spectral resolution. Carbon particle temperatures are determined by fitting Planck's radiation law with black body conditions. Ar* emission lines are identified by reference to the NIST spectroscopic database.

3.0 Gas/Liquid/Carbon Analyses

Microwave (MW) activated natural gas decomposition for H_2 generation produces no CO_2 and requires zero process water, in contrast to the present industrial process – steam reforming. Notably it opens a path for renewable energy storage by coupling electrical energy into chemical bond energy – with H_2 being particularly versatile as fuel or chemical precursor. Moreover, MW activated natural gas decomposition does not require additional carbon as catalyst but rather produces carbon [2]. Therein MW driven decomposition of methane achieves green hydrogen production and production of value-added carbon materials [3]. Here a suite of characterization tools have been applied to the light gases, liquid and solid carbon products. Gas and liquid analyses were performed using gas chromatography-mass spectrometry. For the solid carbon products SEM has been applied to coal chars for cenosphere analysis, and for the carbon products from methane decomposition, scanning electron microsocpy (SEM) for morphology, transmission electron microscopy (TEM) for (nano)structural assessment and identification of different carbon forms (i.e. sp^2 phases), X-ray diffraction (XRD) for evaluation of graphitic structure, thermo-gravimetric analysis (TGA) for bulk determination of oxidative reactivity of the carbons as a means by which to assess graphitic content and gas (N_2) adsorption analyses for texture – i.e. surface area and porosity. Elemental analysis was also performed for HCNS and O by difference. Complementary thermo-gravimetric analyses assess the oxidative reactivity of the carbon products. Optical diagnostics such as multi-wavelength pyrometry relate these material characteristics to reactor conditions and can be applied for process control.

4.0 Summary

H Quest Vanguard is developing broad-spectrum microwave plasma processes targeting conversion of hydrocarbon feedstocks such as coal and natural gas to value-added materials, chemicals and fuels [4,5].

A microwave pyrolysis approach was originally proposed in response to the DARPA initiative for green-house gas (GHG)-emission-free and cost-effective production of US Air Force jet fuel from the domestic coal resources. In H Quest Vanguard's process, natural gas can be used in single-stage reactor as a hydrogen source, eliminating external hydrogen production units and the associated CO_2 production, water consumption, and capital costs, and providing excess hydrogen sufficient for downstream hydro-treating. The MW process produces fast heating resulting in flash devolatilization and pyrolysis followed by fast quenching, preserving the primary pyrolysis products and volatiles' molecular structure.

Since company formation in 2014, H Quest Vanguard, Inc. has developed ample material base and expertise in development of chemical and catalytic processes enhanced by microwave plasma. In particular, H Quest has developed a conversion process that applies microwave energy to rapidly co-pyrolyze solid hydrocarbons (e.g. coal) and natural gas to produce liquid hydrocarbons. A wide range of carbon materials including graphene and ordered carbon blacks have been observed across a wide range of experiments. These forms have potential for high value applications: electrical conductivity additives for plastics, and as electrode material in supercapacitors and Li-batteries. Subsequent and ongoing work has demonstrated feasibility of a microwave driven plasma mediated methane decompostion for H_2 production and nanographene formation as a premium carbon.

5.0 Acknowledgements

H Quest Vanguard, Inc. is a privately held technology company, based in Pittsburgh, Pennsylvania, focused on the development and commercialization of novel hydrocarbon conversion technologies. This material is based on work supported by the Department of Energy, Office of Science through subaward agreement no. 186949 with H Quest Vanguard, Inc. under the Prime Award DE-SC0015895 Phase I SBIR.

6.0 References

[1] Vander Wal, R. L., Gaddam, C. K., and Kulis, M. J. (2014). An Investigation of Micro-Hollow Cathode Glow Discharge Generated Optical Emission Spectroscopy for Hydrocarbon Detection and Differentiation. *Applied spectroscopy*, *68*(6), 649-656.

[2] Dagle, R., Dagle, V., Bearden, M., Holladay, J., Krause, T., and Ahmed, S., (2017). *An Overview of Natural Gas Conversion Technologies for Co-Production of Hydrogen and Value-Added Solid Carbon Products* (No. PNNL-26726; ANL-17/11). Pacific Northwest National Lab. (PNNL), Richland, WA (United States); Argonne National Lab. (ANL), Argonne, IL (United States).

[3] Abbas, H. F., and Daud, W. W. (2010). Hydrogen production by methane decomposition: A review. *International Journal of Hydrogen Energy,* 35(3), 1160-1190.

[4] Skoptsov, George L., and Alan A. Johnson. *Method for processing hydrocarbon fuels using microwave* energy. U.S. Patent 9,682,359, issued June 20, 2017.

[5] Strohm, J.J., Linehan, J.C., Roberts, B.Q., McMakin, D.L., Sheen, D.M., Griffin, J.W. and Franz, J.A., Battelle Memorial Institute, 2012. *Heavy Fossil Hydrocarbon Conversion and Upgrading Using Radio-Frequency or Microwave Energy.* U.S. Patent Application 13/401,216.

537027 Hydrocarbon Chain Growth Via a Nonthermal Electrical Plasma Microreactor

Abstract:

Currently, large amounts of methane are emitted or flared to the environment from stranded sources such as landfills and agricultural sources. The infrastructure required to bring these resources to market is cost prohibitive. There is an economic opportunity to convert this methane into a more transportable or value-added product that can be brought to market.

Using a nonthermal electrical plasma, reactions that usually require high temperatures and pressures can actually take place close to ambient conditions. Instead of thermal collisions of molecules causing reactions, high energy electrons collide with molecules in order to disrupt bonds. In this case, carbon-hydrogen bonds are broken which allows for longer hydrocarbons to form. This process therefore transforms electrical energy into chemical energy.

Performing these reactions at the microscale level has the advantage of decreasing the space between the electrodes that create the electrical plasma. By reducing the size, less voltage is required in order to generate these plasmas which reduces the costs of the necessary power electronics.

Recent results of the chemical species produced by the process and the energy efficiency of the plasma will be discussed.

525092 Combined Experimental and Density Functional Theory Studies on the Modified $SrTiO_3$ Catalysts for Oxidative Coupling of Methane

Abstract:

The main component of natural and shale gas, methane, attracted global attention because of the richness of deposits and significant potential for direct or indirect conversion into value added chemicals and fuels. Oxidative Coupling of Methane (OCM) is one of promising methods of adding value to methane via directly producing ethane and ethylene, vital building blocks in chemical industry. We studied modified $SrTiO_3$ based perovskites with metal dopants which were prepared by polymeric precursor method. $A_xSr_{1-x}TiO_3$ (A = Ba, Ca, La, Y, Li, Na, K, Rb, Cs, Al, Mg, Tl, and Pb, x = 0.2) and $SrB_xTi_{1-x}O_3$ (B = Sn, Al, Mg, Ce, Pr, Nd, Sm, and In, x = 0.2) were investigated. Also, we use density functional theory (DFT) calculations to remove ambiguity on the identification of the catalytically active site and devise descriptors for high-performance catalyst. The volcano curve was observed for C_{2+} selectivity with methyl radical adsorption energy from experimental and DFT results. The correlation indicated that the appropriate strength of methyl radical adsorption is required for improving C_{2+} compounds formation. Based on catalyst characterization using X-ray diffraction (XRD), O_2-pulse injection and temperature programmed desorption (O_2-TPD), and X-ray photoelectron spectroscopy (XPS), the influences on catalytic activity by partial replacement of cations with A- and B-site dopants were investigated.

538208 Process Technology Assessment for Mega Scale Projects

Abstract:

The assessment of process technologies is an important part of integrated process engineering for enterprises to make strategic decisions on new commercialization investment or on the expansion of existing operations with cutting-edge technologies. It becomes even more critical especially when a project requires technologies that are unfamiliar to an organization.

This paper discusses a basic understanding of the work required and the fundamentals in evaluating innovative process technologies for mega scale projects in chemicals manufacturing. The assessment principles, criteria and the processes will be discussed in both technology, economics as well as sustainability perspectives. Coal Gasification technology for Syngas manufacturing is used as an example.

531904 Toward Supply Chain Optimization of Renewable Energy Carriers

Abstract:

A main challenge of renewable energies such as solar and wind is their intermittency. Wind and solar availabilities fluctuate daily, seasonally, and geographically. Moreover, they are often temporally and spatially asynchronous with our energy demands. One proposed solution to the intermittency problem is to store energy during periods and in areas of excess supply and to utilize the stored energy during periods and in areas of excess demand. An option for energy storage is using chemicals such as hydrogen[1], ammonia[2], or methanol[3]. Unlike batteries, chemicals are more energy dense, easily transportable to the end-user, and offer flexibility in terms of end-usage. They act as energy carriers, moving solar and wind energy from one location to another.

While the industrial production of these chemicals is well-established, the integration of renewable energies into the manufacturing process and supply chain is an unresolved task. Toward this latter aim, in this work, we describe the development of a supply chain model framework to systematically analyze the potential of solar and wind to displace energy demand through the deployment of energy carriers. The framework investigates the various tradeoffs and competing options for building an energy carrier infrastructure, and through optimization, the best energy carrier network is elucidated. We will highlight the results on a Texas case study and showcase the feasibility of energy carriers.

[1] Mazloomi, K.; Gomes, C., Hydrogen as an energy carrier: prospects and challenges. Renewable and Sustainable Energy Reviews 2012, 16 (5), 3024-3033.

[2] Klerke, A.; Christensen, C. H.; Nørskov, J. K.; Vegge, T., Ammonia for hydrogen storage: challenges and opportunities. Journal of Materials Chemistry 2008, 18 (20), 2304-2310.

[3] Olah, G. A., Beyond oil and gas: the methanol economy. Angewandte Chemie International Edition 2005, 44 (18), 2636-2639.

527237 Methods for Quantitative Consideration of Ecosystem Services in Supply Chain Design and Optimization

Abstract:

Methods for Quantitative Consideration of Ecosystem Services in Supply Chain Design and Operations

Daniel J. Garcia[a], Fengqi You[b]*

1. Department of Chemical and Biological Engineering, Northwestern University, Evanston, IL, 60626
2. Department of Chemical and Biomolecular Engineering, Cornell University, Ithaca, New York, 14853

*Email: Fengqi.you@cornell.edu

Submitted for consideration in session 23C00 Design, Analysis, and Optimization of Sustainable Energy Systems and Supply Chains

Ecosystems provide essential services – termed ecosystem services (ES) – to humanity that make our lives possible, such as nutrient cycling within a forest or flood management from mangroves.[1] However, despite these critical services that ecosystems provide, many ecosystems around the world are either degraded or destroyed in the name of economic growth.[1] Usually, the inherent value of ecosystems that will be degraded or destroyed is not considered in supply chain design and operations. When they are, they are treated relatively qualitatively or only a few ES are considered at a time.[2-4] In addition to replacing ecosystems entirely, current industrial facilities or enterprises that have already displaced natural ecosystems could further diminish ES provided to their local communities. For example, air quality near a significant polluter of total suspended particulates in the air is worse than air quality in areas not near such a polluter.[5] This may seem obvious, but the underlying concept that enterprises could diminish local ES – in this case clean air provisioning – is important to consider and quantify in supply chain operations, regardless of the enterprise under study or the set of affected ES. With better understanding of how an enterprise or supply chain produces or diminishes economic and ecological value, more sustainable supply chains and operations can be achieved. However, there are many different ecosystems, ES, and methods for valuing ES, making quantitative integration of ES values into decision-making frameworks difficult. On the other hand, many of the individual ES valuation methods are now quite sophisticated, and the ES research community has recently emphasized quantitative application of these methods as soon as possible.[6]

In this work, we integrate ES values into a supply chain design and operations framework utilizing three different evaluation strategies for ES. To the best of our knowledge, this work is the first to demonstrate quantitative application of several ES valuation approaches in a single decision-making modeling framework. We consider ES valuation by the hedonic method, the avoided cost method, and the aggregated benefit transfer method. In the hedonic valuation method, changes in nearby housing values after the construction of a nearby industrial enterprise or reduction of existing enterprises' impacts on local ES, subject to stringent statistical tests, are considered to be proxies for the decrease or increase, respectively, in local ES.[7,8] The avoided cost method considers negative economic costs that would have been avoided if the ecosystem had been undisturbed.[9] Finally, the aggregated benefit transfer approach identifies the overall value of all ES in a particular ecosystem, and applies that value to other, similar ecosystems.[10] There are drawbacks to each method, but they represent a range of valuation techniques often used in ecological economics to influence policy and make land use decisions.[11]

We formulate a multiobjective mixed integer linear program (MILP) that considers ES degeneration or regeneration from the three aforementioned methods. Two objectives of overall cost and Green GDP (nominal GDP plus sum of all ES values in a region) are considered.[12] We apply the model to a case study on bioenergy production in the US state of Illinois, where the state would like to boost renewable energy production while simultaneously improving the environment its people live in. 160 PJ of bioenergy (approximate production of corn ethanol in Illinois in 2016) can be produced as bioethanol or biogas. For bioethanol production, we assume existing forests and/or prairie are repurposed as cropland for input feedstocks, losing the existing ES. We model these ES impacts with the aggregated benefit transfer method. Bioenergy may also be produced by biogas from swine manure, which in turn regenerates local ES lost from on-site storage and spilling of the manure, modeled with the hedonic method and avoided cost method. Spatial data for distribution of ecosystems in Illinois were analyzed in ArcGIS 10.5 and integrated directly into the MILP model.[13] Data from the USDA and US Census Bureau were also integrated directly into the model. Traditional supply chain and process construction costs are considered in the overall cost objective, and ES values lost or regenerated in addition to the value of ethanol and electricity produced are considered in the Green GDP objective.

Green GDP ranges from -\$125M/y to \$132M/y with corresponding costs ranging from \$107M/y to \$260M/y. Solutions with low Green GDP do not often convert manure to biogas and only produce ethanol only from corn. As Green GDP increases, more manure is converted to biogas, and a transition to switchgrass over corn emerges to produce ethanol. We thus conclude that considering ES values significantly influences supply chain design and operations decision-making. With our more holistic approach, more cost-effective and ecologically-friendly solutions are identified.

References

1. Assessment ME. Ecosystems and Human Well-being: Biodiversity 26. *Synthesis. World Resources Institute, Washington, DC.* 2005.
2. Guo M, Richter GM, Holland RA, Eigenbrod F, Taylor G, Shah N. Implementing land-use and ecosystem service effects into an integrated bioenergy value chain optimisation framework. *Computers & Chemical Engineering.* 2016;91:392-406.
3. Hanes RJ, Gopalakrishnan V, Bakshi BR. Synergies and trade-offs in renewable energy landscapes: Balancing energy production with economics and ecosystem services. *Applied Energy.* 2017/08/01/ 2017;199:25-44.
4. Gopalakrishnan V, Bakshi BR, Ziv G. Assessing the capacity of local ecosystems to meet industrial demand for ecosystem services. *AIChE Journal.* 2016;62(9):3319-3333.
5. Chay KY, Greenstone M. Does air quality matter? Evidence from the housing market. *Journal of political Economy.* 2005;113(2):376-424.
6. Polasky S, Tallis H, Reyers B. Setting the bar: Standards for ecosystem services. *Proceedings of the National Academy of Sciences.* 2015;112(24):7356-7361.
7. Seok Lim J, Missios P. Does size really matter? Landfill scale impacts on property values. *Applied Economics Letters.* 2007;14(10):719-723.
8. Herriges JA, Secchi S, Babcock BA. Living with hogs in Iowa: the impact of livestock facilities on rural residential property values. *Land Economics.* 2005;81(4):530-545.
9. de Groot RS, Wilson MA, Boumans RMJ. A typology for the classification, description and valuation of ecosystem functions, goods and services. *Ecological Economics.* 6// 2002;41(3):393-408.
10. Costanza R, de Groot R, Sutton P, et al. Changes in the global value of ecosystem services. *Global Environmental Change.* 5// 2014;26:152-158.

11. Plummer ML. Assessing benefit transfer for the valuation of ecosystem services. *Frontiers in Ecology and the Environment.* 2009;7(1):38-45.
12. Boyd J. Nonmarket benefits of nature: What should be counted in green GDP? *Ecological economics.* 2007;61(4):716-723.
13. *ArcGIS Desktop: Release 10.5* [computer program]. Redlands, CA: Environmental Systems Research Institute 2017.

538655 Superstructure-Based Optimization of Carbon Dioxide Conversion and Utilization Via Syngas Intermediate

Abstract:

CO_2 is one of the primary components in greenhouse gas (GHG) emissions due to human activities [1]. Global levels of CO_2 emissions have been steadily increasing over the decades reaching an unprecedented high of 45 Gt in 2017. An effective way to mitigate the rising GHG emissions is to convert the CO_2 present in feedstocks such as flue gas, natural gas and biogas to syngas [2]. In chemical industries, syngas is prevalently used as an intermediate for manufacturing several value-added products including methanol, dimethyl ether, olefins and gasoline [3,4]. Most of the previous optimization studies on CO_2 conversion and utilization have focused on producing specific fuels or chemicals, thereby resulting in sub-optimal process flowsheet routes. This creates a need for developing a comprehensive superstructure encompassing viable alternatives to obtain optimal CO_2 utilization and process costs [5].

In this work, we postulate a superstructure consisting of multiple layers to form process intermediates and products. To achieve a desired conversion, each layer is comprised of necessary reaction, separation and product upgradation steps along with heating and compression steps which produce auxiliary emissions. The first layer of the superstructure consists of seven process alternatives (dry reforming, partial oxidation, steam methane reforming, auto-thermal reforming and their hybrid combinations) for syngas formation starting from flue gas, natural gas and biogas. The second layer forms intermediates (e.g., methanol, dimethyl ether, heavy olefins) which can further be processed to produce hydrogen-based fuels and chemicals in subsequent layers. The overall superstructure is then used to identify an optimal CO_2 conversion and utilization route in GAMS environment with two separate objectives – maximization of net CO_2 utilization and minimization of total annualized cost. Specifically, we obtain a maximum net CO_2 utilization of 57% for feedstock conversion to syngas with consideration of the auxiliary emissions. The presentation would also cover further conversion of syngas to subsequent value-added products.

References

[1] M. M. F. Hasan, F. Boukouvala, E. L. First, and C. A. Floudas, "Nationwide, regional, and statewide CO2 capture, utilization, and sequestration supply chain network optimization," *Ind. Eng. Chem. Res.*, vol. 53, no. 18, pp. 7489–7506, 2014.

[2] S. S. Iyer, I. Bajaj, P. Balasubramanian, and M. M. F. Hasan, "Integrated Carbon Capture and Conversion To Produce Syngas: Novel Process Design, Intensification, and Optimization," *Ind. Eng. Chem. Res.*, vol. 56, no. 30, pp. 8622–8648, 2017.

[3] K. Roh, R. Frauzem, T. B. H. Nguyen, R. Gani, and J. H. Lee, "A methodology for the sustainable design and implementation strategy of CO2 utilization processes," *Comput. Chem. Eng.*, vol. 91, pp. 407–421,

2016.

[4] Z. Yuan, M. R. Eden, and R. Gani, "Toward the development and deployment of large-scale carbon dioxide capture and conversion processes," *Ind. Eng. Chem. Res.*, vol. 55, no. 12, pp. 3383–3419, 2015.

[5] P. Balasubramanian, I. Bajaj, and M. M. F. Hasan, "Simulation and optimization of reforming reactors for carbon dioxide utilization using both rigorous and reduced models," *J. CO2 Util.*, vol. 23, pp. 80–104, 2018.

523641 Heuristic Algorithm Utilizing Mixed-Integer Linear Programming to Schedule Electric Vehicles for Reduced Cost and Energy Use

Abstract:

A heuristic algorithm based on Mixed-Integer Linear Programming (MILP) is proposed to obtain economical schedules for routing electric vehicles with multiple depots. Electric vehicles are an opportunity to handle delivery or geographically distributed activities in a sustainable way. However, besides the original problem of routing vehicles, special attention must be given to attributes of these technologies, especially recharging. Assignment and scheduling are not solely to be taken into account, but significant travelling times and costs may arise. The goal is the minimization of total cost and use of energy while performing the most activities possible, which involves keeping travelled distances, and hence environmental impacts low. A novel MILP model is proposed, which is capable of handling several factors that may arise in electric vehicle management problems, including time windows. A heuristic greedy algorithm is developed, which assigns activities one by one to the available electric vehiclesâ€™ daily schedules based on the MILP. This two-level approach is tested over medium scale problem instances, proving that computational requirements are acceptable.

526092 Integrated Power Systems Capacity and Transmission Planning with High Spatial and Temporal Resolution

Abstract:

Integrated modelling can combine different scales with regards to time and space. For power systems such integrated planning is crucial to optimal investment and operational decision-making. Driven by a growing level of network interconnections, electrication (e.g., in transport and heating), and digitalisation, there is increasing interest to model and investigate future energy systems at a high spatial and temporal resolution.

To explore spatially-dependent future power systems pathways, we have developed the Electricity Systems Optimisation Model with high temporal and spatial granularity (ESONE). This contribution is based on previous work by Heuberger et al. [1, 2] and open source models available at https://zenodo.org/record/1212298. ESONE is a mixed-integer linear program that minimises total system cost subject to operational constraints, system-wide requirements and carbon emissions targets. On a planning time horizon (years) we perform optimal capacity expansion and grid reinforcement, while on an operational time horizon (hours) we determine the optimal dispatch and unit commitment schedule for up to 2000 units of 16 different types of power generation and energy storage technologies. Optimal power flow is modelled as simplified direct current considering energy loss as a function of transmission volume and distance of zone centres, i.e., 29 nodes for a case study on the United Kingdom (UK). Additionally, we account for system ancillary services in form of spinning reserve, as well as an approximation of frequency and voltage control.

In order to enable model feasibility we develop a multi-dimensional k-means clustering approach and apply this to the spatially and temporally dependent data (i.e., solar, onshore/offshore wind availability [3, 4], electricity demand, power import price [5]). In addition, we introduce a convex hull reformulation [6] for the integer scheduling constraints and apply an iterative solution method implemented in GAMS 25.0.3/CPLEX 12.8. This results in a reduced model size of 2.15 10^6 compared to 8.6 10^6 variables. We parametrise the ESONE model to the UK's power system and explore future scenarios from 2015 to 2050 based on regional resource availability, technical, economic, and environmental drivers. We highlight the role of Carbon Capture and Storage (CCS) equipped power plants, including bio-energy and CCS, in decarbonising future energy systems. While we see a push for offshore wind in the northern half of the UK, we also see a signficant usage of CCS in the south and northeast, with gas-fired CCS plants coming online in 2025 and bio-energy CCS plants appearing in 2030. This contribution presents the methodology behind the ESONE model and showcases its capabilities for the UK power system.

[1] Heuberger, C. F., Rubin, E. S., Staffell, I., Shah, N., Mac Dowell, N. Power Capacity Expansion Planning Considering Endogenous Technology Cost Learning, Applied Energy 204, 831-845 (2017)
[2] Heuberger, C. F., Staffell, I., Shah, N., Mac Dowell, N. A systems approach to quantifying the value of power generation and energy storage technologies in future electricity networks. Computers & Chemical Engineering 107, 247256 (2017)
[3] Pfenninger, S., Staffell, I. Long-term patterns of European PV output using 30 years of validated hourly reanalysis and satellite data. Energy 114, 1251-1265 (2016)
[4] Staffell, I., Pfenninger, S. Using Bias-Corrected Reanalysis to Simulate Current and Future Wind Power Output. Energy 114, 1224-1239 (2016)
[5] National Grid, UK. Data Explorer - Real Time Demand Data. http://tinyurl.com/kbe65ca, (2015)
[6] Hua, B., Baldick, R., Wang, J. Representing Operational Flexibility in Generation Expansion Planning Through Convex Relaxation of Unit Commitment, IEEE Transactions on Power Systems 33(2), 2272-2281 (2017)

527791 The Potential of Cooperative Game Theory for the Design of Chemical Supply Chain Networks Under the Carbon Trading Scheme

Abstract:

Abstract

Nowadays, business decisions are dominated by the globalization of markets and increased competitiveness among firms. In particular, decision problems in supply chain management can be enhanced by taking into account that they are not only influenced by a single decision maker but for several. Here is where cooperation strategies come into play. In the literature, several types of cooperation are distinguished. Interorganizational cooperation (cooperation of economically and legally independent units) versus intra-organizational cooperation (cooperating business units belonging all to the same company). Another classification differentiates between vertical (buyer-seller) and horizontal (buyer-buyer, seller-seller) cooperation. In the latter case, companies that belong to the same supply chain stage jointly work to produce or trade the same product (they are competing and are part of the same industry). The most prominent reasons for horizontal cooperation are the synergistic effects (economies of scale, scope, and speed) and strengthening the competitiveness[1]. Nevertheless, horizontal cooperation is not exempt from drawbacks, such as the creasing complexity of the purchasing process; the loss of flexibility and control[2].

As the interactions among single companies in a supply network directly lead to a multi-decision maker framework, game theory predicts the rational strategic behavior of individuals in conflicting or cooperating situations seems to provide an adequate modeling basis for problems in supply chain management. The field of game theory may be divided roughly in two parts, namely non-cooperative game theory and cooperative game theory. Models in non-cooperative game theory assume that each player in the game (e.g., a company in a supply chain) optimizes its own objective and does not care for the effect of its decisions on others. The focus is on finding optimal strategies for each player. Binding agreements among the players are not allowed. One of the main points of concern is whether the proposed mechanism provides a solution that maximizes the total supply chain profit under Nash equilibrium. Alternatively, cooperative game theory assumes that players can make binding agreements. One of the main questions when applying cooperative game theory to supply chains is whether cooperation is stable, that is, whether there exists an allocation of the joint profit among all the parties in the supply chain such that no group of them can do better on its own.

As well as other topics related to chemical engineering, supply chain networks demand the application of the sustainability principles during their process design. One major motivation for this trend is the pressure placed on governments and regulatory agencies to tighten environmental regulations (e.g. on reducing greenhouse gas, GHG). In particular, capping GHG emissions and establishing a price through trading on them is an essential foundation for climate change policy[3]. The idea behind this cap-and-trading scheme is to set a price tag on carbon emissions and in consequence, a financial incentive to decrease them. After a cap is fixed on emissions, companies are allowed to buy or sell from each other the allowances to emit GHGs. Firms exceeding their emissions cap must buy extra credits to cover the excess. Meanwhile, those that do not use all their allowances can sell them, providing the least-polluting firms with an extra revenue [4]. There are already active carbon markets for GHG emissions such as the European Union Emissions Trading Scheme (EU ETS), the New Zeland Emissions Trading Scheme (NZ ETS), the Chicago Climate Exchange, and the Montreal Climate Exchange [5]. Applying this carbon trading scheme, governments seek for reducing the overall GHG emissions in an industry sector, and here is where the cooperation among companies (i.e., players) fits perfectly. Models based on cooperative game theory can not only improve the economic performance of the whole set of companies but also the overall environmental impact for all the firms, which is the aim of the governments.

The aim of this work is to show the applicability of Cooperative Game Theory as a methodology for analyzing sustainable supply chains. The success and sustainability of cooperation depends on the stability as constituting element. In this way, the solution concept based on this theory provides us under what allocation of the costs would the players want to form the grand coalition (i.e., all the players cooperate in unique coalition). Under the perspective of the game theory, the supply chain design problem we pose, is a transferable utility game, which is a pair consisting of a finite set of players (i.e., plants in the supply chain) and a characteristic function, which measures the cost of every coalition of players (i.e., any subset of all the players) through a real valued mapping We applied our methodology to a three-echelon petrochemical supply chain (production-storage-market), which satisfies the game property of being subadditive (i.e., it is always beneficial for two disjoint coalitions to cooperate and form a larger coalition). The environmental performance of each company in the SC is assessed according to the principles of Life Cycle Assessment (LCA) using the global warming potential (GWP) indicator. CO_2 emissions trading is modeled by introducing a balance equation that relates the total equivalent CO_2 emission for each company in the SC to be equal to the free allowance emissions cap plus the extra rights bought to emit minus the sold rights[6]. First, we illustrate the main concepts of cooperative game theory with a small motivation example with only 3 companies. To this purpose we introduce the concept of the non-emptiness of the core, that means that there exists at least one stable allocation of the total cost such that no group of players has an incentive to leave. This first case study can be solved geometrically (see Figure 1). Then, we extend the proposed framework to a case study with 8 companies cooperating. As the number of coalitions rises exponentially with an increasing number of players, here we use an algorithm that finds an element in the core without testing the constraints for all possible coalitions. This solution strategy is based on a row generation procedure, whose key idea is to start with a relaxed version of the original problem and add missing relaxed constraints over several iterations. The solution to these case studies shows that the Cooperative Game Theory provides a holistic solution for the design of sustainable supply chains with environmental concerns and, can be included as another useful mathematical tool for Chemical Engineers.

Figure 1. Geometric interpretation of the core in a cooperative game with 3 players.

References

(1) Drechsel, J.; SpringerLink (Online service), Cooperative lot sizing games in supply chains. In *Lecture notes in economics and mathematical systems 644*, Springer-Verlag,: Berlin ; Heidelberg, 2010; pp xiv, 167 p.

(2) Schotanus, F. Horizontal Cooperative Purchasing (Ph.D. Thesis). University of Twente, Enschede, the Netherlands, 2007.

(3) Stern, N. Stern review on the economics of climate change. London,UK: HM Treasury. http://www.sternreview.org.uk/

(4) Young, T. The beginners' guide to the UK's carbon trading schemes. Business Green. http://www.businessgreen.com/bg/analysis/1805900/the-beginners-guide-uks-carbon-trading-schemes

(5) Chaabane, A.; Ramudhin, A.; Paquet, M., Design of sustainable supply chains under the emission trading scheme. *Int. J. Prod. Econ.* **2010, doi: 10.1016/j.ijpe.2010.10.025**.

(6) Ruiz-Femenia, R.; Salcedo-Díaz, R.; Guillén-Gosálbez, G.; Caballero, J. A.; Jiménez, L., Incorporating CO 2 emission trading in the optimal design and planning of chemical supply chain networks under uncertainty. *Computer Aided Chemical Engineering* **2012,** 30, 127-131.

525237 Metal-Modified Transition Metal Nitride Electrocatalysts for Oer, HER, and Other Reactions

Abstract:

Transition metal nitrides are attractive electrocatalytic support materials for similar reasons as their extensively studied carbide counterparts â€“ they are particularly useful for reducing precious metal loading, compared to a carbon support. Nitrides, however, have the distinct advantage of avoiding the carbonaceous overlayer resulting from typical carbide synthesis techniques. The bare nitride surface allows ad-metals to directly contact nitrides, which can improve performance and strengthen correlations with density functional theory calculations on model surfaces. This work utilizes these advantages to develop low-cost and effective electrocatalysts for various fuel cell and electrolysis reactions. Specifically, (1) iridium-metal nitride core-shell particles are used to reduce Ir loading for acidic OER by 50 percent, (2) monolayer platinum on niobium nitride and tungsten nitride thin films are demonstrated as effective HER catalysts in alkaline and acidic electrolytes, and (3) systematic stability studies are conducted on nitride thin films to evaluate their utility for various reactions across wide potential-pH regimes.

525521 Combining Electrochemistry and Surface Science to Identify Electrocatalytic Structure-Property Relationships

Abstract:

The oxygen evolution reaction (OER) is the anodic process associated with the electrochemical CO_2 reduction reaction (CO_2RR). State-of-the-art OER electrocatalysts contain precious metals, such as Iridium, which introduce substantial cost-barriers to large-scale CO_2RR deployment. Current efforts have focused on identifying inexpensive, earth-abundant transition metals with satisfactory OER activity. Nickel, cobalt, iron and bimetallic CoFe and NiFe electrocatalysts have shown remarkable OER activity in basic conditions, but atomic-level understanding of how catalyst shape, size, and composition are still lacking in the literature.

Here, we present an approach that couples surface science and electrochemistry to synthesize, characterize and evaluate well-defined transition metal electrocatalysts. We demonstrate this technique with well-defined Fe_2O_3 nanocatalysts that show outstanding, structure-dependent mass activity with values exceeding 10^5 A/g nanocatalysts were grown on Au(111) supports in ultra-high vacuum (UHV) and characterized with scanning tunneling microscopy (STM) and x-ray photoelectron spectroscopy (XPS). Our synthetic approach provided extremely precise structural control and we varied catalyst structures from near monolayer coverage to discrete, isolated nanoparticles. High-resolution STM imaging quantified the nanocatalyst morphology, including size, thickness, total surface area, and perimeter-to-surface-area ratios. Samples were translated from UHV and tested for electrochemical OER in 0.1M KOH solutions. Post-reaction STM and XPS characterization revealed structural stability during OER and we identified a relationship between particle size, density of perimeter Fe-oxide(hydroxide) sites, and OER activity. This represents some of the clearest identification of reactive site location in any electrocatalyst system to date, and the results confirm long-held hypothesis that undercoordinated or â€œedgeâ€ sites are responsible for catalytic activity. Ongoing computational efforts within Density Functional Theory (DFT) will provide atomic-level details about the reaction energetics and mechanisms at various active sites to explain experimentally observed activity trends. Future efforts will be directed towards more complex OER systems to probe how catalyst structure, composition, and active site location impact electrocatalytic activity.

534739 Active Structures and Species of Modified Oxide Catalysts for the Oxygen Evolution Reaction (OER)

Abstract:

A major cause of energy losses in photocatalytic systems for solar fuels, and also in a number of other emergent technologies such as rechargeable metal-air batteries, water splitting devices, and unitized regenerative fuel cells, is the oxygen evolution reaction (OER). The low energy efficiencies of OER are the result of its sluggish reaction kinetics and large overpotentials. Significant improvements to the OER activity of transition metal oxides (TMOs) have been made by tailoring the morphology and crystal structure of the catalysts, incorporating dopants, as well as using conductive supports. However, clear structure-activity correlations remain elusive because of the complex composition and structure of TMO catalysts. Further understanding of these relationships for these promising catalysts for oxygen evolution may lead to additional developments that will enable TMOs to replace precious metal-based OER catalysts (e.g. IrO_x and RuO_x).

Insights from two recent examples of our studies will be discussed. We have utilized a range of spectroscopic techniques for characterization of Ce-modified copper oxide (CuO_x) [1] and Ni-modified cobalt (oxy)hydroxides [2] to reveal the OER active species and structures of these catalysts. In the case of Ce-modified CuO_x, Ce incorporation (6.9 at%) into CuO_x led to 3.3 times greater OER activity compared to pure CuO_x and this was coincident with significant structural changes due to an increasing amount of disorder. By combining X-ray photoelectron and Raman spectroscopy techniques, a strong correlation between OER performance with tetravalent Ce (Ce^{4+}) ions was observed up to a concentration corresponding to CeO_2 phase segregation. We propose a strong promoting effect of Ce^{4+} for OER in this system. In the case of Ni-modified CoO_xH_y, operando Raman spectroscopy was used to reveal a drastic transformation of a spinel Co_3O_4-like structure into a more active (oxy)hydroxide structure under applied potential. Such a transformation was only observed in the presence of uniformly distributed Ni ions. These two examples, i.e. the promoting effect of Ce^{4+} and the formation of active OER structures in Ni-modified CoO_xH_y, reveal the importance of chemical state and local structure considerations for the rational design of improved oxide-based OER catalysts.

[1] Z. Chen, C.X. Kronawitter, X. Yang, Y.-W. Yeh, N. Yao, and B.E. Koel, *Phys. Chem. Chem. Phys.,* **2017**, *19*, 31545-31552.

[2] Z. Chen, L. Cai, X. Yang, C. Kronawitter, L. Guo, S. Shen, B. E. Koel, *ACS Catal.*, **2018**, *8*, 1238â^'1247.

537033 Lowering the Charge Overpotentials in Li-O_2 Battery By Tailoring the Oxygen Reduction and Evolution Reaction Energetics Using Non-Precious Metal Oxide Electrocatalysts

Abstract:

Lithium-oxygen (Li-O_2) batteries are considered as the next generation energy storage technology with vast potential to contribute to electrification of the transportation sector. Although these systems are promising, several factors have resulted in performance far below the theoretical potential, including sluggish oxygen evolution reaction (OER) kinetics leading to high overpotential losses during charging.[1] We report a decrease in these overpotential losses achieved by perturbing the energetics associated with oxygen reduction reaction (ORR) and OER in aprotic Li-O_2 cathodes containing non-stoichiometric mixed metal oxide electrocatalysts with mixed electronic and ionic conducting properties.[2-3] A combination of well controlled synthesis, electrochemical studies and detailed characterization by Raman spectroscopy, titration, and differential electrochemical mass spectrometry show that these non-stoichiometric mixed metal oxide nanostructures change the energetics associated with ORR as compared to carbon-only electrodes by stabilizing Li-deficient intermediate species (> 50% more LiO_2 is detected in the presence of the catalyst), which consequently leads to lower OER overpotentials during charge. In addition, the presence of the catalyst also reduces the charge overpotential losses by lowering the potential required to dissociate Li_2O_2 during charge. These insights are key in providing new avenues for improving the efficiency of Li-O_2 batteries via tailoring the ORR/OER energetics using well-defined nanostructured, non-precious metal oxide electrocatalysts.

References:

(1) Lu, J.; Li, L.; Park, J. B.; Sun, Y. K.; Wu, F.; Amine, K., Aprotic and Aqueous Li-O_2 Batteries. *Chem. Rev.* **2014,** *114*, 5611-5640.

(2) Gu, X.-K.; Samira, S.; Nikolla, E., Oxygen Sponges for Electrocatalysis: Oxygen Reduction/Evolution on Non-stoichiometric, Mixed Metal Oxides. *Chem. Mater.* **2018,** DOI:10.1021/acs.chemmater.8b00694.

(3) Nacy, A.; Ma, X. F.; Nikolla, E., Nanostructured Nickelate Oxides as Efficient and Stable Cathode Electrocatalysts for Li-O-2 Batteries. *Top. Catal.* **2015,** *58*, 513-521.

538815 Density Functional Theory Study of Oxygen Evolution Reaction on Specific Terminated Facets of Perovskite Oxides

Abstract:

Developing efficient oxygen evolution reaction (OER) electrocatalysts is vital to the new generation of electrochemical storage and conversion devices such as electrolyzer and metal-air batteries. In order to address the longstanding problems with high cost and poor stability of noble metal electrocatalyst, this study selects perovskite oxides (ABO_3) as a model system because of their low-cost and high versatility in compositions and tunable electronic structures. In this talk, we will present the fundamental mechanism whereby the electrocatalytic activity on various facets of perovskite oxides in oxygen evolution reaction using the ab initio density functional theory calculations. Elemental doping on A-site and therefore, the fundamental role and effect of doping on the performance of OER reactions on the various perovskite surfaces will also be discussed. Free energy of electrochemical reactions will also be discussed by referencing to the computational hydrogen electrode model. Some of the results will also be compared with the experimental findings.

539677 Core-Shell Nanoparticles for Efficient Oxygen Evolution Electrocatalysis in Alkaline and Acidic Media

Abstract:

By storing excess renewable energy in chemical bonds, water electrolyzers provide a promising means of renewable energy storage and clean hydrogen fuel generation toward a sustainable energy infrastructure. Currently, the slow kinetics of the oxygen evolution reaction (OER) occurring at the electrolyzer anode limit device efficiencies, necessitating the development of high performance electrocatalysts. Core-shell and surface-leached nanostructures provide promising strategies to achieve enhanced OER catalysis through electronic and geometric structure alterations. This presentation will explore these strategies in alkaline and acidic electrolyte through the discussion of two model systems: $Au@M_xO_y$ (where M=Ni, Co, Fe, and CoFe) core@shell nanoparticles for alkaline OER and surface-leached crystalline strontium iridate nanostructures for acidic OER. In alkaline media, the beneficial effects of Au-supported catalysts were translated to core-shell $Au@M_xO_y$ nanoparticles, demonstrating a universal activity enhancement with the inclusion of the Au-core. The highest activity particles, $Au@CoFeO_x$, demonstrated an overpotential of 328 ± 3 mV over a 2 h stability test at 10 mA cm^{-2}, illustrating that strategically combining Au support and mixed metal oxide effects in a core-shell nanoparticle architecture is a promising avenue for practical catalyst design. In acidic media, the effect of crystal structure and stoichiometry on the activity and stability of crystalline strontium iridate (e.g. Sr_2IrO_4, $SrIrO_3$, and Sr_4IrO_6) nanostructures is evaluated in comparison to commercial Ir/C nanoparticles. In-situ leaching results in an IrO_x-rich surface on a strontium iridate core. Detailed microscopy and spectroscopic evaluation before and after electrochemistry is provided to elucidate performance trends and highlight structure-activity-stability relationships. These two systems demonstrate the potential catalytic benefits of core-shell motifs resulting from both direct chemical synthesis and post-synthetic *in-situ* modification towards the OER in alkaline and acidic electrolyte.

528872 Carbon Footprint Analysis of Gasoline and Diesel from Forest Residues and Algae Using Integrated Hydropyrolysis and Hydroconversion Plus Fischer-Tropsch

Abstract:

Carbon Footprint Analysis of Gasoline and Diesel from Forest Residues and Algae using Integrated Hydropyrolysis and Hydroconversion Plus Fischer-Tropsch (IH$^{2®}$ Plus cool GTL™).

Olumide Winjobi[1], Hossein Tavakoli[2], Bethany Klemenstrud[3], Robert Handler[1], Terry Marker[4], Michael Roberts[4], and David Shonnard[1].

[1] Department of Chemical Engineering and the Sustainable Futures Institute, Michigan Technological University, 1400 Townsend Drive, Houghton, MI 49931, USA

[2] Department of Civil and Environmental Engineering, Michigan Technological University, 1400 Townsend Drive, Houghton, MI 49931, USA

[3] Department of Chemical Engineering, University of North Dakota, 243 Centennial Dr. Stop 7101, Grand Forks, ND 58202, USA

[4] Gas Technology Institute, 1700 S Mount Prospect Road, Des Plaines, Illinois 60018, USA

To address the dual challenge of meeting the rising energy demand while reducing environmental impacts associated with the production and consumption of fossil fuels, it is vital also to understand the environmental impacts from the various alternative transportation fuel products and processes being developed to complement the current fossil-derived fuels. This study evaluated the greenhouse gas (GHG) emissions and the cumulative energy demand (CED) associated with the production of 'drop-in' renewable fuels produced by the IH^2® Plus cool GTL™ using forest residues and algal biomass augmented with natural gas as feedstock. The IH^2® Plus cool GTL™ developed by GTI is a modification to the previous IH^2® process also developed by GTI, to achieve a higher yield of biofuels relative to the IH^2® process. The IH^2® Plus cool GTL™ is a hydropyrolysis, hydroconversion combined with Fischer-Tropsch technology for converting a broad range of biomass types into liquid hydrocarbon transportation fuels spanning the range of diesel and gasoline.

The previous IH2® process uses the biogenic C1-C3 gases produced in the process as feedstock within the steam methane reformer to generate the hydrogen required in the process. However, in the modified IH^2® Plus cool GTL™, the biogenic C1-C3 gases are initially processed through a dry reformer to generate syngas that is subsequently processed through a Fischer Tropsch reactor to generate liquid biofuel products,

increasing the gasoline and diesel yield from 26 to 38%. Due to the use of the biogenic C1-C3 gases to boost the yield of biofuels in the IH^2® Plus cool GTL™, the hydrogen required in the process is from an external source increasing; the amount of fossil fuel used within the process relative to the previous IH^2®. This LCA compared the IH^2® Plus cool GTL™ to the original base case studied previously by our group (Fan et al. 2015). This work also evaluated the ability of the biofuel produced from the IH^2® Plus cool GTL™ to still qualify for a renewable identification number with the addition of fossil methane being added as feedstock.

From our study, a life cycle GHG emission savings of 63.3% was estimated for the optimized IH^2® Plus cool GTL™ relative to fossil-derived fuel for forest residue feedstock. Higher emission relative to fossil-derived fuel was estimated for algal feedstock for the IH^2® and IH^2® Plus cool GTL™ processes. However, GHG savings of about 42% can be potentially achieved if the IH^2® Plus cool GTL™ process is in an area with low GHG intensity electricity grid.

Reference

Fan, Jiqing, John Gephart, Terry Marker, Daniel Stover, Ben Updike, and David R. Shonnard. "Carbon footprint analysis of gasoline and diesel from forest residues and corn stover using integrated hydropyrolysis and hydroconversion." ***ACS Sustainable Chemistry & Engineering*** 4, no. 1 (2015): 284-290.

529431 Life Cycle Assessment of Bioenergy Oilseed Crops Produced in Rotation with Dryland Cereals in the Inland Pacific Northwest

Abstract:

Purpose: Oilseed crops are expected to become an important feedstock for production of renewable jet fuel. The objective of this study is to determine the life cycle energy and greenhouse gas (GHG) emissions of several 2- and 3-yr crop rotations with cereals and bioenergy oilseeds in a low precipitation environment of the inland Pacific Northwest. The purpose is to ascertain whether cropping intensification could improve energy efficiency and reduce GHG emissions.

Methods: A life cycle assessment (LCA) was carried out to evaluate the fossil energy demand and carbon footprint of nine cropping systems characterized by different inputs applied to spring carinata [*Brassica carinata* (A.) Braun] and winter canola (*B. napus* L.) in rotation with wheat (*Triticum aevistum* L.) and other cereal crops. Grain yield and field activity data from cropping systems were acquired from a field experiment over a five year period. Soil gas emissions were measured weekly over two years using static chamber methodology and laboratory gas chromatography. Inputs for the LCA regarding fertilizers, machinery fuel use, and pesticides were from the field trials and literature for fuel use.

Results and discussion: Life cycle emission results of winter wheat (WW) rotations are between 300-400 g CO_2 eq. kg^{-1} WW, in the range for U.S. average WW cropping emissions (i.e., 300-600 g CO_2 eq. kg^{-1} WW). Reduced tillage fallow (RTF)-Winter oilseed (WO)-RTF-WW and summer fallow (SF)-WW rotation were the most promising, from a trade-off of GHG emissions versus total crop sales over six years per hectare with low emissions and high sales. The best oilseed result was 660 g CO_2 eq. kg^{-1} for canola following RTF. Highest yields were observed when cereal or oilseed crops were planted following RTF. Efficiency in terms of Energy Return on Energy Investment was 3.85 for winter oilseed yields 1,338.9 kg ha^{-1} and 1.6 for spring oilseed yields 552.2 kg ha^{-1}.

Conclusions: Compared to conventional SF-WW, bioenergy oilseed cultivation may increase CO_2 equivalent emissions in 3-yr cereal-based rotations due to increased inputs with inclusion of fallow-substitution cultivation. Fossil energy inputs required to produce oilseed crops were smaller than the total energy in final seed and thus oilseeds have the potential to reduce reliance on fossil fuels. Improving energy efficiency and encouraging adoption by growers will depend on ability to enhance agronomic performance with higher yielding, drought and cold tolerant oilseed varieties.

535305 Evaluation of Environmental Tradeoffs of ProducingÂ Renewable Jet Fuel and Polyisoprene from Biomass

Abstract:

Polyisoprene is a commercially valuable material that is used in the manufacture of more than 40,000 products. The two main resources used for its production are natural rubber trees and petroleum. This study examines the environmental performance of a new bioconversion method for producing polyisoprene from biomass and compares it to the two main alternative production routes. In this process, following pretreatment and hydrolysis, which release the fermentable sugars in biomass, we convert them to methyl butenol, which is further converted to isoprene in a dehydrogenation reactor and then polymerized into polyisoprene. In the same process, methyl butenol can be co-converted to dimethyl cyclooctadiene, a high-density jet fuel blend, in a dimerization reactor.

We develop a life cycle assessment (LCA) model based on experiments and chemical process simulations formulated with experimental measurement. To generate mass and energy balances required for the LCA, the energy and chemical input requirements for process stages are calculated using Aspen PLUS and Aspen HYSYS simulations, along with stoichiometry, thermodynamics, and kinematics equations.

The metrics studied are land use intensity as measured in land occupation per kg of polyisoprene produced and the 100-year global warming potential (GWP100), as a measure of climate change, measured in kg CO_2 equivalent per kg of polyisoprene. We found that the GWP of polyisoprene produced from both corn stover and forest residue is negative while that produced from petroleum and natural rubber tree is in the range of 0.5 to 15.9 kg CO_2/kg isoprene. Also, while the land use intensity of biomass-based polyisoprene is higher than that of petroleum, which is very close to 0, it is much lower than the land use intensity of natural rubber based polyisoprene.

538293 Life Cycle Assessment of Forest Biomass Energy Pathways in the Northeast US

Abstract:

Forest biomass is anticipated to represent a significant portion of the developing supply of bioenergy feedstocks. Increasing interest to utilize renewable and sustainable energy systems with forest biomass feedstocks necessitates an understanding of potential environmental impacts arising from these systems. Life cycle assessment (LCA) is a method used to evaluate the potential environmental impacts of a process, product, or system over its life cycle. This method is used to determine the life cycle climate change impacts, water use, and eutrophication potential of utilizing a forest biomass feedstock in electrical generation, combined heat and power (CHP), and wood chip heating systems typical to the Northeast United States. The LCA scenarios are modeled in Python code with primary data from harvest sites and energy systems in the Northeast US to define the harvest, transportation, energy conversion, and ash disposal processes. The silviculture processes are based on carbon cycling and partitioning modeling specific to hardwood forest growth and harvesting practices in New York State. The functional unit for the wood chip heating system LCA is 1 MJ of heat delivered. The functional unit for the electrical generation and CHP systems is 1 kWh, allowing for comparison to other forms of electricity generation. System expansion is used to account for the space heat coproduct of the CHP system. Coproducts of forest biomass, including sawtimber, pulpwood, and firewood, are accounted for through mass and economic allocation methods. Sensitivity and uncertainty analyses are conducted to investigate the impacts of variability in forest growth, species of harvested trees, market conditions, harvest systems, transportation distances, and energy conversion efficiencies on the life cycle environmental impacts of the studied bioenergy pathways. The life cycle environmental impacts of the forest bioenergy pathways are compared against those of conventional energy systems, and recommendations are made for system improvements for greater sustainability.

535326 Life Cycle Analysis As an Assessment Tool to Compare Process Alternatives

Abstract:

Bio-fuels have come to be recognized as environmentally friendly alternatives to fossil fuels. This work studies the Bio-ethanol production process complementing a sugar mill.

The process flowsheet examined includes three major processing stages typical of distillery installations for fuel-grade production in sugar-cane processing facilities: (1) fermentation, (2) separation, and (3) dehydration stages.

Using plant data from a bio-ethanol pilot plant in the Northwestern of Argentina (NOA) a kinetic model is formulated for a fed-batch molasses fermentation process. The model for this unit, along with a user-defined separation module, is then integrated into a commercial process simulator. The distillation step is optimized by performing rigorous energy analysis and heat integration on the process utilizing the resources made available by the sugar mill, while pervaporation is used in the dehydration step. Different schemes meeting equivalent performance standards in terms of purity and recovery are examined in this paper.

Particular attention is given to energetic integration by performing a pinch-analysis to each of the fundamental stages to find a global optimum. Optimization is achieved by minimizing a given objective function that accounts for bio-ethanol recovery, separation efficiency, and minimization of effluents. Life Cycle Analysis is used to complete the assessment of operating parameters/conditions on environmental effects.

527461 Carbon Dioxide Introduced for Green Synthesis of Thiuram Disulfides Vulcanization Accelerators Under Mild Conditions

Abstract:

Thiuram disulfides have been widely used in rubber industry as rubber vulcanization ultra-accelerators. The conventional industrial process for the production of thiuram disulfide included reacting a secondary amine with carbon disulfide in an aqueous solution of sodium hydroxide to form a sodium salt of dialkyldithiocarbamate acid, then adding the mixed solution of sulfuric acid and hydrogen peroxide to neutralize and oxidize sodium dialkyldithiocarbamate. The precipitate, thiuram disulfide, can be obtained by filtering, washing, drying. While this method uses the large quantities of sulfuric acid and generates huge amount of salt by-product, sodium sulfate. Thus, an eco-friendly method for oxidation of sodium dialkyl dithiocarbamate to the corresponding thiuram disulfides has been developed using carbon dioxide as neutralizing agent in the presence of H_2O_2 as the oxidant and water as the solvent. After optimizing the reaction conditions such as temperature, concentration of reactants, flow rate of carbon dioxide, dropping speed of hydrogen peroxide, aging time and so on, this method could provide green mild reaction conditions, reduce CO_2 emission and H_2SO_4 consumption remarkably, and reach the almost closely high yields and purity of products as traditional method. With the HPLC-MS analysis of mother liquor, it is proved that the main side reaction is the peroxidation of dialkyldithiocarbamate acid which produced corresponding sulfoxide and sulfone. By comparing the alkalinity of sodium carbonate, sodium bicarbonate and sodium dialkyldithiocarbamate, doing related experiments and reading relevant references, the reaction path has been proposed.

530545 Plastic Waste Accumulation Problem and Emerging Solutions

Abstract:

More than eight billion tons of plastic waste has been accumulated over the past 50 years. The majority (90%) of the worldâ€™s plastic waste goes directly into landfills and 3% ends up in the oceans. At the current rate, the planet will have 30 billion tons of plastic waste by 2050, and the oceans will have 250 million tons of plastics by 2025, or â€œone pound of plastics for every three pounds of fishâ€. Plastics are persistent in environment and degrade very slowly (~100 years timescale), while releasing toxic microplastics and chemicals into the landfills and oceans. This poses serious threats to our ecosystems, drinking water, and food supply. Our analysis shows that to remove the plastic waste from landfills is estimated to cost about \$0.5/kg, or \$4 trillion globally. To remove the plastic waste from the oceans using the state of the art technology is estimated to cost about $\$10^{18}$, about 10,000 times the global GDP.

Polyolefin waste accounts for about 63% of the total accumulated plastic waste. About \$100 billion of post-consumer polyolefin products are consigned to landfills annually. Converting polyolefin waste into useful products can reduce the accumulated waste and associated risk to the environment. In this study we compare incineration, which is the most mature technology of reducing plastic waste, with two emerging solutions, converting polyolefin waste into (1) pristine polymers or (2) clean fuels. The results show that both solutions potentially are more profitable than incineration. Sorted polyolefin waste can be converted into pristine polymers with the highest potential profit. Mixed polyolefin waste can be converted into clean fuels with the second highest profit. The supplies of the sorted waste, however, are limited presently. Developing more efficient sorting technologies or changing the current waste management system to collect more sorted polyolefin waste should have a major positive impact on reducing the polyolefin waste accumulation.

Process Systems Engineering: Limits of performance of the "cyclic" waste plastic economy

James Alistair Fox, Baraka Celestin Sempuga

Institute for the Development of Energy for African Sustainability, University of South Africa, Florida Science Campus, Johannesburg, South Africa

Abstract

The rapid accumulation of waste plastics in the environment has led to recent media and public attention focused on an impending environmental crisis. One potential solution is the idea of a "cyclic plastic economy". It should be possible to maintain an entirely closed and independent cycle, with negligible environmental impact. Old plastics enter the cycle while new plastics leave the cycle. How realistic is this, though? Just how "cyclic" is the cycle? This work seeks to investigate this question by applying high-level systems engineering to identify the limits of performance for such a cycle.

It was found that, if conventional fossil fuel based power generation methods were used, the cyclic economy would have a maximum carbon efficiency of 97.5%. While this is a very high efficiency, for the cycle to be completely closed, a carbon efficiency of 100% is required. An efficiency of 100% can be achieved if renewable energy sources are used instead.

While a real process will always perform worse than this idealized result, the analysis shows that a cyclic economy will never be feasible if non-renewable energy is used. However, it can be conceptually feasible with the use of renewable energy.

Introduction

The plastics industry has seen consistent growth for many decades, with the global value of the industry predicted to reach USD 654-million by 2020[1]. This growth has been accompanied with significant growth in waste generation. It is estimated that some 8-million tons of plastic are disposed of in the oceans every year[2]. With waste plastics taking anything from 100 to 1000 years to degrade naturally, plastic waste is rapidly accumulating in the environment.

China had been importing as much as 51% [3] of the worlds waste plastics, leading many nations and economies to adopt an "out of sight-out of mind" view on their waste production. However, China ceased their imports at the end of 2017, leaving those same economies with vast amounts of waste plastic to deal with and garnering significant media and public attention. This has led to an upsurge in new research in how to deal with plastic waste.

Much of the attention has been focused on the idea of a "cyclic plastic economy"; the principle of this idea is that plastics are first mechanically recycled (washed, repaired and reused) as much as possible until their condition has deteriorated to a point where the plastic needs to be chemically broken into their chemical intermediates. This transformation is primarily accomplished using pyrolysis, which breaks the plastics down to alkene molecules. These alkenes can then be re-polymerized into new plastic products.

In principle, it should be possible to maintain an entirely closed and independent cycle, with negligible environmental impact. Old plastics enter the cycle while new plastics leave the cycle. How realistic is this, though? Just how "cyclic" is the cycle? This work seeks to investigate this question by applying high-level systems engineering to identify the limits of performance for such a cycle.

The proposed system to investigate this question is comprised of two major sections:

1. Pyrolysis, which is an endothermic reaction that involves using heat to split the polymer chains into predominantly alkene components that is typically carried out at temperatures in the region of 823K.
2. Polymerization, which is an exothermic process that is carried out at temperatures around 373K. Normally, coordination polymerization of ethylene monomers is used. However, pyrolysis does not produce ethylene alone but rather a range of alkene products. The re-polymerization of pyrolysis products for this system would require reversing the pyrolysis reaction using something analogous to radical polymerization. However, for this work, the mechanism of polymerization is not important but only that it is the reverse of pyrolysis.

Idealised thermodynamics for the cyclic economy

In order to begin an analysis of the hypothetical cyclic economy it is necessary to represent the processes involved with their respective material balances.

Due to limiting hydrogen, pyrolysis primarily produces unsaturated alkene hydrocarbons of varying carbon chain lengths. The relative amounts of each alkene depends on the temperature at which the pyrolysis is carried out. Most empirical research [4–6] carried out on plastic pyrolysis reports the product distribution in mass fractions of gases, liquid fraction and heavy hydrocarbons. At a temperature of 873K, a typical pyrolysis yield is given as 25% gases, 45% liquids and 30% heavies.

For the purposes of a complete material balance, it would be better to have a full distribution of all the alkene products but such detailed information is not available in the literature. Instead, each fraction will be represented by a "representative component", the component that comprises a significant proportion of a given fraction. The gas fraction will be represented by propylene, the liquid fraction by heptadecene and the heavy fraction by triacontene. With the given information it is now possible to write the material balance for polyethylene pyrolysis.

$$(C_2H_4)_n \rightarrow 0.17nC_3H_6 + 0.05nC_{17}H_{34} + 0.02nC_{30}H_{60}$$

Here "n" is the number of ethylene monomers present in a single mole of polyethylene. Since this value can vary over a wide range, the monomer number will be treated as an unknown variable. It is important to note that this does not represent a reaction for pyrolysis; it represents a material balance for pyrolysis as a whole. In addition, this material only applies to pyrolysis conducted at 823K. Investigation of a different temperature would require data on the product distribution at that specific temperature. With sparse data on plastic pyrolysis available, this work will stick to using the data at 823K.

The goal of the cyclic economy would be to re-polymerize the alkenes back into polyethylene; this would need to be done perfectly to ensure that the loop is closed. The mechanism for the re-polymerization would probably need to be something akin to radical polymerization. This analysis will not be investigating the mechanism but rather the system as a whole. The material balance for re-polymerization would be represented as the reverse of the material balance for pyrolysis.

Returning to the first step of the system, pyrolysis, the limit of performance would be reversible pyrolysis. A hypothetical pyrolysis process that would operate reversibly is represented in Figure 1

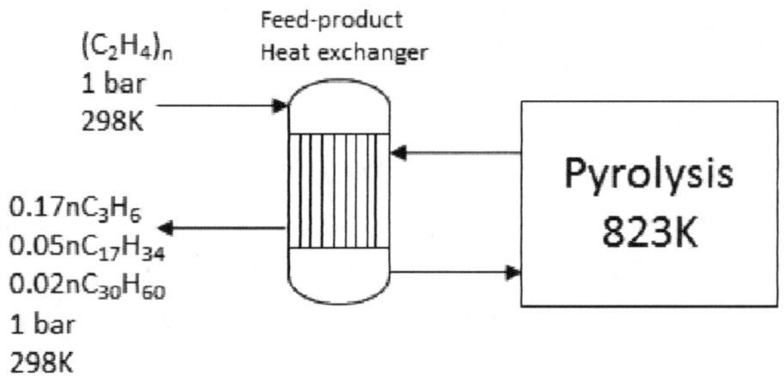

Figure 1: Representation of a reversible pyrolysis process

The process in Figure 1 is not a practical one. One of the most obvious problems is that the feed-product heat exchanger would need to have an infinite surface area. The purpose of this idealized analysis is to find the limit of performance and provide a basis for comparison. No real process can achieve what is represented in Figure 1 but if the cyclic economy does not work under idealized conditions, it will *never* work under realistic conditions and that applies for all systems.

Applying the first law of thermodynamics to Figure 1 reveals that, for every mole of polyethylene, pyrolysis needs 72.6n kW of heat supplied at a temperature of 823K. This energy needs to come from somewhere. One of the major premises of the cyclic economy is that this energy can come from the exothermic re-polymerization. This is simply the process in Figure 1 with the arrows reversed but there is one important difference, the polymerization takes place at a lower temperature (around 373K) than pyrolysis. This immediately implies a violation of the second law of thermodynamics, since the heat produced in the re-polymerization would need to transfer to pyrolysis against the temperature gradient, but it is possible to move heat against the gradient with heat pumps.

The goal is now to devise a method of "upgrading" the heat produced by re-polymerization and then transfer that heat to pyrolysis. Consider the ideal/reversible re-polymerization process, Figure 2

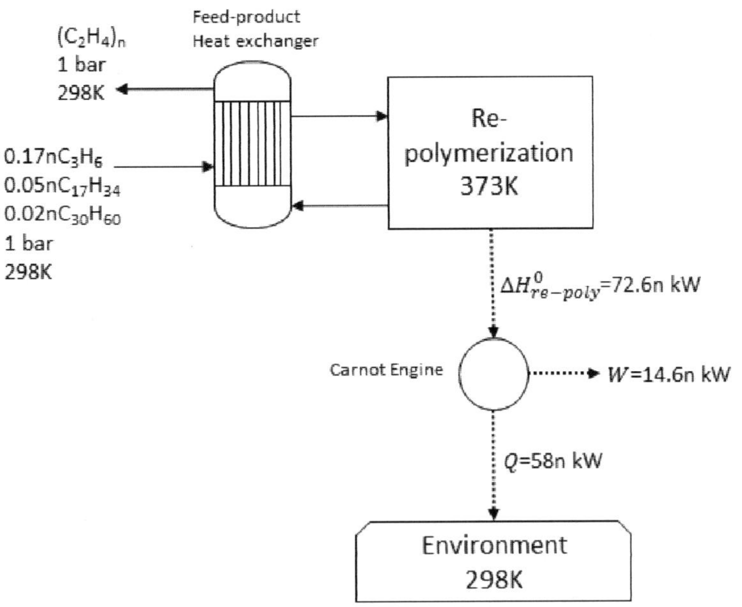

Figure 2: Ideal re-polymerization with work recovery using a Carnot engine. Solid arrows indicate mass flows; dotted arrows indicate energy flow

In Figure 2, the heat generated by the re-polymerization is sent to a Carnot engine where work is recovered by the engine and remaining heat expelled to the surroundings at ambient conditions. The well-known Carnot's Theorem calculates for the work recovery

$$W = Q_h\left(1 - \frac{298K}{T}\right)$$

Where:

- W is the work recovered by the engine
- Q_h is the heat entering the engine from the hot reservoir. In this case, $Q_h = \Delta H$.
- T is the temperature of the hot reservoir.

The re-polymerization is able to provide 14.6n kW of work towards the upgrading of energy.

This same idea can then be applied to the process in Figure 1,

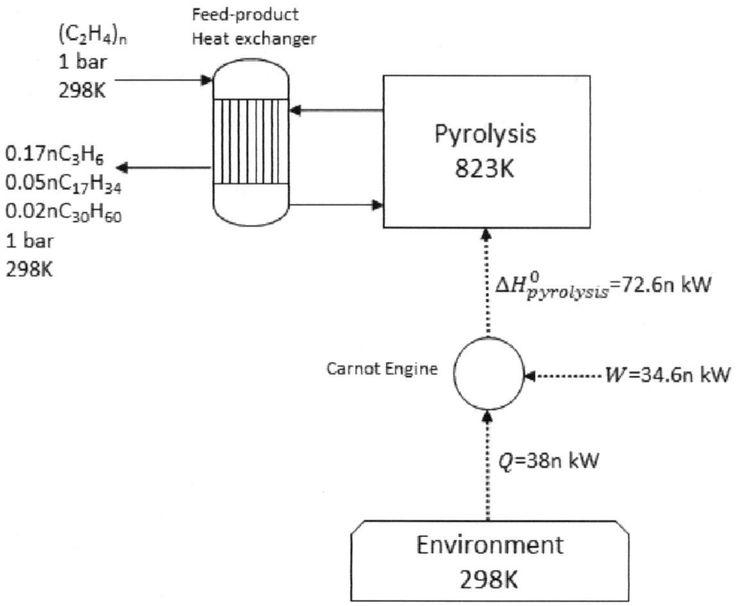

Figure 3: Ideal pyrolysis with work addition using a Carnot heat pump. Solid arrows indicate mass flows; dotted arrows indicate energy flows

Figure 3 shows that 34.6n kW of work is needed in order to upgrade the heat from the environment, but the re-polymerization is only capable of supplying 14.6n kW. It is clear that there is a deficit of 24n kW of work that needs to be supplied from an external source.

This result already shows that the cyclic economy is not a completely closed cycle, even under ideal circumstances. The question is now: how far off is the cycle?

In order to complete the construction of the idealized chemical waste plastic recycling process, a source of energy is needed to supply the outstanding 24n kW of work. The easiest way to do this is to use electricity. The material balances for how this electricity is produced then need to be taken into account. Previous work by one of the authors involves the design of a process that can convert waste plastic into electricity. The initial estimates show that waste plastic could produce up to 34 GJ of power per ton of waste polyethylene. The material balance that allows this process to provide 24n kW of electricity is given by:

$$0.025(C_2H_4)_n + 2nAir \rightarrow 0.05nCO_2 + 0.1nH_2O + 1.94nAir$$

A portion of the waste polyethylene now needs to be diverted to a process that can convert the polyethylene into electricity. In addition, this process now produces carbon dioxide, a different type of pollutant. However, it is now possible to re-polymerize the waste polyethylene into new "virgin" polyethylene.

It is now possible to construct a diagram by combining Figure 2, Figure 3 and the electricity production.

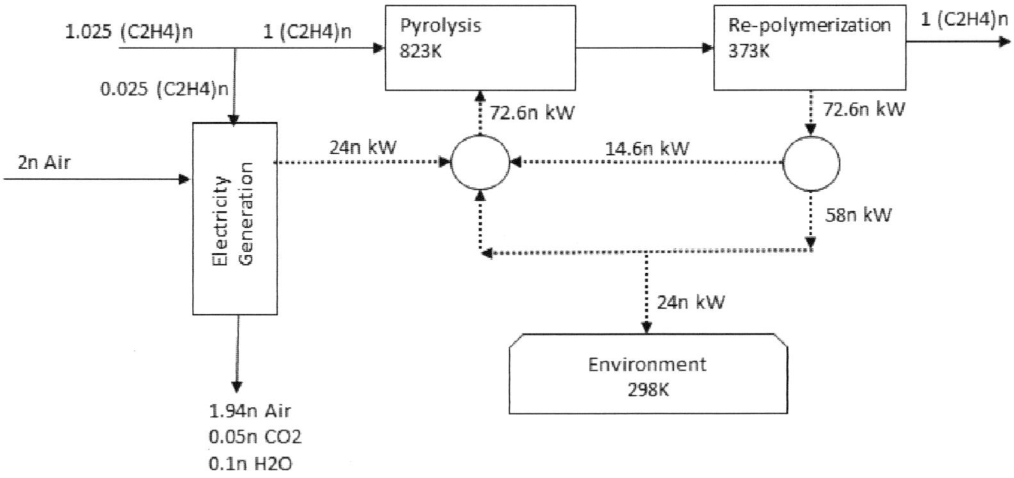

Figure 4: Supplying the energy deficit through electricity via "polyethylene to power" process. Solid arrows indicate mass flow and dotted arrows indicate energy flows.

The system shown in Figure 4 is now overall exothermic. Additional work (in the form of electricity) is now being added to account for the work deficit. In order to maintain the conservation of energy, an identical amount of heat now needs to be rejected from the system. However, this heat has the same temperature as the surroundings and thus no potential is being lost.

To assess the efficiency of the system, or how close to "closed" the cycle is, the carbon efficiency will be used. The carbon efficiency is defined as the amount of carbon in the desired products, relative to the amount of carbon in the feed.

$$C_{eff} = \frac{C_{products}}{C_{feed}} = \frac{2\ mol/s}{2.05\ mol/s} = 0.975$$

Where:

- C_{eff} is the carbon efficiency
- $C_{products}$ is the amount of carbon in the desired product. Here, the desired product is polyethylene. Carbon dioxide is not a desired product.
- C_{feed} is the amount of carbon in the feed.

While 97.5% is a very high carbon efficiency, a carbon efficiency of 100% is required in order for the system to be completely cyclic. In addition, this efficiency is achieved by generating electricity from waste polyethylene, which is not a widely used technology. If this electricity was instead provided by a more typical fossil fuel, like coal, the efficiency of the system would be even lower.

This shows that the cyclic economy will never be an entirely closed and sustainable system unless some way can be found to provide the energy deficit by some method that does not result in the emission of other types of pollutant. This leads to the possibility of using renewable energy.

A closed cyclic economy loop using renewable energy

Wind and solar power offers an opportunity to supply the work deficit of the waste plastic recycling system without emissions that will reduce the efficiency of the system.

The costs of renewable energy installations are constantly changing and have a strong dependence on the region and current technology, but overnight costs can be approximate from data available in the literature [7]. An industrial wind turbine rated at 1.5MW would have an overnight cost of approximately 2.5-million USD. The amount of waste plastic that could be chemically recycled by this turbine is given by,

$$Recycled = \frac{P_{rated}\mu}{W_{deficit}}.M_{C_2H_4}$$

Where:

- Recycled is the mass of waste plastic that can be chemically recycled
- P_{rated} is the rated power of the renewable energy source. 1500 kW for the wind turbine/
- μ is the efficiency of the wind turbine, assumed to be an average of 30%
- $W_{deficit}$ is the work deficit required by the chemical recycling system, 24n kW
- M_{C2H4} is the molar mass of polyethylene, 0.028n kg/mol

The amount of waste polyethylene that could be chemically recycled by a 2.5-million USD, 1.5 MW wind turbine thus comes to 0.525kg/s or 45.4 tons/day.

Using the same reasoning a 1000 kW Concentrated Solar Power (CSP) array, at a cost of 6300 USD/kW [8] and an assumed efficiency of 40%, could chemically recycle 40 tons/day of waste polyethylene.

Taking the market value of High-density polyethylene as around USD 670/ton and assuming the chemically recycled plastics could be sold at the same value, a rough payback period for a renewable energy installation would be between 85 and 160 days. This number only considers the overnight cost of the renewable energy itself, with no operating expenses or the capital for the other units. Even so, such a payback period is not unreasonable.

The system could now be represented as,

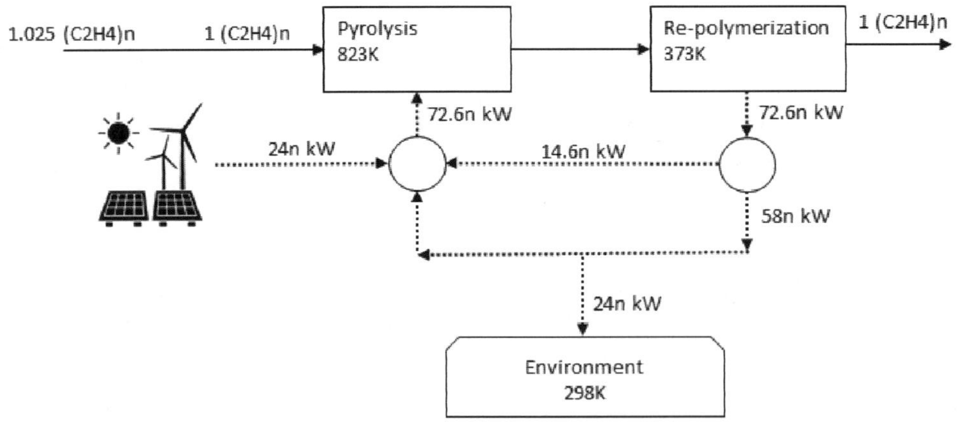

Figure 5: Conceptual closed loop chemical waste plastic recycling system

Instead of using a carbon dioxide emitting energy source, the system in Figure 5 gains the 24n kW of work deficit from renewable energy. This system does achieve the required carbon efficiency of 100%. This shows that, provided a renewable energy source or non-emitting energy source is available, the "cyclic economy" is expensive but conceptually viable.

Conclusion

Conventional power production invariably requires some form of combustion but the use of renewable energy avoids this issue and presents the possibility of achieving the required 100% carbon efficiency needed for a truly cyclic system. It was found that a 1.5 MW wind turbine, operating at 30% efficiency could provide the energy requirements for a cyclic system processing 45 tons/day of waste polyethylene. A 1000 kW concentrated solar power installation, operating at 40% efficiency, could process 40 tons/day. These renewable energy sources carry substantial capital costs of between 2.5 and 6 million US-dollars, for the renewable energy units alone. However, they do achieve the required 100% carbon efficiency needed for the cyclic system.

All of these values represent an idealized limit of performance, the very highest efficiency any process could hope to achieve. Real processes would always perform worse than this limit, general energy loses and fluctuations in renewable energy power supply are just some of inefficiencies present in real systems.

The implication posed by this analysis is that polymers contain a significant amount of chemical potential locked away in the polymer structure. There could be great potential in viewing waste plastics not as a "waste" but rather as an opportunity, a potential new fuel source waiting to be unlocked.

References

1. Grand View Research: Plastics Market Size Analysis, Trends | Global Industry Report, 2020.
2. Jambeck, J.R., Geyer, R., Wilcox, C., Siegler, T.R., Perryman, M., Andrady, A., Narayan, R., Law, K.L.: Marine pollution. Plastic waste inputs from land into the ocean. Science. 347, 768–71 (2015). doi:10.1126/science.1260352
3. Reuters: China's plastic demand to rise as foreign garbage ban to curb recycled supply, https://www.reuters.com/article/us-china-plastics-demand/chinas-plastic-demand-to-rise-as-foreign-garbage-ban-to-curb-recycled-supply-idUSKCN1BO0J8
4. Almeida, D., Marques, M. de F., Almeida, D., Marques, M. de F.: Thermal and catalytic pyrolysis of plastic waste. Polímeros. 26, 44–51 (2016). doi:10.1590/0104-1428.2100
5. Kumar, S., Singh, R.K.: Thermolysis of High-Density Polyethylene to Petroleum Products. J. Pet. Eng. 2013, 1–7 (2013). doi:10.1155/2013/987568
6. Yan, G., Jing, X., Wen, H., Xiang, S.: Thermal Cracking of Virgin and Waste Plastics of PP and LDPE in a Semibatch Reactor under Atmospheric Pressure. Energy & Fuels. 29, 2289–2298 (2015). doi:10.1021/ef502919f
7. Energy Information Administration, U.: Cost and Performance Characteristics of New Generating Technologies, Annual Energy Outlook 2018. (2012)

8. Renewable Energy Agency, I.: RENEWABLE ENERGY TECHNOLOGIES: COST ANALYSIS SERIES Concentrating Solar Power Volume 1: Power Sector Issue 2/5 Acknowledgement. (2012)

528218 Discovery of Electronics Cooling Fluids

Abstract:

Modern miniaturization trends have enabled the development of power intensive electronic devices and resulted in a dramatic increase in electronics cooling requirements that traditional cooling approaches are unable to supply. Efforts to identify working fluids for electronics equipment could enhance cooling performance and meet environmental and safety requirements. Among all cooling approaches, two-phase cooling systems can achieve high heat flux as they rely on phase change of the liquid. Micro-channel heat sinks have become one of the most promising two-phase cooling systems as they allow the combined action of heat conduction, latent heat absorption, and forced convection, leading to sufficient heat dissipation while maintaining a safe device temperature. However, heat transfer performance of micro-channel heat sinks has been plagued by the poor heat transfer characteristics of available cooling fluids. In this work, we aim to identify an efficient replacement of the common industrial cooling liquid HFE-7200.

We build on molecular design methodology [1] that addresses the challenge of large combinatorial search spaces in computer-aided molecular design through decomposition. The approach decomposes the complex CAMD problems into three subproblems: composition design, structure generation, and extended design. This three-phase approach has been implemented in the software AMODEO, which is an ideal tool to solve the coolant design problem. We looked closely into the physical and chemical property constraints that new fluids should satisfy. We used group contribution methods as the major property estimation tool along with additional accurate property models. The resulting candidate molecules were ranked by a heat transfer per unit pressure drop metric, developed specifically for the micro-channel heat sinks regime. Five organic families with the least environmental impact were identified, including hydrofluoroethers (HFEs), hydrofluoroolefins (HFOs), fluorinated esters, ethers, and aliphatics. Most candidate molecules are novel compounds, suggesting promising directions for developing safe, biodegradable, low ozone depleting, and low global warming potential coolants.

[1] A. Samudra and N. V. Sahinidis., Optimization-based framework for computer-aided molecular design., AIChE Journal, 59:3686â€“3701, 2013

535465 Computer-Aided Design of Products Derived from Biomass Pyrolysis

Abstract:

The use of biomass to produce fuels and chemicals required for the modern world has emerged as a promising alternative to reduce our dependency on the fossil fuels which are used as the main feedstocks for the petrochemical industry. Among chemical biomass-to-fuel conversion processes, pyrolysis is considered as the cheapest conversion pathway towards liquid biofuels and it is gaining increased attention as a sustainable option for the production of bio-oil and biochemicals (Galadima and Muraza, 2015). In addition to its significant economic potential, pyrolysis technology has a high degree of flexibility as any type of biomass can be potentially used as a feedstock. Despite its advantages, the use of biomass pyrolysis to produce renewable fuels poses several challenges. The difficult characterization of highly diverse biomass resources, reaction intermediates and final products makes the commercialization of the process quite challenging. Although extensive work has been done on different pretreatment technologies (Hosseini et al., 2010) and reaction pathways of several feedstocks (Papari and Hawboldt, 2015), the upgrading of pyrolysis products into a viable product mix remains a challenge that requires further investigation.

In a recent study, Sharifzadeh et al. (2017) addressed the challenges of modeling the hydrothermal upgrading of pyrolysis oil in order to deoxygenate it and make it usable in the current energy infrastructure. In this work we develop a computer-based methodology for identifying an optimal product mix given information on the effluent from a pyrolyzer. A set of possible products is specified based on their properties and prices, and we formulate an optimization problem to identify the most valuable product portfolio. The proposed methodology builds on a general computer-aided model for designing mixtures and product blends, in which the number, identity and composition of different chemicals in the final product are optimized simultaneously using advanced optimization techniques (Jonuzaj et al., 2016, 2018). The modeling approach is demonstrated though the determination of optimal product blends from predefined feedstock.

References

Galadima, A., Muraza, O., 2015. In situ fast pyrolysis of biomass with zeolite catalysts for bioaromatics/gasoline production: A review. *Energy Conversion and Management 105*, 338–354.

Hosseini, S.A., Lambert, R., Kucherenko, S., Shah, N., 2010. Multiscale modeling of hydrothermal pretreatment: from hemicellulose hydrolysis to biomass size optimization. *Energy Fuels 24*, 4673–4380.

Jonuzaj, S., Akula, P.T., Kleniati, P.M., Adjiman, C.S., 2016. The formulation of optimal mixtures with Generalized Disjunctive Programming: A solvent design case study. *AIChE Journal 62*, 1616–1633.

Jonuzaj, S., Gupta, A., Adjiman, C.S., 2018. The design of optimal mixtures from atom groups using Generalized Disjunctive Programming. *Computers & Chemical Engineering.* URL: *https://doi.org/10.1016/j.compchemeng.2018.01.016.*

Papari, S., Hawboldt, K., 2015. A review on the pyrolysis of woody biomass to bio-oil: Focus on kinetic models. *Renewable and Sustainable Energy Reviews 52*, 1580–1595.

Sharifzadeh, M., Richard, C.J., Shah, N., 2017. Modelling the kinetics of pyrolysis oil hydrothermal upgrading based on the connectivity of oxygen atoms, quantified by ^{31}P-NMR. *Biomass and Bioenergy 98*, 272–290.

524516 Reactive Distillation Combined with Pervaporation for Biobased By-Product Recovery

Abstract:

In downstream processing the biobased industry often struggles with complex multicomponent mixtures. By-products in effluent with low concentration may complicate the wastewater treatment and limit complete usage of biomass. In pulping, an effluent containing mainly acetic acid and formic acid is produced from black liquor evaporation. State of the art process for effluent treatment is reactive extraction followed by several distillation steps for solvent recovery and product isolation. Reactive extraction combined with alternative separation methods were investigated to reduce the high energy consumption and apparatus demand for solvent regeneration and isolation of acetic acid and formic acid from black liquor evaporation condensate. The aqueous mixture of acetic acid and formic acid forms inseparable azeotropes.

Reactive distillation and pervaporation are perfectly suited for separation of azeotropic mixtures. Reactive distillation overcomes azeotropes by changing the substance properties via chemical conversion. With carefully chosen esterification the market value of the by-products may be increased, and the significant change of the physico-chemical properties and product separation decrease the cost of product isolation. For separation of acetic acid, formic acid and water reactive distillation with methanol as esterifying agent was successfully performed, but with low reaction rates without catalyst. For successful solvent regeneration high reaction rates and high separation efficiency of the carboxylic acids are needed. The reaction rates were raised by admixture of 5 wt.% 4-DBSA the catalyst 4-dodecylbenzenesulfonic acid (4-DBSA) to the solvent. Batch reactive distillation experiments of laden solvent showed complete conversion of formic acid and a methyl acetate yield of 88 % in the distillate. The feasibility of catalyst admixture to the solvent phase was successfully tested. The catalyst has a minor impact on the liquid-liquid equilibrium. The distillate contains the low boiling methyl esters and excess methanol. Water is not transferred to the distillate. This product mixture exhibits a low boiling azeotrope of methyl acetate and methanol.

Pervaporation is able to isolate azeotropic mixtures because the separation does not depend on the vapor-liquid equilibrium; it is the result from solution-diffusion properties of the components in the membrane. Therefore, pervaporation with hydrophilic membranes was investigated for the separation of methanol from the methyl esters. Pervaporation experiments with methanol/methyl acetate mixtures, the bottleneck in the system, were performed with the commercial membrane Pervap^{TM} 4155-70. With the ternary mixture, the viability of separating methanol from both methyl esters was confirmed.

The implementation of alternative separation tools shows a high potential for raising the product yield and for reducing energy demand. Simplification of process concepts satisfies the specific needs of the biobased industry.

538047 Analysis of a Circular Economy: From Food Waste to Foods

Abstract:

The US-EPA estimated that 38 million tons of food waste went to landfill in 2014 and in 2010 66.5 million tons of food didnâ€™t make it to consumers. The food waste streams total to around a billion pounds of food waste, where the majority go to landfill. Publicly owned waste water treatment plants in U.S. process approximately 14 billion kilograms of dry solids per year.[1] Turning waste streams into valuable products is the goal of the circular economy.[2]

Hydrothermal carbonization (HTC) process is proposed to be a solution to the large streams of food waste and biosolids from waste water treatment plants that creates a highly valuable, multiuse product. HTC is a thermochemical process that converts wet biomass materials to a coal-like product. Under sub-critical conditions, water in the HTC reactor stays in liquid form and acts as a reaction medium to promote the breakdown and cleavage of chemical bonds in the solid biomass. The absence of liquid-to-vapor phase change of the water makes the process significantly less energy intensive compared to a process that involves water vaporization or drying. The temperature range for HTC is from 180-240°C producing solid hydrochar as the main product.

The solid product of hydrothermal carbonization, hydrochar, has multiple applications ranging from a combustion fuel source, a catalyst for anaerobic digestion that increases the purity of methane production, a soil amendment, and the base for other carbon product such as activated carbon.

In this study we will explore the circular economy of utilizing organic wastes into foods. The steps involved in the circular economy system include: 1) collecting organic wastes (specifically food wastes and bio-solids), 2) converting the organic waste streams to hydrochar via HTC process, 3) growing food produces by indoor gardening, and 4) using hydrochar from step 2) as fuels to power the indoor gardens.

Utilizing organic waste streams in cities such as Chicago or Philadelphia, where there are a multitude of abandoned warehouses, provides an opportunity to produce locally grown produces that would typically come from across the country from California or Mexico. Not only will this circular economy reduce environmental impacts, but it will increase positive societal impacts by benefitting the local economy, providing educational opportunities of farming that would not typically be available in cities, and serving fresher produces to the local community.

A comparative life cycle assessment of our proposed process will be applied to compare conventional farming to our circular economy proposal. We hope to prove a reduction of environmental impacts, such as global warming potential, fossil fuel depletion, natural resource depletion. The life cycle assessment results will aid in a full system analysis which evaluate the proposed circular economy idea from societal, technical, environmental, economic, and political perspectives.

- https://www.epa.gov/sustainable-management-food/sustainable-management-food-basics
- https://www.sciencedirect.com/science/article/pii/S0301479717303808

*Contact: Dr. Justinus Satrio (justinus.satrio@villanova.edu)

538097 Re-Wiring the Domestic Food Trade for Reducing Irrigation Impacts in the United States

Abstract:

The food-energy-water(FEW) nexus concept recognizes the interconnected nature of the three systems and advocates managing them together to avoid sub-optimal outcomes and unintended consequences. The FEW nexus challenges differ for developed agro-economies such as the United States(U.S.) from agriculture centric developing nations. For U.S., irrigation provides a crucial link to study interconnections between food-water-energy resources. Additionally, food trade is an integral part of food system as it bridges the growing disconnect between production and consumption centers. Depending on the origin of food production, trade can potentially reduce environmental impact of food consumption. However, environmental impacts of agriculture are not directly considered when forming trade connections and may carry risk associated with suboptimal FEW system management (e.g. sourcing food from water-depleted regions). Structural optimization of food trade provides an opportunity to assess whether FEW resources can be managed together with respect to land-water-energy constraints. This work focuses on environmental impacts associated with FEW systems, namely- water required for irrigation, greenhouse gas(GHG) emissions from commodity transport, and pumping irrigation water for food production.

We leverage publicly available datasets from agriculture census and transportation surveys to create a network of regional food trade. The study combines data on state-wide food production, irrigation statistics, and life cycle inventory databases to convert food transfer network into embedded irrigation water, and GHG emission flows. Specifically, we focus on rice production and trade in the U.S. Rice is one of the staple food crops and produced entirely through irrigation, making it an ideal crop for the irrigation impact analysis. Majority of rice production is concentrated in four regions-Arkansas Grand Prairie, Mississippi Delta, Gulf Coast, and Sacramento Valley. The model optimizes rice trade network by keeping the demand constant and evaluate a series of constraints pertaining to reduction in energy and water usage to understand whether production can be shifted realistically to optimize all three systems. We limit the potential crop shifting to current production regions to account for soil and climatic parameters conducive to crop production. Additionally, we consider type of rice grown in each region as crop physiological requirements may differ based on the rice type. Preliminary results of the optimization framework reveals a rewired domestic rice trade network with a 27% reduction in virtual water use (15.1 billion m^3 reduced to 11.1 billion m^3). However, due to shift in production to more water-efficient states, a 10% increase in marginal land use is required. The land results are significant as they demonstrate the resulting sub-optimal solutions when interdependencies are overlooked. On the other hand, minimizing groundwater usage from water stressed areas also reduces overall GHG emissions demonstrating that these are not mutually exclusive goals in many states. The Pareto optimal solutions showing water-GHG emissions tradeoffs resulting from the model will be presented and their implications for regional FEW nexus sustainability will be discussed.

527168 Food, Energy, Fuels and Chemical Feedstocks from Rice Crops: Multi-Objective Optimisation of Multi-Product Value Chains for the Philippines

Abstract:

Rice is an important staple food in the Philippines such that its 20 M metric tons annual production places the country rank 8th among the rice producers in the world. However, there are many improvements in the efficiency and sustainability of the Philippines rice value chains that could be made by addressing various constraints related to technological, environmental, economic and socio-political factors [1]. Consequently, the Philippine government is striving to achieve Philippine rice competitiveness and to transform its value chains to be more profitable, resilient and sustainable; they recognise the importance of research and development as basis for such a laudable goal [2]. From a recent analysis of the Philippine rice value chain commissioned by the government, by-products like rice straw, rice hull and rice bran can be important sources of raw material for other industry sectors. Ninety-five percent of the rice bran is being used as animal feed component while seventy percent of the of the rice hull is being used as an alternative source of energy and fuel. Moreover, the silica-rich rice hull ash is being used as additive in the cement industry [3]. These could not only provide affordable renewable energy, fuels and materials but also much-needed rural development and employment opportunities. However, there are complex issues that need to be addressed: rice production is a water- and energy-intensive activity. The global water footprint from rice production is 1308 Mm^3/year with irrigated rice cultivation consuming the most water. Energy is required for the production of fertilisers, electricity for pumping water and heat for parboiling, drying and other milling activities [4]. Furthermore, the limited land for rice cultivation is being threatened by the rapid and unregulated conversion of agricultural land to other uses [1]. In this project, a comprehensive optimisation model, based on mixed-integer programming [5-6], is being developed that can support complex decision-making related to multi-product rice value chains and can capture the trade-offs between water, energy and land utilisation. The model is used to identify promising rice value chains for sustainable and efficient production of food, energy, fuels and chemical feedstocks. In this conference, we will present the optimisation model and the key findings from the case studies.

References:

[1] B.A. Excode (2016). Rice production and consumption trends in the Philippines from 2000 to 2015. Yuchengco Center, De La University; Philippines.

[2] C.T. Briones (Ed.) (2016). The PhilRice strategic plan 2017-2022. Philippine Rice Research Institute; Philippines.

[3] A.B. Mataia, A.B. Mataia, R.G. Manalili, J.C. Beltran, B.M. Catudan, N.M. Francisco and A.C. Flores (2016). Analysis of the rice value chain in the Philippines. Philippine Rice Research Institute; Philippines

[4] R. Sims, A. Flammini, M. Puri, S. Bracco (2015). Opportunities for Agri-Food Chains to become energy-smart. Food and Agriculture Organization of the United Nations and the United States Agency for International Development; United States of America.

[5] S. Samsatli, N.J. Samsatli (2018). A multi-objective MILP model for the design and operation of future integrated multi-vector energy networks capturing detailed spatio-temporal dependencies. Applied Energy. DOI: 10.1016/j.apenergy.2017.09.055.

[6] S. Samsatli, N.J. Samsatli, N. Shah (2015). BVCM: a comprehensive and flexible toolkit for whole-system biomass value chain analysis and optimisation - mathematical formulation. Applied Energy, 147, pp. 131-160.

541194 Modeling the Impacts of International Food Trade on Contaminant Transport and Human Exposure

Abstract:

Introduction: Quantifying and addressing the implications of the transfer of resources across borders is a hallmark of research on the food, energy and water (FEW) nexus. International trade in goods can result in severe imbalances in the water stress, green house gas emissions, and energy consumption experienced by or attributed to producing and receiving countries[1,2]. Yet an important component of food trade—that is, the potentially hazardous chemicals contained in foods as a result of accumulation from production environments, processing, transport, or storage—is still poorly quantified[3–5]. Here, we investigate the importance of international food trade on the dietary exposure of the Swiss population to polybrominated diphenyl eithers (PBDEs). PBDEs are lipid soluble compounds used as flame retardants in synthetic fibers and molded plastics. Of major concern is PBDE exposure through the diet. Studies have confirmed that fish, meat, and dairy products contribute most to the daily PBDE intake[6]. Like other countries, Switzerland has reported a gradual shift in dietary patterns over the past few decades, with increasing consumption of fish reported, reaching a peak value of 17.6 kg in 2010, and with a shift in the origin of fish. We therefore constructed a model to quantify the exposure of the Swiss population to PBDEs from local and imported fish and seafood.

Methodology:

Model boundaries: Our study focused on Switzerland as a geographical case study. All calculations referred to fish imports for the year 2016. Fish, shrimp/prawn, and mussels were selected as key fish and seafood types to include in the study, as they are among the most highly consumed.

Data collection: An extensive literature review was conducted to derive useful input data for our model, including PBDE concentrations in fish, shrimp and mussels from different countries, and to estimate daily per capita fish consumption in Switzerland. Trade data (imports, exports, and re-exports) for fish, crustaceans (shrimp and prawns) and mollusks (mussels) was extracted from the UN Comtrade Database. Data were collected for fresh, frozen, fresh fillet, and frozen fillet categories of the respective commodities.

Model Structure: The model consists of four primary components. These are: (i) PBDE concentrations in fish from different origins, resulting from bioaccumulation from their environment and/or food; (ii) food trade data for Switzerland; (iii) local fish production; and (iv) dietary intake to the Swiss population via fish and seafood consumption.

Results and Discussion: Based on the difference between average per capita fish consumption, production, imports and re-exports, we estimate that Switzerland imports approximately 66% of fish consumed, while only 34% is locally caught. Of the imported fish, Norway, Denmark and the UK are the largest exporters of salmon to Switzerland. For shrimp and prawn, the majority is from Vietnam. For other fish, Norway, Denmark, the Netherlands, Vietnam, France, Germany, Italy, and China are the largest exporters. Within the locally produced fish, whitefish (*Coregonus sp.)* is the most common fish contributing to 60% of local the catch, followed by perch (19%), pike (7%) and roach (7%)[7]. Using an average of 23g/day fish consumption, results show that from 70000 metric tons fish consumed in Switzerland in 2016, 66 % (46000metric tons) was imported corresponding to 34 % locally caught fish (24000 metric tons).

Based on published studies of PBDEs in fish, the highest concentrations were detected in whitefish (Coregonus spp.) in Switzerland[8]. Norway also reported high levels of PBDEs in whitefish[9]. The US reported PBDEs as high as 3078 pg/g wet weight in salmon[10]. Compilation of country-specific PBDE body burdens in fish continues. In addition, we are employing modeling approaches to estimate PBDE body burdens in key species from exporting countries where data are lacking. These data will aid in forming a comprehensive source-receptor relationship for Swiss population exposure to PBDEs via seafood and fish.

References Cited

(1) Vora, N.; Shah, A.; Bilec, M. M.; Khanna, V. Food–Energy–Water Nexus: Quantifying Embodied Energy and GHG Emissions from Irrigation through Virtual Water Transfers in Food Trade. *ACS Sustain. Chem. Eng.* **2017**, *5* (3), 2119–2128.

(2) Leck, H.; Conway, D.; Bradshaw, M.; Rees, J. Tracing the Water–Energy–Food Nexus: Description, Theory and Practice. *Geogr. Compass* **2015**, *9* (8), 445–460.

(3) Ng, C. A.; von Goetz, N. The Global Food System as a Transport Pathway for Hazardous Chemicals: The Missing Link between Emissions and Exposure. *Environ. Health Perspect.* **2017**, *125* (1), 1–7.

(4) D'Odorico, P.; Carr, J. A.; Laio, F.; Ridolfi, L.; Vandoni, S. Feeding Humanity through Global Food Trade. *Earths Future* **2014**, *2* (9), 2014EF000250.

(5) Ercsey-Ravasz, M.; Toroczkai, Z.; Lakner, Z.; Baranyi, J. Complexity of the International Agro-Food Trade Network and Its Impact on Food Safety. *PLoS ONE* **2012**, *7* (5), 1–7.

(6) Schecter, A.; Haffner, D.; Colacino, J.; Patel, K.; Päpke, O.; Opel, M.; Birnbaum, L. Polybrominated Diphenyl Ethers (PBDEs) and Hexabromocyclodecane (HBCD) in Composite U.S. Food Samples. *Environ. Health Perspect.* **2010**, *118* (3), 357–362.

(7) Swiss fact: around 1.8 million kilograms of fish are caught in Switzerland every year https://lenews.ch/2017/01/04/swiss-fact-around-1-8-million-kilograms-of-fish-are-caught-in-switzerland-every-year/ (accessed May 31, 2018).

(8) Zennegg, M.; Kohler, M.; Gerecke, A. C.; Schmid, P. Polybrominated Diphenyl Ethers in Whitefish from Swiss Lakes and Farmed Rainbow Trout. *Chemosphere* **2003**, *51* (7), 545–553.

(9) Mariussen, E.; Fjeld, E.; Breivik, K.; Steinnes, E.; Borgen, A.; Kjellberg, G.; Schlabach, M. Elevated Levels of Polybrominated Diphenyl Ethers (PBDEs) in Fish from Lake Mjøsa, Norway. *Sci. Total Environ.* **2008**, *390* (1), 132–141.

(10) Schecter, A.; Päpke, O.; Tung, K.-C.; Staskal, D.; Birnbaum, L. Polybrominated Diphenyl Ethers Contamination of United States Food. *Environ. Sci. Technol.* **2004**, *38* (20), 5306–5311.

523949 Using Agricultural Wastes to Recover Rare Earth Elements from End-of-Life Materials

Abstract:

An economical and eco-friendly bench-scale process that relies on microbial production of organic acids has been developed to solubilize and release rare earth elements (REE) from solid recyclable materials (Reed, Hydrometallurgy, 2016, 166:34-40). Techno-economic analysis (TEA) and life cycle analysis (LCA) for bioleaching of REE from fluid catalytic cracking (FCC) catalysts suggested that a bioleaching plant could be profitable, but could be even more profitable if the cost of the sugar (glucose) used for growing the microbes was reduced (Thompson, ACS Sustainable Chemistry & Engineering, 2017, 6:1602-1609). This presentation will focus on our efforts to improve the economics of REE bioleaching using alternative carbon and energy sources as microbial growth substrates.

In our previous work examining the potential for several organic acid producing microorganisms to leach REE from end-of-life waste materials we identified *Gluconobacter oxydans* B58 as a promising industrial strain. Lixiviant produced by *G. oxydans,* containing predominantly gluconic acid, could recover ~50% of the total REE content of FCC catalyst. The biologically produced lixiviant was more effective than an abiotically prepared solution containing a similar concentration of gluconic acid. TEA and LCA showed that bioleaching could be more profitable and environmentally friendly than industrial standard chemical leaching processes, suggesting commercial viability. However, substantial reductions in operating costs could be achieved by replacing the refined glucose used in the original growth medium with a lower cost substrate. Consequently corn stover and potato wastewater were tested as growth substrates for *G. oxydans*. The initial lixiviants generated were tested against FCC and yielded a leaching efficiency 50-90% of the best case leaching efficiency of glucose-derived biolixiviant. Work is ongoing to optimize biolixiviant production from agricultural wastes. The conditions for improving bioleaching using waste-derived biolixiviants, including TEA and LCA evaluations, will be discussed.

526360 Membranes for Nutrient Concentration, Industrial Separations and Applications Beyond

Abstract:

With the increasing importance of the waterâ€“energyâ€“food nexus, it is critical that membranes can be designed for optimal performance and energy efficiency for large varieties of water and food applications. The cost and energy consumption of milk and whey concentration systems can be minimized when one selects a reverse osmosis element that reduces fluid flow requirements while maintaining optimum dewatering efficiency. In addition, superior performance of RO membranes which address the emerging importance of high rejection will be discussed for nutrient concentration, with a properly designed membrane based on the understanding of membrane structure-property relationships. Potentially there are many more applications where RO technology can be used to meet critical customer requirements with high efficiency and favorable economics.

541565 25 by 25: Chemical Engineering in the Next 25 Years

Abstract:

In 2008, the centennial year for AIChE, we asked 25 chemical engineers for their visions of the profession from around the US and the world: Nine faculty, eight from industry, and eight post-docs and graduate students. We posed four questions related to each individual's industrial sector or area of research, asking for extrapolation, new impacts on existing sectors, new sectors, and an optional, more open-ended comment on the future of the profession as a whole.

In this 110th year of AIChE, we have revisited these questions with many of the same people. Ten years has seen many developments in our profession, in society, and in our thought-leaders' experiences. Here we report on their perspectives, comparing them to the 2008 visions.

541542 The Future of Chemical Engineering Itself

Abstract:

Chemical engineeringâ€™s future is bright because it is vital to so much of society and industry -- and its future is complicated by its rapidly expanding universe of activities. It must represent depth in core skills centered on the chemical sciences including biology and materials. At the same time, it must be transdisciplinary; it must exemplify the need and value of reaching across disciplines in borderless ways.

One way is to create and disseminate a compelling vision and strategy for chemical engineering, research, and education for the future. AIChE and the National Academies are working together to that end. In 1988, the National Academies published "Frontiers in Chemical Engineering: Research Needs and Opportunities.â€ Its committee of authors outlined a wide-ranging roadmap to turn promising research opportunities into reality, while calling for researchers to embrace new frontiers.

Since 1988, that has happened -- along with many other transformative changes like the Web, fracking, personalized medicine, and social media. A new nationwide effort, aided by international contributions, will address both the identity of chemical engineering and its future directions. Its goal is to ensure that the people in chemical engineering are primed to meet society's needs and rapid changes.

544708 Accelerating Innovation through Academic-Industrial Partnerships

Abstract:

In 2011, The Dow Chemical Company launched the University Partnership Initiative (UPI) to sponsor collaborative research at leading U.S. Universities. This 10-year commitment aims to address fundamental questions in industrial chemistry and chemical engineering with alignment to Dowâ€™s next-generation process technology and product development efforts. To date, this portfolio of projects has generated an impressive array of intellectual property, publications, and numerous examples of the collaborative developments implemented at Dow. This presentation will introduce UPI program and highlight several contributions from our collaborators and leaders in the chemical engineering profession.

540559 Maximizing Uptime, Efficiency, and Safety of Industrial Operations through Early Risk Detection

Abstract:

Achieving zero incidents while maximizing process uptime and efficiency is a key vision for operating companies. An important building block towards this vision is establishment of a proactive risk mitigation culture, supported by effective systems and a strong management. Studies have shown that most of the incidents and unexpected process failures can be avoided if the operating teams get timely information about developing risks and take preventive actions early on. While advances have been made in process monitoring and automation in the past decades, there still remains significant technological and behavioral gaps that prevent plant operations from proactively mitigating process risks.

A modern industrial plant monitors thousands of parameters, generating upwards of 50-100 million data points every day. Thanks to recent breakthroughs in machine learning and artificial intelligence approaches, today there are autonomous systems that can sift through this data and point out meaningful and timely insight. This can help operating teams ascertain process issues that are hidden in the data, long before process variables reach critical levels.

In this presentation, a new autonomous system, Dynamic Risk Analyzer™ (DRA), will be introduced that points out early indicators of process issues using proprietary machine learning. The risk indicators enable operations team to respond and make required process changes, days and sometimes weeks ahead of any alarm, avoiding late stage (expensive) fixes. To realize these objectives fully, a proactive management workflow must also be in place. The presentation will discuss new initiatives and workflows developed by plant operations to implement a proactive culture. Real-life case studies from plants using DRA technology will be presented on how this culture change resulted in increased uptime, efficiency and safety of plant operations.

531309 Gaussian Processes for Hybridizing Analytical & Data-Driven Decision-Making

Abstract:

Surrogate models are widely appreciated in process systems engineering [1]. The typical setting focuses on expensive-to-evaluate, possibly uncertain functions. Examples include: modular process simulators [2], integrated gasification combined cycle processes [3], a carbon capture absorber [4], and many other applications, e.g. [5-8]. Resources are typically limited, so effective decision making requires data-efficient learning.

The data science and statistical machine learning communities typically focus on models learned solely from observed data. But chemical engineering applications may also require explicit, parametric models, e.g. modeling known process constraints, operations constraints, and cost objectives [9]. So, work has integrated semi-algebraic functions with those learned from data [10] or developed semi-physical modeling techniques [11, 12].

This presentation surveys the state-of-the-art in hybridizing analytical and data-driven decision making [13]. We consider three probabilistic modeling applications to these hybrid situations:

Design of experiments for model discrimination [14]. We bridge the gap between classical, analytical methods [15] and Monte Carlo-based approaches [16]. Classical methods may have difficulty managing non-analytical model functions and data-driven Monte Carlo approaches come at a high computational cost. We replace the original, parametric models with probabilistic, non-parametric Gaussian process surrogates learned from model evaluations. The surrogates are flexible regression tools that extend classical analytical results to non-analytical models, while providing us with model prediction confidence bounds and avoiding the computational complexity of Monte-Carlo approaches.

Multi-objective optimization [17, 18]. We make novel extensions to Bayesian multi-objective optimization in the case of one analytical objective function and one black-box, i.e. simulation-based, objective function. The resulting method has been applied to a bone neotissue application [19] and a more general test suite.

Scheduling plant operations under uncertainty. For processes with equipment degradation, we use Gaussian processes to approximate large-scale, mixed-integer optimization problems.

We close by offering a broad outlook on applying probabilistic surrogate models to chemical engineering. Statistical machine learning has recently attracted significant interest in process systems engineering [20]. Here we show that state-of-the-art research in Gaussian processes [21, 22] and probabilistic modeling more generally [23] can have a big impact on chemical engineering.

References

[1] Bhosekar A, Ierapetritou M. Advances in surrogate based modeling, feasibility analysis and and optimization: A

review. *Comput Chem Eng*. 108: 250-267, **2018**.

[2] Caballero JA, Grossmann IE. An algorithm for the use of surrogate models in modular flowsheet optimization. *AIChE J*. 54(10):2633-2650, **2008**.

[3] Lang Y, Zitney SE, Biegler LT. Optimization of IGCC processes with reduced order CFD models. *Comput Chem Eng*. 35(9):1705-1717, **2011**.

[4] Cozad A, Sahinidis NV, Miller DC. Learning surrogate models for simulation-based optim-ization. *AIChE J*. 60(6):2211-2227, **2017**.

[5] Boukouvala F, Ierapetritou M. Surrogate-based optimization of expensive flowsheet modeling for continuous pharmaceutical manufacturing. *J Pharm Innov*. 8: 131, **2013**.

[6] Soepyan FB, Cremaschi S, Sarica C, and others. Estimation of percentiles using the Kriging method for uncertainty propagation. *Comput Chem Eng*. 93:143-59, **2016**.

[7] Shokry A, Ardakani MH, Escudero G, Graells M, Espuña A. Dynamic Kriging-based fault detection and diagnosis approach for nonlinear noisy dynamic processes. *Comput Chem Eng*. 106:758-76, **2017**.

[8] Tran AP, Georgakis C. On the estimation of high-dimensional surrogate models of steady-state of plant-wide processes characteristics. *Comput Chem Eng*. DOI 10.1016/j.comp-chemeng.2018.02.014, **2018**.

[9] Boukouvala F, Hasan MF, Floudas CA. Global optimization of general constrained grey-box models: new method and its application to constrained PDEs for pressure swing adsorption. *J Glob Optim*. 67(1-2): 3-42, **2017**.

[10] Boukouvala F, Floudas CA. ARGONAUT: AlgoRithms for Global Optimization of coNstrAined grey-box compUTational problems. *Optim Lett*. 11(5): 895-913, **2017**.

[11] Pearson RK, Pottmann M. Gray-box identification of block-oriented nonlinear models. *J Process Contr*. 10(4):301-315, **2000**.

[12] Cozad A, Sahinidis NV, Miller DC. A combined first-principles and data-driven approach to model building. *Comput Chem Eng*. 73: 116-27, **2015**.

[13] Boukouvala F, Misener R, Floudas CA. Global optimization advances in mixed-integer nonlinear programming, MINLP, and constrained derivative-free optimization, CDFO. *Eur J Oper Res*. 252: 701-727, **2016**.

[14] Olofsson S, Deisenroth MP, Misener R. Design of Experiments for Model Discrimination Hybridising Analytical and Data-Driven Approaches. *arXiv:1802.04170*. **2018**.

[15] Buzzi-Ferraris G, Forzatti P, Emig G, Hofmann H. Sequential experimental design for model discrimination in the case of multiple responses. *Chem Eng Sci*. 39(1): 81-85, **1984**.

[16] Vanlier J, Tiemann CA, Hilbers PAJ, van Riel NAW. Optimal experiment design for model selection in biochemical networks. *BMC Syst Biol*, 8(20), **2014**.

[17] Olofsson S, Mehrian M, Geris L, Calandra R, Deisenroth MP, Misener R. Bayesian Multi-Objective Optimisation of Neotissue Growth in a Perfusion Bioreactor Set-Up. *Computer Aided Chemical Engineering*. 40: 2155-2160, **2017**.

[18] Beykal B, Boukouvala F, Floudas CA, Pistikopoulos EN. Optimal design of energy systems using constrained grey-box multi-objective optimization. *Comput Chem Eng*. DOI 10.1016/j.compchemeng.2018.02.017, **2018**.

[19] Mehrian M, Guyot Y, Papantoniou I, Olofsson S, Sonnaert M, Misener R, Geris L. Maximizing neotissue growth kinetics in a perfusion bioreactor: An *in silico* strategy using model reduction and Bayesian optimization. *Biotechnol Bioeng*. 115(3):617-29, **2018**.

[20] Lee JH, Shin J, Realff MJ. Machine learning: Overview of the recent progresses and implications for the process systems engineering field. *Comput Chem Eng*. DOI 10.1016/j.compchemeng.2017.10.008, **2017**.

[21] Damianou A, Lawrence N. Deep Gaussian processes. In *Artificial Intelligence and Statistics*. 207-215, **2013**.

[22] Deisenroth MP, Ng JW. Distributed Gaussian processes. In *32nd International Conference on International Conference on Machine Learning*. 37: 1481-1490, **2015**.

[23] Ghahramani Z. Bayesian non-parametrics and the probabilistic approach to modelling. *Phil Trans R Soc A*. 371(1984):20110553, **2013**.

537126 Cost-Competitive Electrolysis-Based Hydrogen Under Current U.S. Electric Utility Rates

Abstract:

Hydrogen has the potential to facilitate the integration of disparate energy systems. Electrolysis-based hydrogen production enables a key link between the electricity sector and end uses for hydrogen, including transportation, heating, agricultural, and industrial applications [1–4]. Hydrogen cost-competitiveness analyses are usually oversimplified by using annual average electricity cost by state or region. This approach does not allow for location-specific value assessments. Using 7,182 industrial and commercial U.S. retail electric utility rates [5], this study addresses a lack of understanding about the electric utility rates, the costs to produce hydrogen, and it addresses the potential to reduce production costs and support electric grid needs by operating electrolyzers more dynamically. Results show that hydrogen production from a 1 MW electrolysis unit is already cost-competitive under a large number of current electricity rate structures. Sensitivities show that the ideal yearly capacity factor for an electrolyzer, which is a highly flexible load, is between 90%–95%, depending on the structure of the utility rate and the electrolyzer capital cost. Based on the operation of electrolysis equipment, hydrogen production costs range from U.S. \$2.6 kg-1 to U.S. \$12.3 kg-1, and costs are less for electrolyzers that operate flexibly according to the changing rate structure. Our analysis shows that electrolysis-based hydrogen production costs in some locations are already less than the U.S. Department of Energy target of U.S. \$4 kg^{-1} for 81 utility rates. Thus, hydrogen is already cost-competitive in the U.S. energy sector and has a number of interesting possible roles to play in future energy and transportation systems.

References

[1] International Energy Agency. Technology Roadmap: Hydrogen and Fuel Cells. Paris, France: International Energy Agency; 2015. doi:10.1787/9789264239760-en.

[2] Stevens J, Rustagi N, Pivovar B, Ruth M, Boardman R, Gilroy N, et al. " H2 @ Scale " – an emerging cross- sector opportunity in the USA. Gas Energy 2017:22–7.

[3] Dodds PE, Staffell I, Hawkes AD, Li F, Grünewald P, McDowall W, et al. Hydrogen and fuel cell technologies for heating: A review. Int J Hydrogen Energy 2015;40:2065–83. doi:10.1016/j.ijhydene.2014.11.059.

[4] Staffell I, Dodds P. The role of hydrogen and fuel cells in future energy systems. 2017.

[5] National Renewable Energy Laboratory. Utility Rate Database (URDB) 2017. https://openei.org/wiki/Gateway:Utilities (accessed October 13, 2017).

521790 Environmental and Economic Analysis for Sustainable Management of Livestock Waste

Abstract:

Livestock waste may cause air quality degradation from ammonia and methane emissions, soil quality detriment from the in-excess nutrients and acidification, and water pollution issues from nutrient and pathogens runoff to the water bodies, leading to eutrophication, algal blooms, and hypoxia [1-2]. Despite the significant environmental benefits of performing pollution management for these organic materials, the recovery of value-added products from livestock waste is not a current practice due to the high investment costs required and the low market values being offered for the products that are recovered. In addition, the deployment of waste treatment technologies is hindered by the difficulty to attribute an economic value to environmental benefits.

In this talk, we will present a supply chain design framework [3] to conduct simultaneous economic and environmental analysis of post-livestock organic material to value-added products. The proposed framework captures techno-economic and logistical issues and can accommodate diverse types of policy incentives obtained at federal and state levels, allowing stakeholders to conduct systematic studies on the effect of incentives on the economic and environmental viability of diverse technologies. We apply the framework to a case study for dairy farms in the State of Wisconsin (U.S.), where we consider the placement of anaerobic digestion in combination with nutrient recovery and biogas upgrading technologies. The framework reveals that, from a purely economic perspective, products recovered from dairy waste are not competitive at current market prices. We also find that incorporating current and potential U.S. government incentives in the form of Renewable Identification Numbers (RINs) [4] and phosphorus credits [5] can achieve economic viability of the recovery of liquefied biomethane and nutrient-rich products. On the other hand, current incentives for electricity generation (Renewable Energy Credits or RECs [6]) cannot achieve economic viability. The analysis also reveals that the best strategy to manage waste is to synergize the deployment of technologies that conduct simultaneous recovery of liquefied biomethane and nutrients [7].

References:

[1] Burkholder, J.; Libra, B.; Weyer, P.; Heathcote, S.; Kolpin, D.; Thorne, P. S.; Wichman, M. Impacts of waste from concentrated animal feeding operations on water quality. *Environmental Health Perspectives*115, no. 2 (2007): 308.

[2] Aguirre-Villegas, H. A.; Larson, R. A. Evaluating greenhouse gas emissions from dairy manure management practices using survey data and lifecycle tools. *Journal of Cleaner Production* 143 (2017): 169-179.

[3] Sampat, A. M.; Martín, E.; Martín, M.; Zavala, V. M. Optimization formulations for multiproduct supply chain networks. *Computers & Chemical Engineering* 104 (2017): 296-310.

[4] Schnepf R.; Yacobucci B. D. Renewable Fuel Standard (RFS): overview and issues. *In CRS Report for Congress* **2010** Jul 14 (No. R40155).

[5] Trading and Offsets in the Chesapeake Bay Watershed. *U.S. Environmental Protection Agency available at* https://www.epa.gov/chesapeake-bay-tmdl/trading-and-offsetschesapeake-bay-watershed, [Online; accessed 06-December-2017].

[6] Renewable Energy Certificate (REC) Arbitrage; Technical Report; U.S. Environmental Protection Agency, 2017.

[7] Sampat, A. M.; Gerardo J. Ruiz-Mercado, and Victor Zavala. "Economic and Environmental Analysis for Advancing Sustainable Management of Livestock Waste: A Wisconsin Case Study." *ACS Sustainable Chemistry & Engineering* (2018). In Press.

534071 Multi-Objective Planning Approach for Supply Chains Based on Biomass Processing By a Sequential Methodology Via Geographic Information System and Mathematical Programming

Abstract:

Biomass is a renewable resource with attractive characteristics to produce energy and biofuels. One of main reasons to use biomass as renewable source for energy production is that biomass can capture a significant amount of CO_2 emissions during growing. In this respect, diverse studies have stated that biomass used for biofuels and energy production can contribute partially to solve the energy demand problem and mitigate the increasing of CO_2 emissions. Most works have focused on developing new technologies, processes and processing system based on biomass. In addition, several works have focused in determining locations, raw materials and processing technologies. Nevertheless, most of them have based the options to select potential locations on their own criteria, neglecting important issues associated to viability; such as distance to roads for transportation or distance to water bodies to avoid affecting rivers and lakes.

Therefore, models based on Geographic Information Systems (GIS) are useful tools for data management and processing in order to determine viable locations in this type of systems. GIS models have not been widely used to assess the economic performance of a production system. For that reason, GIS methodologies can be complemented using mathematical programming strategies. Therefore, this paper proposes a sequential and multi-objective approach for the optimal planning of a residual biomass processing system. The methodology considers the selection of potential locations through a multi-criteria methodology based on GIS; which considered the uncertainty associated to the amount of available biomass by diverse scenarios generation. Additionally, the approach uses the viable locations obtained by the GIS methodology and then selects locations larger than a minimum size in order to define where processing facilities can be installed. Also, chosen locations by the GIS methodology can receive raw material from specific harvesting sites; which should be into a radius of 161 km from the centroid of processing facilities. Subsequently, data of potential locations for processing facilities and harvesting sites are fed in a mathematical formulation to determine the optimal supply chain topology considering various points of view. The mathematical programming approach consists of mass balances to define the interconnections between the different supply chain nodes, as well as, constraints to model the considered technologies. Moreover, mathematical model involves two objective functions. Environmental objective is modeled via the eco-efficiency while the economic performance is associated to net annual profit; which produce a multi-objective programming problem. It should be noted that the proposed approach takes into account raw material production variations, since several probable values for available biomass amount were taken into account when scenarios were generated. The methodology was tested in a case study for Mexico considering 8 types of residual biomass (corn bagasse, sugar cane bagasse, wheat straw, barley straw, sorghum straw, rice straw, pecan nut shell and agave residue), 8 types of products (ammonia, ethanol, methanol, acetaldehyde and 4 types of Fisher Tropsch liquids), 9 processing routes, as well as different number of potential processing plants according to the raw material. The number of potential processing facilities were 257 for corn straw, 57 for sugar cane bagasse, 3 for wheat straw, 0 for barley straw, 54 for sorghum straw, 47 for rice straw, 8 for pecan nut shell and 6 for agave residue. The results present the trade-off between both objectives through a Pareto curve, where each of points represent a different production system based on biomass. Also, the results illustrate that the proposed methodology can be used to determine the supply chain topology considering diverse initial highly viable locations and several objectives. Furthermore, results show that scale and number of considered locations have a strong influence over the transportation, capital and operating costs.

537572 Accounting for Spatial Variability of Ecosystem Services in Sustainable Supply Chain Design

Abstract:

Supply chain design problems determine environmental impact by performing life cycle assessment as an addendum to the framework. Moreover, these frameworks never consider the limits of ecosystems to provide environmental impact remediating services or supply resources. Not including ecosystems results in these designs eventually exceeding nature's capacity resulting in ecological degradation. Previous work (1) has demonstrated the benefits of including ecosystems in process design. The key benefits are that designs with ecosystems can be ecologically and economically superior to conventional techno-centric designs. However, the spatial variability of ecosystems has not been considered in these studies.

Objective:

In this research work, spatial variability of ecosystem services is considered while designing supply chains as well as manufacturing processes. With a goal to perform sustainable process design of any manufacturing process as well as determine upstream material and service pathways to that process, we develop a matrix based mathematical framework that uses upstream process based life cycle network as a platform for supply chain design The framework also allows the incorporation of ecosystem services as unit operations during optimization as a measure of environmental sustainability and designing of ecological systems.

Background:

Sustainable Process Design (SPD) is used to design manufacturing processes while accounting for environmental impacts as an optimization objective along with conventional objectives such as profit or production quantity. (2) Process to Planet (P2P) framework is applied to perform SPD which calculates environmental impact using a hybrid approach to life cycle assessment. (3)

Techno-Ecological Synergy: Techno-Ecological Synergy(TES) framework (4) considers supply of ecosystem services while satisfying the environmental sustainability condition of maintaining that demand of ecosystem services by technological systems do not exceed the capacity of ecosystems to provide the same.

Approach:

This research combines process design, supply chain design (SPD), technology choices, hybrid life cycle assessment (LCA) as well as incorporation of ecosystem services in a comprehensive optimization model. It is achieved by development of the P2P-TES framework that combined the sustainable process design and hybrid LCA characteristics of P2P with the ecosystem service integration of TES.

Results:

We perform a sustainable supply chain design study to determine the distribution of biorefineries within a spatial region, design the bioethanol conversion process and simultaneously determine upstream raw material network while considering tradeoff between food – energy demands. Farming practice choice of till versus no till was explored as well as effect of ecosystem services such as wetlands for mitigation of N runoff. Other ecosystem services being explored are food provisioning, C sequestration and air quality regulation by forests, etc. For this presentation, we apply this optimization framework to the watershed region of Maumee river in Northern Ohio to analyze spatial distribution of biorefineries and corn farming to reduce P output – a leading cause for eutrophication in Lake Erie.

References:

(1) Gopalakrishnan, Varsha, Bhavik R. Bakshi, and Guy Ziv. "Assessing the capacity of local ecosystems to meet industrial demand for ecosystem services." *AIChE Journal* 62.9 (2016): 3319-3333.

(2) Bakshi, B. R., "Methods and tools for sustainable process design, "Current Opinion in Chemical Engineering, vol. 6, pp. 69–74, 2014. [Online]. Available: http://linkinghub.elsevier.com/retrieve/pii/S2211339814000768 \\\vspace*{0.3cm}

(3) Hanes, Rebecca J., and Bhavik R. Bakshi. "Sustainable process design by the process to planet framework." AIChE Journal 61.10 (2015): 3320-3331.

(4) Bakshi, Bhavik R., Guy Ziv, and Michael D. Lepech. "Techno-ecological synergy: A framework for sustainable engineering." Environmental Science and Technology Letters 49.3 (2015): 1752-1760.

(5) Nurjanni, K.P., Carvalho, M.S. and Costa, L., 2017. Green supply chain design: A mathematical modeling approach based on a multi-objective optimization model. *International Journal of Production Economics*, *183*, pp.421-432.

539349 Techno-Economic Analysis and Life-Cycle Assessment for Gas Phase Catalytic Oxidation of Lignin to Produce Phenolic Compounds

Abstract:

Displacing fossil fuels with biofuels is an effective way to sustainably mitigate greenhouse gas (GHG) emissions and natural resource consumption. As DOEâ€™s only national laboratory solely dedicated to researching and developing clean energy and energy efficiency technologies, NREL has been developing conceptual biomass-to-fuels conversion processes. To improve the economic feasibility of biofuels production from lignocellulosic biomass, NRELâ€™s researchers have been exploring various design options. Our team has developed a vapor phase selective oxidation process that will convert lignin streams from biofuels production into high-value chemicals and thereby improve the economics and GHG emissions of biofuels plants. This approach is directed towards valorizing lignin, which is often treated as waste or burnt for low value process heat. For techno-economic analysis (TEA) and life-cycle assessment (LCA), we have developed and modeled this hybrid process: 1) biochemically converting C5/C6 sugars from biomass to a hydrocarbon blendstock, and 2) thermochemically converting lignin from biomass, via selective partial oxidation of lignin pyrolysis vapor, to phenols. Results from the TEA and LCA for this process will be presented and discussed. The lignin valorization readily improves the biofuel production cost and reduces the life cycle carbon footprint. Even though more electricity is required as lignin is used for coproduct production instead of for heat and power generation, the GHG coproduct credit from phenols is more than enough to compensate for the GHG associated with the additional import electricity use. These analyses show that our process to produce phenol has significant economic and environmental impacts, which will enable biofuel production.

539815 Kinetics of Pyrolysis-GCMS and Thermogravimetric Analysis of Oil-Laden Biomass Intermediate from Flashed-Hydrolyzed Microalgae

Abstract:

The continuous increase in price of petroleum-derived fuel has led to the need for an attractive and renewable-based feed stock such as biomass for energy generation. One of such energy sources is the prolific microalgae that has considerable amount of lipid, which can be converted to liquid hydrocarbon fuels(LHF). Not only can this feedstock be used to produce liquid fuel, but it can also be used as solid fuel or converted to gaseous fuel. To increase the usefulness of microalgae as solid or gaseous fuel feed stock, flash hydrolysis (chemical-free method) has been used to upgrade or concentrate the lipid content (10wt%) in the fresh microalgae to obtain biofuel intermediate(BI), which has 52wt% fatty acids. With this upgrade, the BI can be used directly as a solid fuel or converted to gaseous form via thermochemical conversion process such as pyrolysis, gasification, or liquefaction.

In the present study, the devolatilization kinetics of oil-laden BI pyrolysis using thermogravimetric analysis(TGA) was investigated. To further investigate the pyrolysis, a pyro-GCMS facility was used to analyze the products and the temperature at which these products were produced. Non-isothermal pyrolysis was performed by considering heating rate, gas flowrate, temperature range, and activation energy evaluation. SEM, BET, FT-IR, and XRD methods were used to characterize both fresh BI and its char.

533432 Framework for Multi-Scale Modeling and Dynamic Simulation of a Biorefinery

Abstract:

Biorefineries are a promising approach to produce chemicals and fuels from sustainable natural resources in order to reduce carbon dioxide emissions and the dependency on fossil raw materials. A variety of process steps, e.g. mechanical pretreatment, enzymatic hydrolysis, microbial conversion and downstream processing, is necessary to convert biomass into desired products. Techniques from computer-aided process design play an important role in increasing competitiveness of bio-based processes compared with existing conventional processes [1-5]. In that context, plant-wide dynamic models of a biorefinery help to improve the process understanding and allow model-based process operation and control. To this end, we developed a Modelica library with replaceable building blocks that allow dynamic modeling of an entire biorefinery.

As an example, we consider a biorefinery based on lignocellulosic raw material. The biomass is pretreated via the OrganoCat process [6] yielding process streams containing carbon sources for microbial conversion. For modeling the microbial conversion, we built on the dynamic flux balance analysis (DFBA) approach to formulate process models for the simulation of cellular metabolism under changing environmental conditions, typical for bioreactor-based production processes. Classical flux balance analysis (FBA) is based on the stoichiometry of all intracellular reactions within defined network boundaries and additional constraints on some fluxes to result in a linear equation system [7]. In most cases, the system of equations is underdetermined because there are more unknown reaction fluxes than metabolites such that there are infinite solutions. Formulation of a cellular optimization criterion, e.g., maximization of growth or ATP production, allows choosing among the set of solutions in order to determine one particular flux distribution. The DFBA model is obtained by coupling the FBA model to the time-dependent behavior of further process variables related to the extracellular environment of a bioreactor such as biomass, substrate and product concentration [8]. This leads to a system of differential algebraic equations with embedded optimization criteria. The DFBA approach is successfully applied to model reaction networks of different size and focus [9-12]. Our framework couples DFBA process models to entire plant simulation. The framework comprises a Modelica library with replaceable building blocks for dynamic modeling of typical biorefinery process steps, a converter to embed DFBA process models formulated in Omix [13] and an external program to solve the embedded optimization problem allowing simulation from available software, e.g. Dymola. We demonstrate the applicability of our framework by coupling a dynamic model of the OrganoCat pretreatment process [6] to a DFBA process model describing the growth of *Corynebacterium glutamicum* on substrates derived from lignocellulosic biomass.

References:

[1] W. Marquardt, A. Harwardt, M. Hechinger, K. Kraemer, J. Viell, and A. Voll. The biorenewables opportunity - toward next generation process and product systems. AIChE Journal, 56(9):2228–2235, 2010.

[2] A. C. Kokossis and A. Yang. On the use of systems technologies and a systematic approach for the synthesis and the design of future biorefineries. Computers & Chemical Engineering, 34(9):1397–1405, 2010.

[3] M. Martín and I. E. Grossmann. On the systematic synthesis of sustainable biorefineries. Industrial & Engineering Chemistry Research, 52(9):3044–3064, 2013.

[4] R. T. Ng and C. T. Maravelias. Design of biofuel supply chains with variable regional depot and biorefinery locations. Renewable Energy, 100:90–102, 2017.

[5] B. Bao, D. K. Ng, D. H. Tay, A. Jiménez-Gutiérrez, and M. M. El-Halwagi. A shortcut method for the preliminary synthesis of process-technology pathways: An optimization approach and application for the conceptual design of integrated biorefineries. Computers & Chemical Engineering, 35(8):1374–1383, 2011.

[6] J. Viell, A. Harwardt, J. Seiler, and W. Marquardt. Is biomass fractionation by organosolv-like processes economically viable? A conceptual design study. Bioresource Technology, 150:89–97, 2013.

[7] Palsson BØ. 2006. Systems biology: Properties of reconstructed networks. New York, NY: Cambridge University Press.

[8] R. Mahadevan, J. S. Edwards, and F. J. Doyle. Dynamic flux balance analysis of diauxic growth in escherichia coli. Biophysical Journal, 83(3):1331-1340, 2002.

[9] Hanly and M. A. Henson. Dynamic flux balance modeling of microbial co-cultures for efficient batch fermentation of glucose and xylose mixtures. Biotechnology and bioengineering, 108(2):376–385, 2011.

[10] J. L. Hjersted and M. A. Henson. Optimization of fed-batch saccharomyces cerevisiae fermentation using dynamic flux balance models. Biotechnology progress, 22(5):1239–1248, 2006.

[11] K. Höffner, S. M. Harwood, and P. I. Barton. A reliable simulator for dynamic flux balance analysis. Biotechnology and Bioengineering, 110(3):792-802, 2013.

[12] X. Zhao, S. Noack, W. Wiechert, and E. v. Lieres. Dynamic flux balance analysis with nonlinear objective function. Journal of Mathematical Biology, 147(1):761, 2017.

[13] P. Droste, S. Miebach, S. Niedenführ, W. Wiechert, and K. Nöh. Visualizing multi-omics data in metabolic networks with the software Omix: A case study. Bio Systems, 105(2):154–161, 2011.

521084 Solar Powered Biomass Pyrolysis: A Carbon Neutral Pathway for Producing Fuels and Chemicals

Abstract:

A solar powered pyrolysis scheme was developed for both fixed and fluidized bed reactor. A cinema projector was converted to a 5KW solar simulator. A Gardon-type gauge was used to calibrate the solar simulator. The solar simulator can produce peak intensity of up to 700 suns to heat the system above 1000 Â°C. Pine and corn stover was pyrolyzed using this simulator. The maximum bio-oil production occurred at 750 Â°C measured at the focus of the simulator. The effect of Fe_2O_3 catalyst on solar power pyrolysis was also studied. The resulting bio-oil composition was analyzed using â€˜Mass spectroscopyâ€™ for different solar light intensity. Solar light can enhance the production of furan derivatives and phenolic compounds from biomass. The scheme can be used for production of carbon neutral biofuel.

535362 A Total Site Synthesis (TSS) Model for the Selection, Integration and Planning of Multiple-Process and Multiple-Feedstock Biorefineries

Abstract:

This paper presents a systematic approach to screen biorefinery value chain paths by combining process integration thermodynamics with mathematical programming. The incentives of this work originate in lignocellulosic biorefineries, where numerous candidate valorization paths (value chains) are based on the use of C5/C6 sugars and lignin intermediates. Once minimum energy consumption is always the case, energy integration should be applied (beyond energy targeting) to systematically evaluate the infinite process combinations that emerge through value chains and may get integrate across the biorefinery site. The work introduces systems representations (i) to map all synthesis routes along value chains simultaneously coping with biomass seasonality issues and (ii) to mathematically formulate Total Site energy integration of multiple-process sites considering processes as additional degrees of freedom. The process synthesis representation is based on the use of a bipartite graph that follows the properties of petri-nets, while the integration representation is constructed based on a properly extended transshipment model that accounts for both direct and indirect integration of involved processes.

As competitive paths branch along the value chains, critical questions raise about which process portfolios increase the biorefinery efficiencies. The candidate processes may get integrated in various ways changing energy efficiencies each time. The integration problem requires for energy complementary process combinations that optimally utilize generated steam and minimize energy cost. A Total Site heat Cascade is introduced describing heat exchange options among hot-cold streams serving with additional options for process-to-process integration. Heat contribution to the proposed cascade is adjusted by process selections made by the process synthesis representation. Once the problem applies for lignocellulosic biorefineries, additional challenges emerge related with seasonal availability of biomass varieties. Since biomass varieties feature with different intermediates (sugars and lignin) compositions, process and energy efficiencies differentiate at the use of multiple feedstocks over the year. The proposed process synthesis and integration model is used to make selections about processes and products portfolios and plan multiple-feedstock biorefinery operations so as to minimize annual energy cost of the under-construction biorefinery.

The proposed methodology has been applied in contexts of a real-life biorefinery case that involves 21 candidate processes and 6 biomass varieties (miscanthus, wheat, rise, barley, poplar, birch). The model has been used (i) to highlight energy promising biorefinery solutions including the production of xylitol, butanol and poly-urethanes by using miscanthus and wheat mixtures and resulting up to 24% steam cost savings from process-to-process integration and (ii) to examine preferences on biomass varieties that improve biorefinery efficiencies.

532742 Optimal Bio-Refinery Configuration Using Economic Metrics and Environmental Impacts Considering Supply, Demand and Process Uncertainties

Abstract:

Optimal Bio-refinery Configuration using Economic Metrics and Environmental Impacts Considering Supply, Demand and Process Uncertainties

Abhay Athaley, Yue Zhang and Marianthi Ierapetritou

Department of Chemical and Biochemical Engineering, Rutgers - The State University of New Jersey

This work principally concentrates on a multi-objective optimization approach for the integrated design of bio-refineries that deals with uncertainties in raw material supply, product demand and process parameters. The idea of bio-refinery has been proposed to combine the use of different technologies to produce multiple products including fuels and chemicals from biomass. The success of bio-refinery is highly dependent on the flexibility to produce not only high volume low cost fuels but also to produce low volume high cost chemicals[1]. Different simulation based approaches and mathematical programming models have been proposed in the literature for the optimal synthesis and design of complex bio-refineries. However the consideration of uncertainties in the supply of raw materials, production processing alternative technologies, and demand of products results in economic losses and increased emissions[2].

In our previous work, we investigated a novel hydrolysis process[3] for the production of sugars and the subsequent production of different chemicals and fuels. We considered the production of p-Xylene[4], butadiene, jet fuel and surfactants and carried out economic analysis to calculate the minimum selling price of the produced chemicals as well as performed life cycle analysis to analyze carbon emissions and water consumption[4]. In our present work, we design an optimal configuration for the production of multiple high value chemicals and fuels taking into consideration fluctuations in demand and supply of biomass resources and optimizing both economic and environmental metrics.

This work uses an optimization approach to design and determine an optimum configuration of a bio-refinery considering various biomass resources to produce different fuels, chemicals and energy via alternative production routes. The detailed flowsheet is first developed for production of p-Xylene, ethanol, surfactants, butadiene, jet fuels and energy via both enzymatic and thermochemical production pathways from a variety of different sources of biomass feedtstock using Aspen plus. After this step an economic analysis is performed to calculate the minimum selling price of each product. Life cycle analysis is also performed using SimaPro® to compute carbon emission and water consumption of each process. Multi-objective optimization is carried out to generate a pareto curve identifying optimal solutions for maximizing profit and minimizing carbon dioxide generation and water consumption. Finally, robust optimization approach is developed to determine the optimum configuration and capacity with consideration to biomass supply, product demand and process uncertainties.

References:

1. Lin, Z.J., V. Nikolakis, and M. Ierapetritou, *Alternative Approaches for p-Xylene Production from Starch: Techno-Economic Analysis.* Industrial & Engineering Chemistry Research, 2014. **53**(26): p. 10688-10699.

2. Sy, C.L., et al., *Multi-objective target oriented robust optimization for the design of an integrated biorefinery.* Journal of Cleaner Production, 2018. **170**: p. 496-509.

3. Sadula, S., et al., *Process Intensification for Cellulosic Biorefineries.* Chemsuschem, 2017. **10**(12): p. 2566-2572.

4. Athaley, A., et al., *Techno-Economic and Life Cycle Analysis of Different Types of Hydrolysis Process for the Production of p-Xylene.* In review, 2018.

530331 A Model-Based Life Cycle Analysis of Hydrotreated Renewable Jet Fuel (HRJ) from Oilseed Feedstocks Replacing Fallow in the U.S. Northern Great Plains

Abstract:

This life cycle assessment (LCA) of hydrotreated renewable jet (HRJ) fuel produced in the U.S. Northern Great Plains regions from rotation of biofuel rapeseed with grain crops uses a model-based approach. The utilized models to obtain life cycle input data include Environmental Policy Integrated Climate (EPIC), Alternative Fuel Transportation Optimization Tool (AFTOT), tier 1 models from the Intergovernmental Panel on Climate Change (IPCC), and other process models. Life cycle inputs are obtained using geospatial data on soils and weather and spatially explicit information on bioenergy rapeseed replacement of fallow in food crop-fallow rotations for different price points of rapeseed based on relative profitability of growing options. A cradle-to-grave LCA of the entire chemical conversion and utilization pathway was developed considering emissions of greenhouse gases (GHGs) and consumption of total primary, renewable, and fossil energy. Fertilizer utilization was predicted and crop management of the different production options was evaluated using EPIC. Different strategies for allocation of environmental impacts among the planned co-products was investigated. Comparisons is made relative to petroleum jet fuel products to determine savings of GHG emissions and fossil energy demand. LCA results show that introducing fuel oilseeds to existing crop rotations to replace fallow have significant advantages in terms of GHG emissions. With consideration of soil carbon sequestration from replacing fallow period, the GHG emissions can be as low as -235 g CO2 eq/MJ jet fuel to -349 g CO2 eq/MJ jet fuel using displacement allocation, and -55 g CO2eq/MJ jet fuel to -107 g CO2 eq/MJ jet fuel using energy allocation, over the price points from \$400 to \$800/Mg rapeseed. Soil organic carbon and N2O emissions vary along the price points, and are influenced by the fertilizer application, tillage system, crop rotations, etc. The cumulative energy demand for production of rapeseed HRJ is larger than for fossil jet fuel, however, most of the energy inputs are from renewable biomass and at all price points requires less fossil energy comparing to fossil jet, with net energy return on fossil energy invested of up to 4 to 1. Total net revenues at lower prices of \$400 are \$1.6 million per year, and reaches \$56.2 million per year at the highest price of \$800. These results provide insights on the potential impacts of expanded rapeseed biofuel production systems in regional contexts compared to the current cropping systems and answered the questions of what is the best agricultural practice to enhance the sustainability of HRJ fuel production.

537866 Towards Systematic Design on Life Cycle Assessment Models By Accounting for Uncertainty and Network Complexity

Abstract:

Background:

Life cycle assessment(LCA) methods are used for studying environmental impacts of products and manufacturing processes from their "cradle to grave". LCA is plagued with several serious drawbacks such as subjective system boundary selection, uncertainty in data sources and aggregation. Neglecting activities from the inventory can result in exclusion of significant environmental impacts as negligibility of these processes is difficult to guarantee. Hybrid LCA methods were introduced for filling boundary 'gaps' not accounted for by traditional process system LCA using national input-output LCA data. Although hybrid approaches try to address the system boundary problem, location of boundaries between the different scales of hybrid LCA were arbitrary. From a broad literature review on LCA studies, it is found that the boundary selection as well as delineation of boundaries between different scales in multiscale LCA models have an impact on the validity of the final results and there is a compelling need to address this problem through a mathematically rigorous method directed towards removing subjectivity in these decisions. Very few studies have addressed the problem of boundary selection and delineation between multiple scales in LCA models without determining the uncertainty in result estimation.

Objective:

This presentation describes a methodology using Structural Path Analysis(SPA) for exploration of the life cycle network of an activity and prioritize the LCA inventory according to the amount of environmental impact. An algorithm is developed in this study that generalizes the construction of multiscale life cycle models and determine model quality based on uncertainty analysis using analytical error propagation. It can be applied to single scale models, hybrid models as well as the Process to Planet framework (1). This work also combines two distinct modules to accomplish the model generation goal - network theory and uncertainty analysis and extends analytical error propagation approach to multiscale LCA models for performing uncertainty analysis to determine model quality. Another important addition of this work is the introduction of parameters for determining the complexity of life cycle models using price based or data aggregation granularity information. Overall, this algorithm reduces the subjective-ness in determining system boundary in LCA models and introduces parameters that guide the model generation process.

Results:

The algorithm is applied for the design of a biodiesel refinery life cycle network. First using SPA, the life cycle network information is extracted from economic models and then used to create a life cycle model using this algorithm. The important questions such as selection of activities to be included in the system boundary and the determination of confidence in the results are all obtained through this algorithm. Significant challenges still exist in terms of availability of uncertainty and complexity data at different scales of a life cycle model. The algorithm is built into a software with a graphical user interface for ease of use for LCA practitioners who without going into the intricacies of the algorithmâ€™s functions can use it for building LCA models. We intend to provide a small demo of the software along with the presentation.

References:

1. Hanes, Rebecca J., and Bhavik R. Bakshi. "Sustainable process design by the process to planet framework." AIChE Journal 61.10 (2015): 3320-3331.

541078 Designing with the System in Mind: Life Cycle Assessment of Nano-Enabled Agrochemicals

Abstract:

Engineered nanomaterials (ENMs) used as fertilizers, pesticides and growth regulators will involve direct application of large quantities of ENMs to the environment and products intended for human consumption. Assessing their life cycle environmental impacts to mitigate unintended consequences poses several challenges in areas such as (1) definition of functional units that represent the function provided by nano-enabled agrochemicals (2) availability of comprehensive data necessary to inform life cycle material flow for inventory development specific to this application, (3) human and environmental exposure and effects data relevant to the agricultural context for impact assessment models, (4) identification of spatial and temporal dependent components that can affect results, and (5) data uncertainties and the possibility of their reduction through collaborative efforts between life cycle practitioners and experimental researchers using anticipatory decision-based models. While several of these challenges are experienced in LCA of emerging technologies generally, there is heightened relevance with respect to the utilization of ENMs in agriculture. Early application of LCA to this emerging technology can assist in identifying areas that can be further improved to enhance overall environmental benefits. One of the proposed benefits of nano-enabled agrochemicals is reducing our reliance on conventional fertilizers. A life cycle energy analysis of common ENMs used as agricultural supplements recommends the need for lower ENM doses than what the literature suggests, to at least break even with the current fertilization methods on an energy stand point. A more immediate advantage realized through lowering fertilizer use is reduction in eutrophication potential, however, a net benefit is only attainable if releases of nitrogen and phosphorous from upstream production stages do not outweigh the downstream savings. As research on the agricultural application of ENMs progresses, more information on their behavior and effects becomes available, allowing for assessment of other environmental impact categories. Findings from these analyses can inform research prioritization to ensure safe and sustainable design, development, and implementation of nano-enabled agrochemicals.

527180 Environmental Life Cycle Assessment of Peracetic Acid Application in the Pulp and Paper Industry

Abstract:

Environmental Life Cycle Assessment of Peracetic Acid Application in the Pulp and Paper Industry

Darlene Echeverria, Richard Venditti, Hasan Jameel and Yuan Yao

Department of Forest Biomaterials, NC State University.

Raleigh, NC 27607, USA

ABSTRACT

The global pulp and paper industry reached a new production high of 406.5 million tons of paper and paperboard in 2016, with China and the United States being the worldâ€™s leading producers (RISI |, 2015). The large-scale production of pulp and paper generates a significant amount of waste and emissions, and is one of the most environmentally regulated sectors in the U.S., driving the development of new technologies to reduce the environmental footprints. As the bleaching process is one of the largest contributors to a mill's water consumption and wastewater discharge, intensive efforts have been made by the industry and academia in past decades to improve this process. Two main bleaching technologies dominate pulp and paper production. As of 2012, 93% of the worldwide chemical pulp was produced with the Elemental Chlorine-Free (ECF) bleaching process which utilizes chlorine dioxide as its main bleaching agent, while about 5% was produced with the Total Chlorine-Free (TCF) which uses Peracetic Acid (PAA) (AET, 2012). Compared to other agents, PAA does not produce toxic by-products while still having a high selectivity in removing lignin, and brightens pulp to industry requirements. Though PAA is regarded as a more environmentally friendly bleaching agents because of these characteristics, few studies have quantified the overall environmental impacts of PAA production and its application in a bleaching process.

This work quantifies the environmental impacts of PAA as a bleaching agent by developing two comparative Life Cycle Assessment (LCA)s. The first LCA assesses the cradle-to-gate environmental impacts of PAA produced by different technology pathways and feedstocks. The second LCA compares the cradle-to-gate environmental impacts of pulp production using PAA bleaching and conventional ECF process. Experimental work is conducted to test the performance of PAA as a bleaching agent in achieving the same brightness of pulp that are produced by traditional bleaching agents.

The results of this study will enhance decision-making towards more sustainable PAA production and its use in bleaching pulp. From an end-user perspective, a quantitative analysis of the potential environmental benefits of utilizing PAA may foster partnerships between PAA suppliers and the pulp and paper industry. This may further result in the development of more sustainable production methods and supply chains. Finally, the results of this work may shed light on the potential and practical applications of PAA for a reduction of environmental impacts in other industries beyond pulp and paper.

REFERENCES

AET, (Alliance for Environmental Technology). (2012). *Trends in World Bleached Pulp Chemical Pulp Production 1990-2012.* Retrieved from http://www.aet.org/science_of_ecf/eco_risk/2013_pulp.html

RISI |. (2015). Pulp and Paper Industry Intelligence. Retrieved from https://www.risiinfo.com/press-release/global-production-of-paper-and-board-hit-record-levels-in-2014/

526755 Transforming the Circular Economy with the Value Web Model – a Multi-Objective Spatio-Temporal MILP Model for Planning, Design & Operation of Integrated Circular Value Chains

Abstract:

Chemical engineering deals with designing systems that can convert low-value resources to high-value products at a sufficiently large scale to be profitable. As well as designing reactors and unit operations, this also involves determining how and where to source raw materials, where to locate processing facilities, how to deliver products to customers and how to manage inventory levels to satisfy demands at all times. Coordinating all of these activities and designing the network are known as supply chain management. Nearly all current supply chain models are linear and the ones that are closed-loop are quite limited in terms of dealing with problems in the circular economy. More efficient, flexible and resilient supply chains can result from integrating supply chains from different sectors thus forming a value web.

The Value Web Model (VWM) is a multi-objective spatio-temporal optimisation model, based on mixed integer linear programming, that can simultaneously determine the design and operation of any integrated multi-vector energy networks. It has been developed to answer variants of the following questions: "What is the most effective way, in terms of cost, value/profit and/or emissions, of designing and operating the integrated multi-vector energy networks that utilise a variety of primary energy sources to deliver different energy services, such as heat, electricity and mobility, given the availability of primary resources and the levels of demands and their distribution across space and time? When to invest in technologies, where to locate them; what resources should be used, where, when and how to convert them to the energy services required; how to transport the resources and manage storage inventory?"

The many different possible pathways from all possible primary energy sources to all energy services via any number of energy vectors form a complex web of interactions between all of the elements of the system, including circular pathways. As value is being added at each transformation from primary resource to energy service, the whole system can be viewed as a value web. The system needs to be robust with respect to uncertainties such as the availability of primary energy sources, levels and timings of demands, costs and prices, policy changes and other technological uncertainties (e.g. appearance of a disruptive technology). The VWM can be applied to a wide variety of systems such as integrated energy systems, water networks, chemicals and pharmaceutical supply chains. In this presentation the focus will be on the application of VWM for strategic design and tactical operation of integrated energy value chains with various energy carriers such as hydrogen, electricity, syngas, methanol, biofuels/bioenergy and so on.

References:

[1] S. Samsatli, N.J. Samsatli (2018). A multi-objective MILP model for the design and operation of future integrated multi-vector energy networks capturing detailed spatio-temporal dependencies. Applied Energy. DOI: 10.1016/j.apenergy.2017.09.055.

[2] S. Samsatli, I. Staffell, N.J. Samsatli (2016). Optimal design and operation of integrated wind-hydrogen-electricity networks for decarbonising the domestic transport sector in Great Britain. Int. J. of Hydrogen Energy, 41, 447-475.

[3] S. Samsatli, N.J. Samsatli (2015). A general spatio-temporal model of energy systems with a detailed account of transport and storage. Computers and Chemical Engineering, 80, 155-176.

[4] S.M. Jarvis, S. Samsatli (2018). Technologies and infrastructures underpinning future CO2 value chains: a comprehensive review and comparative analysis. Renewable & Sustainable Energy Reviews, 85, 46-68.

[5] S. Samsatli, N.J. Samsatli, N. Shah (2015). BVCM: a comprehensive and flexible toolkit for whole-system biomass value chain analysis and optimisation - mathematical formulation. Applied Energy, 147, pp. 131-160.

543554 Gas Capture by Ionic liquids

Abstract:

With the rapid development of the global industrialization process, a large number of gases, such as CO_2, SO_2, NH_3 and H_2S, have been produced. Among of them, CO_2 and SO_2 have caused air pollution problems, resulting in a serious harm to the environment and human health. On the other hand, CO_2 and SO_2 are also important source for synthesizing chemicals. Recently, ionic liquids (ILs) have been regarded as one of the most promising alternative solvents for capturing CO_2 or SO_2 due to their special physicochemical properties, such as negligible vapor pressure and designable character. For example, it was found that N-butylpyridinium thiocyanate ([C_4Py][SCN]) can absorb 0.841 gSO_2/g IL at 293 K and 0.10 MPa[1].

In this work, the mechanism of CO_2/SO_2 absorption in different kinds of ILs were systemically studied by molecular simulations. The microscopic structure, interaction energy and dynamic properties of the mixtures of ILs and CO_2/SO_2 were investigated. The results of interaction energy and fractional free volume of ILs showed that unlike non-polar gas CO_2, the interaction energy of cation-anion and free volume do not play a dominant role in SO_2 absorption. The solvation of SO_2 in ILs is controlled by enthalpy. Both the cations and anions play an important role in SO_2 solubility. Based on the above studies, a serial of ILs were synthesized for capturing CO_2/SO_2. Moreover, the energy and economic analysis of whole process of CO_2 capture with ILs were simulated. It was found that the IL process saved about 15% regeneration heat energy consumption and 11.5% total capture cost compared to the MEA process.

Acknowledgment

This work was supported by the National Natural Scientific Fund of China 51674234.

Reference

1. J. Zeng, H. S. Gao, X. C. Zhang, H. F. Dong, X. P. Zhang, S. J. Zhang, *Chem. Eng. J*, 2014, 251, 248-256.

533203 Thermoeconomic Optimization of Reverse Brayton Cycle Based Cryocooler for HTS Power Transmission Cable

Abstract:

In the era of the global sustainable energy crisis, an efficient power transmission system is required to meet the increasing power demand. After the discovery of superconductivity in 1911, many materials with zero resistance at a temperature below 25 K, known as Low-Temperature Superconductor (LTS), were discovered and a possible alternative for efficient power transmission. However, the cost of these materials and cooling system to maintain them in the superconducting state was very high. Thus, it is practically not feasible to use such materials for the transmission system. In 1986, the discovery of High Temperature Superconductors (HTS), which can be maintained in the superconducting state using liquid nitrogen, made the practical and economically feasible for use HTS power cable. These cables have to be maintained in a temperature range of 65 to 80 K with a cooling load of 2 to 5 kW for 1 km long transmission cable. Reverse Brayton cycle based Cryocooler (RBC) is the most suitable cooling system because of a wide range of cooling capacity at different temperature level. However, the capital and operational costs of RBC is high compared to the cryogenic roadmap (Gouge, 2004). The current cost of refrigeration for HTS devices is five to ten times of cryogenic roadmap ($25 per watt of refrigeration at 65-80 K). Hence, a thermoeconomic optimisation of RBC has been carried out to enhance the performance of the system from the thermodynamic and economic point of view.

Fig. 1 Reverse Brayton Cycle based cryocooler for HTS power transmission cable.

In the present study, a reverse Brayton cryocooler consists of a compressor, main heat exchanger, turboexpander and load heat exchange as shown in Fig.1, has been investigated and optimised for 10 kW cooling load at 65 K. A commercial process simulator Aspen HYSYS V8.6 has been used for the study. Helium, more efficient than neon (Dhillon, 2017) has been used as working fluid for the cryocooler. The thermodynamic and economic models have been developed for analysing of the cryocooler. It is observed that the exergy efficiency of the system has maxima at 1.9 pressure ratio whereas cost per hour has manima at 2.3 pressure ratio as shown in Fig. 2. The inlet pressure and pressure ratio of the compressor have been varied to optimise the system process parameter. The effect of design parameters of equipment on the thermoeconomic performance of the system has been studied. The results obtained for RBC are presented in the present work and may be helpful for engineers and scientist to reduce the cost of RBC cryocooler without compromising the performance of the system.

Fig. 2 Variation of cost per hour and exergy efficiency of RBC with different pressure ratios

1. Gouge, M. Cryogenic cooling technology for HTS electric machinery, 2004. IEEE Power Engineering Society General Meeting, 1-3, DOI: 10.1109/PES.2004.1373239.

2. Dhillon, A. K., Dutta R. and Ghosh P. Exergetic analysis of reverse Brayton cryocooler for different cooling loads at 65 k for HTS cables, 2017. 14th Cryogenics 2017 IIR International Conference, DOI: 10.18462/iir.cryo.2017.027

530055 Technology Readiness Levels As a Framework for Techno-Economic Assessment

Abstract:

Techno-economic assessment (TEA) is an important basis for decisions about research, development and deployment (RD&D). Methods applied in TEA depend heavily on the availability of data which is closely linked to the development progress. Technology readiness analysis and especially its rating scheme, technology readiness levels (TRL), are popular concepts for rating the maturity of RD&D according to available data. TRLs can therefore serve as a framework that facilitates the choice of methods along RD&D by sorting them according to their required input data.

TRL rating is widely adopted in the evaluation of RD&D projects by government (*e.g.* policy makers) [1,2], industry and academia. However, as the TRL concept was developed for space exploration [3], a domain that integrates a variety of disciplines from mechanical, electrical and chemical engineering to aviation and medicine, TRL guidelines and criteria remain unspecific to technologies, which can lead to misinterpretation and overly optimistic evaluation. A variety of TRL definitions evolved over the years, for example for biomanufacturing [4] or for the defense sector [5], defining specific TRL criteria for their respective domains. The need for a specification of TRL analysis for the chemical industry was identified in a report of the European association of research and technology organizations (EARTO) [6].

In this contribution, we summarize existing TRL concepts and propose adaptions and extensions to criteria for TRL rating in the chemical industry: Firstly, TRLs are characterized by the description of the project phase, its tangible deliverables and other qualitative criteria. Secondly, for a more objective and comprehensible TRL, specific indicators representing known information of various aspects of chemical engineering such as reaction engineering or process development are proposed. In addition, quantitative indicators such as the production capacity are often used and therefore discussed for different types of processes in this contribution.

Furthermore, we outline how methods of (capital) cost estimation and profitability calculation can be sorted by TRL in order to give guidance for sound TEA.

We gratefully acknowledge the financial support by the European Institute of Innovation and Technology (EIT) and the German Federal Ministry of Education and Research (BMBF).

[1] European Commission. EN HORIZON 2020 WORK PROGRAMME 2016 â€“ 2017, General Annexes (European Commission Decision C (2017) 2468 of 24 April 2017), Annex G, Technology Readiness Levels (TRL); 2017.

[2] BMBF, Definition des Technologischen Reifegrades [â€¦] accessed on 08.02.2017 from https://www.ptj.de/lw_resource/datapool/_items/item_5502/definition_des_technologischen_reifegrades.pdf

[3] S. R. Sadin, F. Povinelli, R. Rosen, contribution at 39thIAF Congress, Bangalore, India, Oct. 8-15, 1988

[4] NSF Engineering Research Center for Biorenewable Chemicals (CBiRC). Technology Readiness Levels https://www.cbirc.iastate.edu/industry/technology-readiness-levels/ (accessed Oct 31, 2017).

[5] U.S. DEPARTMENT OF DEFENCE. Technology Readiness Levels in the Department of Defense (DoD) https://www.army.mil/e2/c/downloads/404585.pdf (accessed Oct 31, 2017)

[6] EARTO, The TRL Scale as a Research & Innovation Policy Tool, EARTO Recommendations, 2014

536017 Techno-Economic Modeling and Optimization of Catalytic Reactive Distillation for Bio-Oil Upgradation

Abstract:

The most significant benefit of pyrolysis bio-fuel is its availability from virtually every waste biological materials, which provides a double edge in sustainable waste management in addition to meeting our energy requirements. However, unfavorable characteristics of crude bio-oils received from pyrolysis process compared to petroleum fuels means that it must undergo upgradation before using in combustion engines. Reactive distillation (RD) is emerging as a powerful technology for these chemical reactive separation processes. RD, which has tremendous advantages over traditional reaction and separation unit operations including handling of unfavorable chemical and phase interactions, enhancing conversion and selectivity, circumventing azeotropic mixtures and reducing capital and energy costs, has been exploited for the reactive separation of various equilibrium controlled chemical reaction systems. Pyrolysis bio-oil upgradation is usually carried out by transesterification and esterification reactions with an alcohol and RD is suited pefectly to these types of equilibrium controlled reactions.

Here, equilibrium RD process simulation and economic evaluation was carried out for the esterification of n-butanol with complex mixtures of several fatty acids, phenolics and water (represent pyrolysis crude bio-oils) using Aspen PLUS process simulator and Aspen Process Economic Analyzer, respectively at both atmospheric and high pressure (10 bar). Complete design and optimization was carried out using evolutionary techniques from simplest system (ordinary distillation) to the most rigorous one (complex RD) with introduction of progressive complexities. RD simulation was performed using UNIQUAC as base-property method in Aspen RADFRAC module by minimizing Gibbs free energy. Prior to distillation design, esterification reaction kinetics and binary/ternary interactions were analyzed using Aspen RGibbs reactor module and Aspen Property PLUS, respectively. Simple distillation column was designed and optimized to obtain the initial trial parameters to be used in RD simulation, which was initially optimized only for acetic acid-water (75% acetic acid-25% water) mixture before RD design for complex fatty acid mixtures based on techno-economic feasibility and optimization. The reaction of phenol with fatty acids was incorporated and phenol was found to react with formic acid entirely, which increases the reaction conversion 5-10% due to its favorable effects. The conversions for the esterification reactions were found to be 88-99% for various simulated bio-oils with phenol composition 5-20% for butanol:acid (B:A) ratio of 2.5, and water was almost completely separated from the ester products which should increase the fuel quality. Higher water percentage in bio-oil feed was found to reduce acid conversion and ester separation. A reflux ratio of 0.95, distillate to feed ratio 0.505 and total 18 stages (3 rectifying, 9 reactive and 4 stripping) with n-butanol and bio-oil feed stage of 4 and 13, respectively, was found to be optimum at 1 bar based on reaction conversion, ester separation, heat duties and capital and operating costs. Effects of B:A feeding ratio and water percent in bio-oil is also analyzed and B:A = 2.25 was found to be optimum for 25% water in bio-oil. These works and optimized parameters can serve as a design platform for pyrolysis bio-oil upgradation to transportation fuels.

Reference:

1. Khan, M.A., Adewuyi, Y.G. High pressure reactive distillation simulation and optimization for the esterification of pyrolysis bio-oil. Process Engineering Journal, 1 (2017) 73-85.
2. Khan, M.A., Adewuyi, Y.G. Simulation of reactive distillation for esterification of pyrolysis bio-oil. The Proceeding of AIChE Annual Meeting, 2013, San Francisco, California, USA.

536658 Superstructure Optimization of Membrane-Based Carbon Capture Systems

Abstract:

This work presents the optimal design, configuration and operating conditions of a membrane-based post-combustion CO_2 capture process using computationally identified mixed matrix membranes. A new mixed integer nonlinear model is presented, which includes discrete and continuous decisions to minimize the cost of electricity (COE). The simultaneous optimization of process configuration, membrane materials, and operating conditions ensures rigorous exploration of the entire feasible design space.

Membrane systems often consist of multiple stages of membranes and compressors with intercooling. Flue gas source plays a very important role, since the CO_2 concentration highly affects the final plant configuration. While a simple multi-stage configuration may be optimal for concentrated CO
integrated with the boiler is often more effective for coal fired power plants.[1] Given the importance of finding the right configuration, two approaches have been used to optimize these systems. The first approach uses simulation-based optimization that leverages commercial process simulators such as Aspen Custom Modeler or gPROMS to develop rigorous models[2,3,4]. Previous work has used the Framework for Optimization, Quantification of Uncertainty and Surrogates (FOQUS)[2] to optimize the design and operating conditions of specific process configurations. In these approaches, the number of potential configurations that can be analyzed is limited since each must be set up individually.[3,4]

The second approach focuses on superstructure-based mathematical optimization of multi-stage membrane systems, which often requires simplified models for the equipment.[5,6] The advantage of the second approach is that the configuration as well as the design and operating conditions may be optimized simultaneously, resulting in more cost-effective plants.

Our work leverages efforts at NETL for developing new materials for membrane-based post-combustion capture processes.[2,3,4,7] An advanced superstructure-based process optimization model is developed using rigorous mathematical models for compressors, pumps, mixers, and heat exchangers and reduced order models for liquefier columns and property calculations. The current work makes use of results from a recent study[4], which analyzed a large number of hypothetical mixed matrix membrane materials for carbon capture. This work addresses the aforementioned limitations by analyzing multiple potential advanced process configurations and the selection of the membrane material from a wide set of potential options while optimizing the plant operation.

[1] Merkel T. C., Lin H., Wei X., Baker R. (2010). Power plant post-combustion carbon dioxide capture: An opportunity for membranes. Journal of Membrane Science. 359, 126-139.

[2] Miller D. C., Agarwal D., Bhattacharyya D., Boverhof J., Cheah Y. W., Chen Y., Eslick J., Leek J., Ma J., Mahapatra P., Ng B., Sahinidis N. V., Tong C., Zitney S. E. (2016). Innovative computational tools and models for the design, optimization and control of carbon capture processes, Computer Aided Chemical Engineering, 38, 2391-2396.

[3] Morinelly J. E. M., David C., 2012, Post-Combustion Gas Permeation Carbon Capture System Models, AIChE Annual Meeting, Pittsburgh, PA, October 28-November 2, 2012.

[4] Budhathoki S., Ajayi O., Steckel J.A., Wilmer C.E. (2018). High-throughput computational prediction of the cost of carbon capture using mixed matrix membranes. In preparation.

[5] Hasan M. M., Baliban R. C., Elia J. A., Floudas C. A. (2012). Modeling, Simulation, and Optimization of Postcombustion CO_2 Capture for Variable Feed Concentration and Flow Rate. 1. Chemical Absorption and Membrane Processes. Industrial and Engineering Chemistry Research. 51, 15642-15664.

[6] Arias A. M., Mussati M. C., Mores P. L., Scenna N. J. (2016). Optimization of multi-stage membrane systems for CO_2 capture from flue gas. International Journal of Greenhouse Gas Control. 53, 371-390.

[7] Zamarripa M. A., Ajayi O., Matuszewski M., Miller D. C., 2017, Superstructure-based optimization of membrane-based carbon capture systems. AIChE Annual Meeting, Minneapolis, MN, October 28-November 2, 2017.

543630 Making Hydrogen from Water with a Protein Organized Electrode: Ultra-High Utilization of Noble Metal in Proton Exchange Membrane Electrolysis for Capital Cost Reduction

Abstract:

Renewable hydrogen could play a critical role in low-carbon energy systems and in fulfilling intermittent energy needs. Proton exchange membrane (PEM) electrolysis technology for hydrogen production is safe, simple, and highly efficient. PEM electrolysis has ability to be integrated in a variety of energy systems and infrastructures, such as transportation, renewable electricity, and biogas upgrading. It also has the ability to address seasonal-variation in long timescales. But, one of the major challenges remains the high cost of noble-metal electrodes in the membrane electrode assembly (MEA). Conventional catalysts like platinum and iridium are required to facilitate the hydrogen evolution reaction (HER) and the oxygen evolution reaction (OER) at electrodes for superior stability in long-term. The contribution of catalyst cost in the MEA becomes a more significant fraction of the capital cost when the production is scaled up.

To reduce the capital cost, we have developed protein-integrated electrodes which are designed to allow lower amounts of noble metal catalyst while maintaining high electrolysis performance. The purpose of our designed protein, elastin-like polypeptides (ELP), is to organize ionomer networks and create efficient pathways between the ionomer and catalyst sites, thereby significantly improving catalyst utilization. A preliminary cost analysis indicates that even at current prices for engineered proteins and using reasonable ratios of protein to catalyst particles (~6.5 metal:protein by mass), a 10X reduction in noble metal could be achieved for the same cost. Considering this technology is expected to lower loadings by more than 10X, and that the future cost to produce these proteins will decrease while the cost of noble metals will increase, the technology has great economic potential. For example, if the cost of protein manufacturing drops by a reasonable 25%, and a 20X reduction in metal loading is achieved, the total electrode cost would be 60% lower. ELP is suitable as electrode organizing molecules for its flexibility in design, modular format, and manufacturability. Our results using a quartz crystal microbalance and a transmission electron microscope show a robust binding of ELP to platinum catalyst and ionomers which affects the resulting ionomer-catalyst structures. This system provides an innovative, multidisciplinary solution for improving catalyst utilization in membrane electrode assemblies.

522132 Design Optimization and Economic Evaluation of LNG/NGL Integrated Processes Under Lean Feed Composition

Abstract:

The economic evaluation of the integrated liquefaction and natural gas liquids (NGL) recovery process for the lean feed is required to meet recent increasing demand for lean gas reservoirs. This paper performs economic analysis for the conventional co-production of liquefied natural gas (LNG) and NGL process compared to the proposed simplified integrated process with consideration of the LNG higher heating value (HHV) specification under the lean feed condition. The dual mixed refrigerant (DMR) cycle was selected for liquefaction part due to high efficiency reported by previous studies. The schemes were built by Aspen HYSYS commercial software and the object function was optimized with a genetic algorithm (GA). The results show that the simplified integrated process has economic advantages compared to the conventional integrated scheme under the lean feed composition, which shows different results in terms of previous studies conducted by normal or rich feed condition. Additionally, the capital cost can be reduced remarkably by simplifying NGL recovery section with almost no loss of liquefaction efficiency in the proposed simplified integrated process.

COMPARISON OF FIRST PRINCIPLES AND PARAMETRIC MODELS FOR THE DESIGN AND OPERATION OF A SPRAY DRYER FOR WHEY PROTEIN PRODUCTION

ANIBAL JOSÉ BARRIOS QUANT* 1, JOHN TELOTTE 2, JOSE ANDRES PEREZ MENDOZA 1.

1, Chemical engineering department, Universidad del Atlántico, Km 7, Puerto Colombia, Atlántico, Colombia

2, Department of Chemical and Biomedical Engineering, FAMU-FSU College of Engineering, Florida State University, 2525 Pottsdamer St, Tallahassee, FL, USA

*Corresponding author:abarriosquant@fsu.edu; Tel: +57 3053058185

Abstract

In this work we analyze four models for spray dryer design for a whey protein process. The models include both first principles approaches and empirical correlations. For a large set of given set of design specifications (inlet and outlet moisture contents, inlet air conditions and residence time) the physical size of the spray drying system was determined from all correlations and then the designs were compared to both industry heuristics for size ratios and actual dryer performance (where possible) to assess the validity of the models. Specific consideration was given to the correlation between the systems that met design heuristics and those that produced usable product. An economic analysis was developed to compare the prices that could be obtained designing a spray dryer based on the Masters restrictions. Finally a simulation on ANSYS CFD represented the flowfield of the spray dryer was developed.

KEYWORDS: Spray Dryer, design, sizing, dimensional analysis, critical variables, whey protein.

INTRODUCTION

Whey proteins with improved flavour and increased functionality are obtained with new technologies, as spray drying. While choosing a particular whey product, it is essential to match its function to the characteristics we want to achieve. For example, high protein concentrates or isolates are usually produced in spray dryers due with this process is possible to guarantee high quality powder [1,2].

Spray-drying is a rapid, continuous, cost-effective, reproducible and scalable process for the production of dry powders from a fluid material by atomization through an atomizer into a hot drying gas medium, usually air [3]. The design of spray dryers has remained largely empirical; it is still mainly based on the extensive experience and the vast body of operating data which the manufacturers have acquired over the years. To design spray dryers usually designers develop assumptions based on variables that allow an easier design process for spray dryers [4]. The focus of this study is to compare four different design models and their responses over common variables adjusted to evaluate the sizing of a spray dryer for whey protein production.

COMPARATIVE MODELS

The numerical equations who represented a phenomenon is better called design model, and is commonly used as a representation of spray dryers as well as digital simulations of spray drying. In this study the method was applicable to represent the particle movement inside the spray chamber and the dimensional structure of the equipment.

The dimensional structure of the spray dryer was explained by the models of Mujumdar [7], Suryanarayana [8], while the behavior of the drops inside the spray and the atomization process were showed by the models of Gluckert [9] and Masters [10].

The comparative analysis of these models was focused on study each models and evaluate their responses over the same operational conditions. The equations below show the structure of the models represented only by the diameter, length and volume. The complete structure of each model can be found in the supplementary information.

Mujumdar Model [7]

$$D = \sqrt{4\,G / 3600\,\pi\, j\, u}$$
$$L = 4V / \pi D^2 \qquad (1)$$
$$V = t F_2 / H \;\rho_{whey}$$

Suryanarayana Model [8]

$$D_{Spray} = \sqrt{4\,A_c / \pi}$$
$$L = 0.6\, D_{spray} \qquad (2)$$
$$V = G_s V_{avg}\, t_d$$

Gluckert Model [9]

$$D = 2\sqrt{(t_T\, 2.24\, V_{res}\, (b\,(D_r / 2)^{12}} + (D_r / 4)$$
$$L = 1.5\, D_{spray} \qquad (3)$$
$$V_t = G_s V_{avg}\, t_d$$

Masters Model [10]

$$Dvs = 14000\; Q\; 10.24 / n\; h\; 0.12\; N\; d\; 0.6$$
$$L = 1.5\, D_{spray} \qquad (4)$$
$$V = 1/4\;\; \pi D_i\; 2\; h_{spray}$$

The comparison of the four models was evaluated by the variation of key variables of the model in range adjusted by the study based on avoidance of whey protein denaturation. The key variables were selected according to their influence over the four models. The values assigned to the key variables were reported in the literature and are shown in Table 1.

Table 1. Variables and values

Independent Variables (Factors)	Values	
X1= Humidity content (kg/kg db)	0.60	0.90
X2= Inlet air temperature (°C)	85	180
X3= Outlet air temperature (°C)	80	120
X4= Outlet whey humidity (kg/ kg dry)	0.05	0.10
X5= Air flow (m3/h)	123	288

Key variables

The key variables were organized randomly and set to obtain thirty- two responses. The analysis was evaluated by the effect of the variables on the responses obtained in the models and the relation between the size and the different values that were applied to the models.

The *treatments* were organized by considering the main factor of protein denaturation and this terminology was used to explain the thirty-two possible variables used on this study to determine the value of the responses (spray diameter, length and volume) in each particular model.

Economic analysis

The economic analysis was carried out in this study to determine the cost of building and operating a spray dryer over a specific range of operational conditions. Then, a comparative analysis was made to find out which model was the most economically efficient.

The cost estimation for the spray dryer was calculated following the economic considerations of Mujumdar [7]. This determined the total cost of a dryer taking into consideration the operation costs and the equipment costs, based on the quantity on the material and the dimensions of the diameter and the length.

The dryer unit was determined by the cost of wet product ($ / kg wet base) with Equation 5.

$$
\begin{aligned}
Cp &= CTtop\,Fs\,(1 + Xs) \\
Ct &= eCeq + Cop \\
e &= i\,(1 + i)N \,/\, (1 + i)N - 1 \\
Ceq &= D\,An'D + \alpha nZZn'z \qquad (5) \\
Cop &= hPCetop + Z\,CzTop \\
hp &= qN'D\,(M + W') \\
W &= M\,(\,2D^2/\,4) + \pi DL)\,dx
\end{aligned}
$$

Size analysis and CFD simulation

A CFD simulation was carried out to investigate the fundamental flow behaviour in the spray dryers and their modeling. The simulation is performed using a mixed Eulerian/Lagrangian approach, in which the single single phase Reynolds Averaged Navier Stokes (RANS) equations are solved to determine the flowfield. This work used the responses given by the models to employ them as boundary conditions to simulate the spray dimensions and the CFD. The results of this simulations are shown in Figure 2.

DISCUSSION AND RESULTS

Depending on the modeling and boundary conditions of each author, variations in the dimensional parameters (diameter, length and volume) were obtained in the design models. When the models of Mujumdar [7], Suryanarayana [8], Gluckert [9] and Masters [10] were analyzed using the residence time and air velocity levels (high and low) as variables, only four responses of the possible thirty-two were obtained for the models of Mujumdar, Suryanarayana, and Gluckert [7-9] and one of thirty-two for Masters [10].

Model responses

The responses of the models were calculated by the combination of the five variables selected for this study in the two chosen levels (Figure 1). This procedure showed thirty-two responses for each dimensional variable (diameter, length and volume) for every model. The average responses of each model were compared to describe the behavior with the same operational conditions.

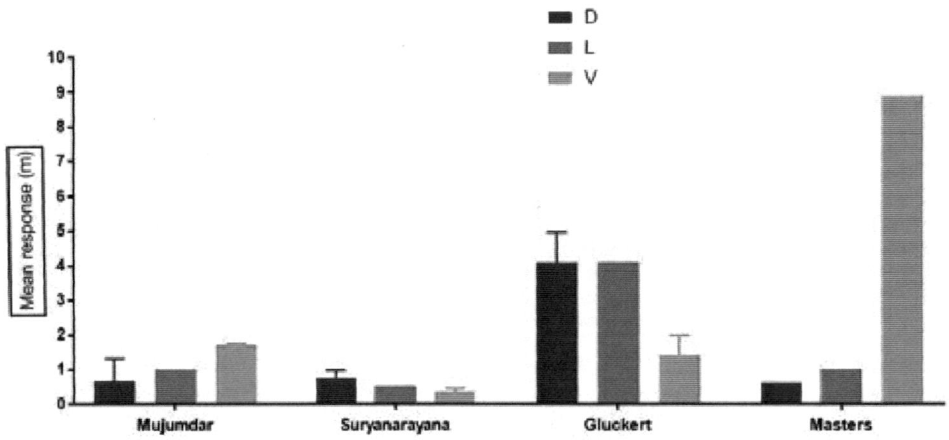

Figure 1. Model responses.
* "T" on top of the bars means average value conquered

The models of Mujumdar, Suryanarayana, Masters and Gluckert did not consider the denaturation conditions of whey protein for the designing of a spray dryer.

Models-comparison

The models comparison was developed with the analysis of the four models on two categories: sizing and modeling and cost estimation, this conditions were adjusted to analyze the scalability of the spray dryer of each model in terms of size of the chamber and cost estimation.

The sizing and modeling of the spray dryers were carried out in AutoCAD Inventor 2019, in this program was simulated the structure of the spray dryers according to the responses obtained on the variables analysis, the results demonstrated that exist appreciable differences on the sizes of the sprays dryers by each model.

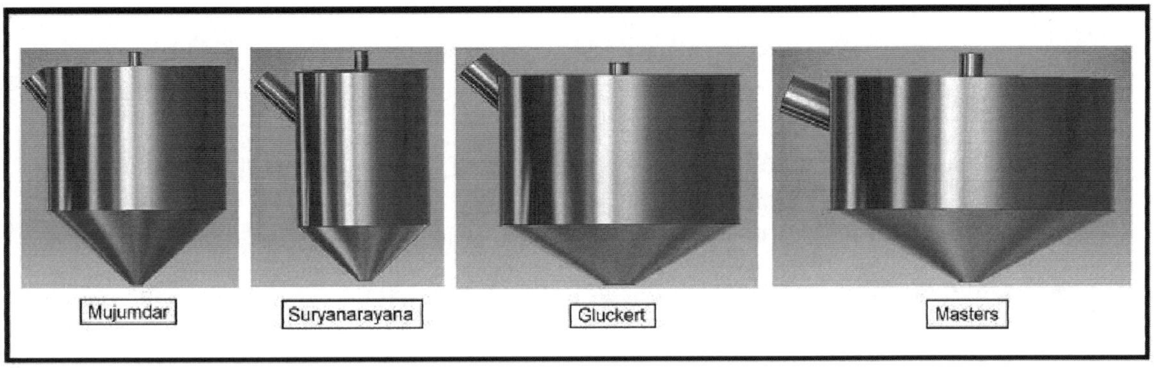

Figure 2. Size analysis

The cost estimation responses were calculated over a set of initial conditions for the 4 models expressed on the Table 2.

The model prices of the spray dryers varied from $6,417 USD to $14,803,215 USD. The lowest-priced option was the Suryanarayana [8] model with dimensional responses (diameter, length and volume) of 0.454 m, 0.313 m and 0.193 m3 respectively, and the highest-priced model was Gluckert [9] with dimensional responses (diameter, length and volume) of 5.301 m, 7.952 m and 0.912 m3 respectively.

Table 2. Cost estimation

Model	**Responses (U.S. Dollars)**
Mujumdar	$18,264 - $205,755
Suryanarayana	$6,417 - $91,540
Gluckert	$2,890,141 - $14,803,215
Masters	$27,711

CFD results

The results showed on the Figure 3 represent the flow patterns inside the spray chamber, the conditions proposed to carry out the simulation were the dimensions obtained on the responses of this work and the values of the temperature and the whey flow were adjusted according the conditions of non-denaturation presented on Table 1.

In this simulation (Figure 3) was possible to observe the changes that suffered the whey liquid particles inside the spray dryer, the particles traveled since the atomizer until the outlet hole, also is appreciable to see the the mass and energy transfer, as well as the the temperature profile in the middle of the spray dryer, represented by the mixture of colors in the center of the Figure 3.

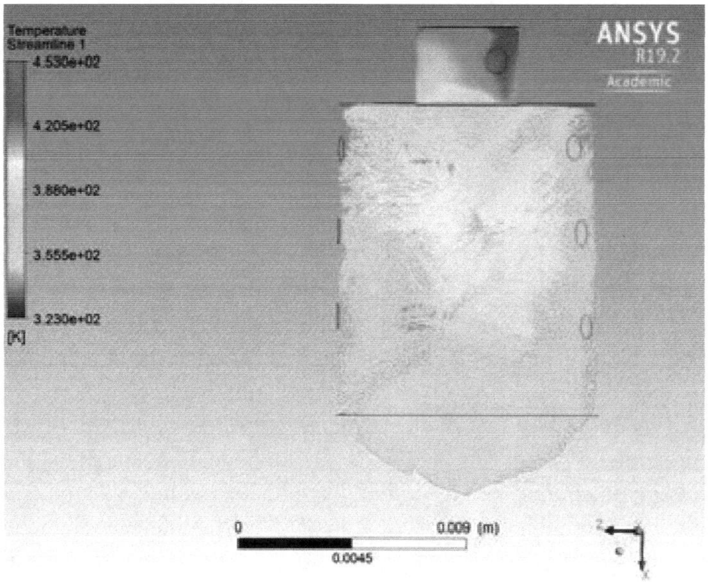

Figure 3. ANSYS CFD simulation

CONCLUSIONS

The models analyzed in this study presented different behaviors when evaluated over the same operational conditions. This is because the creator of each model used different mathematical correlations to describe the same phenomenon. This resulted in different responses in the spray dryer dimensional variables, (diameter, length and volume), that were analyzed by the simulations on ANSYS CFD who showed the behavior of the drop responses inside the spray chamber. Also in this work was studied the economic conditions for design a spray dryer to offer a guidance that allowed to analyze the cost estimation to develop the equipment. The analysis was carried with the key variables and responses adjusted in this research, obtaining a guidance to select a specific model design according the budget.

Acknowledgments

The authors gratefully acknowledge the Universidad del Atlántico in Barranquilla, Colombia for all the support and knowledge delivered. Thank you to Savannah Cayson for editing the writing of this article.

REFERENCES

[1] C. V. Morr, E. Y. W. Ha, Whey protein concentrates and isolates: Processing and functional properties. *Critical Reviews in Food Science and Nutrition*. **29** (2009) 431-476

[2] P. Schuck, R. Jeantet, B. Bhandari, X. D. Chen, Í. T Perrone, A. Fernandes de Carvalho, Recent advances in spray drying relevant to the dairy industry: A comprehensive critical review. *Drying Technol.* **34** (2016) 1773-1790

[3] M. Mezhericher, A. Levy, I. Borde, Spray drying modelling based on advanced droplet drying kinetics. *Chemical Engineering and Processing: Process Intensification*. **49** (2010) 1205-1213

[4] M. Mezhericher, A. Levy, I. Borde, Multi-Scale Multiphase Modeling of Transport Phenomena in Spray-Drying Processes. *Drying Technol.* **33** (2015) 2-23

[5] V. S. Birchal, L. Huang, A. S. Mujumdar, M. S. Passos, Spray Dryers: Modeling and Simulation. *Drying Technol.* **3** (2006) 275-279

[6] H. M. Lisboa, M. E. Duarte, M. E. Cavalcanti-Mata, Modeling of food drying processes in industrial spray dryers. *Food and Bioprod process.* **107** (2018) 49-60

[7] A.S. Mujumdar, *Handbook of industrial Drying.* (2014), 182-200

[8] A. Suryanarayana, *Mass Transfer Operations.* (2002), 571- 578

[9] F.A Gluckert, A theorical correlation of spray-dryer performance. *American Institute of Chemical Engineers*. **22** (1962) 713-724

[10] K. Masters, *Spray Drying Handbook*. (1979), 115-120

[11] M. Tunick, A. Thomas-Gahring, D. Van Hekken, S. Iandola, M. Singh, P. Qi, D. Ukuku, S. Mukhopadhyay, C. Onwulata, P. Tomasula, Physical and chemical changes in whey protein concentrate stored at elevated temperature and humidity. *Journal of Dairy Science*. **99** (2016) 2372-2383

[12] M.A. Haque, P. Aldred, J. Chen, C. Barrow, B. Adhikari, Comparative study of denaturation of whey protein isolate (WPI) in convective air drying and isothermal heat treatment processes. *Food chemistry*. **141** (2013) 702-711

536291 Molecular Thermodynamic Model for Aqueous Na^+–K^+–Mg^{2+}–Ca^{2+}–Cl^-/SO_4^{2-} Quinary Electrolyte Systems

Abstract:

Managing produced water is one of the major challenges in oil and gas production [1]. Produced water can be as much as seven times saline compared to sea water systems, and desalination of this water depends on novel material and process innovations. We present a comprehensive thermodynamic model aimed to support heat and mass balance calculations and predict mineral scale deposition of the saline water. The six major ions that control salinity of the water are: Na^+, K^+, Mg^{2+}, Ca^{2+}, Cl^-, and SO_4^{2-}. Current investigation focuses on establishing a thermodynamic model for the Na^+–K^+–Mg^{2+}–Ca^{2+}–Cl^- and the Na^+–K^+–Mg^{2+}–Ca^{2+}–SO_4^{2-} aqueous quinary electrolyte systems. To model the systems, we adopt electrolyte Non-Random Two Liquid theory (eNRTL) for electrolyte solutions [2] which describes concentration dependency of the solution nonideality with two binary interaction parameters for each water-electrolyte and electrolyte-electrolyte pair. Temperature dependency of each binary parameters is further correlated with a Gibbs-Helmholtz type expression with three temperature coefficients representing Gibbs energy, enthalpy, and heat capacity contributions. With the binary parameters and thermodynamic constants of the precipitating salts regressed from phase equilibria, calorimetric and salt solubility data, the model reliably calculates various thermodynamic properties i.e., salt solubility of the quaternary subsystems as well as quinary systems with temperatures from 273.15 to 473.15 K. The model is being extended to cover additional ions such as Sr^{2+}, Ba^{2+}, and HCO_3^- without losing inherent consistency and it should become a very useful tool for chemical engineering calculations.

Reference:

[1] A. V. García, K. Thomsen and E. H. Stenby, "Prediction of Mineral Scale Formation in Geothermal and Oilfield Operations Using the Extended UNIQUAC Model: Part I. Sulfate Scaling Minerals," Geothermics. 2005, 34(1): 61-97.

[2] Y. Song, C.-C. Chen, "Symmetric Electrolyte Nonrandom Two-Liquid Activity Coefficient Model," Ind. Eng. Chem. Res. 2009, 48:7788-7797.

530194 Numerical Investigation of Heat Transfer in Fixed-Bed Reactors Filled with Complex Particle Shapes Using CFD

Abstract:

Catalytic fixed-bed reactors with a low tube-to-particle diameter ratio are widely used in the chemical and process industry. They are characterized by a heterogenous packing morphology that lead to a strong interplay between local flow effects and the heat and mass transfer within the system. For that reason, simplified pseudo-homogenous models lead to an erroneous numerical description of that reactor type which is why more sophisticated numerical methods are needed.

Particle-resolved computational fluid dynamics (CFD) is a first-principle modeling approach that has been proven to be a reliable and predictive tool to analyze the flow, temperature and species field within fixed-beds in a spatially resolved manner as recently reviewed by *Jurtz et al.* This allows the method to be used as a design tool for new particle shapes to optimize the trade-off between a high active catalytic surface, a low pressure drop and a good heat transfer characteristic.

Within the scope of this presentation a fully automated workflow for the simulation of fluid flow and heat transfer in fixed-beds filled with cylindrical particles, Raschig rings and multi-hole cylinders will be presented. The different particle shapes will be compared regarding their thermal performance and advanced post-processing methods will be used to understand the complex interplay between particle orientation, convective heat transfer within the inner particle voids and the overall thermal performance. It will be shown how effective transport parameters can be extracted from the simulation results that afterwards can be used in simplified numerical models.

Literature
Jurtz, N., Kraume, M. and Wehinger, G. (2018). *Advances in fixed-bed reactor modeling using particle-resolved computational fluid dynamics (CFD).* Reviews in Chemical Engineering. Published: 2018/02/02. Available at: https://doi.org/10.1515/revce-2017-0059 [Epub ahead of print]

534211 Optimal Front-End Crude Scheduling for Refinery with Consideration of Proactive Unit Maintenance

Abstract:

The front-end crude scheduling (FECS) has been well studied and convincingly proven very effective for the scheduling of crude supply process [1,2]. It can be used to tackle crude movement problems from crude importing by vessels, unloading through berths, temporarily storing at port-side storage tanks, transferring via long-distance pipeline, blending at refinery-side charging tanks, to continuously feeding to refineries. Through FECS, optimal crude transfer schedules in terms of profitability and operability can be determined. In previous studies, most works were developed based on the assumption that all processing units were under normal working status without consideration of any possible abnormal performance. In reality, however, the performance of every processing unit will be inevitably decaying along with the processing time and load. Thus, regular maintenance operations have to be scheduled in advance to restore unit performing efficiency and more importantly to prevent potential unit failures [3,4]. Meanwhile, the unit maintenance schedule needs to be proactively integrated with the crude movement schedule to ensure the real operability as well as the minimization of the overall cost. Some research works did deal with maintenance issues by identifying post remedial strategies for unit failures [5,6]. However, the reactive response will unavoidably and stiffly disrupt the original schedule, resulting in higher operating expense and even infeasible operations. Therefore, proactive scheduling of regular unit maintenance is a promising opportunity to further improve and significantly benefit the crude supply process.

In this study, a novel continuous-time FECS model combined with proactive scheduling of unit maintenance has been developed. It involves a large-scale mixed-integer nonlinear programming problem, covering all crude movements from shipping all the way to the refinery processing. Through this model, the optimal schedule for both crude supply and unit maintenance operations will be simultaneously identified for the minimal operating cost. It can not only determine when and how to isolate the designated unit(s) from the process system for the maintenance preparation (i.e., how to discharge the leftover inventory in units), but also identify the best opportunity for the maintenance-completed unit(s) to rejoin the system for continuous production during the scheduling time horizon. In addition, real process requirements, including tank inventory, transfer flowrate, blending specifications are all taken into account to guarantee the obtained schedule is practical. The efficacy and performance of the proposed model has been tested and demonstrated through various unit(s) and maintenance scenarios.

References

[1] Zhang, S., & Xu, Q., (2015) Refinery continuous-time crude scheduling with consideration of long-distance pipeline transportation. Computer and Chemical Engineering, 75, 74-94.

[2] Li, J., Misener, R., & Floudas, C., (2012) Continuous-time modeling and global optimization approach for scheduling of crude oil operations. AIChE Journal, 58, 205-226.

[3] Amaran, S., Sahinidis, N., Sharda, B., Morrison, M., Bury, S., Miller, S., & Wassick, J., (2015) Long-term turnaround planning for integrated chemical sites. Computer and Chemical Engineering, 72, 145-158.

[4] Vieira, M., Varela, T., Povoa, A., (2017) Production and maintenance planning optimization in biopharmaceutical processed under performance decay using a continuous-time formulation: A multi-objective approach. Computer and Chemical Engineering, 107, 111-139.

[5] Zhang, S., Xu, Q., (2014) Reactive scheduling of short-term crude oil operations under uncertainties. Industrial & Engineering Chemistry Research, 53, 12502-12518.

[6] Zhang, S., Wang, S., & Xu, Q., (2015) A new reactive scheduling approach for short-term crude oil operations under tank malfunction. Industrial & Engineering Chemistry Research, 54, 12438-12454.

534540 Thermodynamic Modeling of CO_2 Absorption in Aqueous Amino Acid Salt Solutions with Symmetric Electrolyte NRTL Model

Abstract:

Development and large-scale deployment of renewable energy is the need of the hour, but it is still desirable to mitigate the increased CO_2 emissions as we make a transition from fossil fuel-based energy sources to renewable energy sources. In this perspective, many separation technologies have been designed for the large-scale removal of CO_2 from industrial sources. The most developed technology for CO_2 removal is the solvent based absorption using monoethanolamine (MEA)[1]. However, alkanolamines suffer from volatility, oxidative degradation and toxicity making them less suitable for solvent-based absorption[2,4]. Recently, aqueous solutions of amino acid salts (AAS) have gained considerable attention as alternative solvents for carbon capture because the inherent ionic nature of the system makes it non-volatile[2] which is advantageous at the stripper conditions. Also, their higher oxidative stability is favorable for oxygen-rich flue gas streams[4]. They also have comparable absorption kinetics to alkanolamines[2]. Thus, it is desirable to develop an extensive thermodynamic model of AAS-H_2O-CO_2 system to support process development and optimization. However, scarcity of extensive experimental data on thermodynamic aspects of the binary (AAS-H_2O) and ternary (AAS-H_2O-CO_2) systems[2] as well as the dissociation of salt to zwitterion, anion and cation[3] add to the complexity of modeling the liquid phase chemical equilibrium. Limited work has been reported for the experimental determination of CO_2 solubility in aqueous solutions of potassium salts of Glycine[4] (KGly), Sarcosine[2] (KSar), Proline[5] (KPro) and Taurine[6] (KTau). Prior models using extended UNIQUAC framework[2] and Deshmukh-Mather mehtodology[4] have been developed for potassium salts of sarcosine and glycine respectively.

In this study, we develop a symmetric electrolyte NRTL model to represent liquid phase non-idealities in conjunction with Redlich-Kwong equation of state to model vapor phase fugacity. Solubility data and pH data is used to identify eNRTL binary interaction parameters for one binary system (KSar-H_2O) and three ternary (KSar-H_2O-CO_2), (KGly-H_2O-CO_2), (KPro-H_2O-CO_2) systems. The models accurately represent the VLE behavior and liquid phase speciation in the three AAS-H_2O-CO_2 systems, covering a temperature range of 313-353 K and CO_2 loading of 0.2-1. The models can be further advanced upon the availability of more extensive experimental data on the absorption enthalpy for the three AAS-H_2O-CO_2 systems.

References:

1. Zhang, H.Que, C.-C.-Chen, Fluid Phase Equilibria 311 (2011) 67-75
2. E.Aronu, E.T.Hessen, T H.-Warberg, K.A.Hoff, H.F.Svendsen, Chemical Engineering Science 66 (2011) 2191-2198
3. -C.-Chen, Y.Zhu, Biotechnology Progress, Vol. 5 No. 3 (1989) 111-118
4. F.Portugal, J.M.Sousa, F.D.Magalhaes, A.Mendes, Chemical Engineering Science 64 (2009) 1993-2002
5. E.Majchrowicz, D.W.F.Brilman, Chemical Engineering Science 72 (2012) 35-44
6. S.Kumar, J.A.Hogendoorn, P.H.M.Feron and G.F.Veersteg, J.Chem.Eng.Data 46 (2001) 1357-1361.

525846 Plantwide Process Design with Automatic Column Optimization, Sequencing and Stacking Using a Rigorous Process Simulator

Abstract:

Column sequencing refers to the problem of determining the optimal column sequence of distillation columns for multicomponent separation. For example, for the separation of three components, ABC, there are two possible sequences: the direct split (A/BC followed by B/C) and the indirect split (AB/C followed by B/C). The number of possible sequences grows rapidly as the number of components to be separated grows. The column sequencing problem has been widely studied in the literature [1–11], although most researchers use shortcut equations (e.g. the Fenske-Underwood-Gilliland column design method) for the column design in order to reduce the computational load.

Column stacking refers to the process of adjusting the operating pressure of one or more columns so that the reboiler of one or more columns can be powered by condenser of another column. Depending on the temperature difference between the top and bottom of the columns and the temperatures of the utilities, three or more columns can be stacked on top of each other. This type of heat integration can significantly reduce energy consumption. Column stacking has also been widely studied in the literature [12–16], although again mostly using shortcut models.

Finally, it is well known that for many problems of practical interest, there is a tradeoff between the design of the reactor or reactor network and the cost of separation [17]. Specifically, for many problems of practical interest, selectivity decreases as per-pass conversion increases, so there is a tradeoff between selectivity losses at high conversion and separation cost (for recycle of unconsumed reactant) at low per-pass conversion. Thus in order to determine the optimal plant-wide process design, it is necessary to consider the design of the reactor or reactor network and the separation system simultaneously. For greater accuracy, it would be desirable to use a rigorous process simulator rather than shortcut column design equations.

In the present contribution, these problems are approached using a rigorous process simulator (Aspen Plus) coupled with a high-level procedural programming language (Matlab). The two programs are connected using a client-server architecture, in which Matlab is the client and Aspen Plus is the server. A process flowsheet consisting of a single distillation column is constructed in Aspen Plus, and this flowsheet is used to model each column in every sequence by changing the feed conditions. The reflux rate and boilup rate are manipulated (varied) in order to achieve desired specifications on the top and bottom product distributions. The optimal number of trays and feed tray location are determined by optimization using simulated annealing programmed in Matlab.

The code determines the optimal design of each column in every sequence and adds up the cost to determine the optimal column sequence. Once the optimal sequence is identified, the code automatically implements column stacking by adjusting the operating pressures of the columns within the limits of feasibility (given temperatures for the utilities and a minimum temperature driving force). Finally, this entire process (automatic column design, sequencing and stacking) is coupled with a model of a chemical reactor (in a separate Aspen flowsheet) to determine the optimal plantwide process design by adjusting reactor network design parameters (temperature and residence time) to find the optimal tradeoff between selectivity losses and separation costs using the optimized separation system design in each case.

References

[1] Nishida N.; G. Stephanopoulous; A. W. Westerberg. A review of process synthesis. AIChE Journal 1981, 27, 321-351.

[2] Tedder D. W; Rudd D. F. Parametric Studies in Industrial Distillation Part I. Design Comparisons. AIChE Journal 1978, 24, 303-315.

[3] Stephanopoulos G; Westerberg A. W. Studies in Process Synthesis–II. Evolutionary Synthesis of Optimal Process Flowsheets. Chemical Engineering Science 1976, 31, 195-204.

[4] Malone M. F.; K. Glinos; F. E. Marquez; J. M. Douglas. Simple analytical criterion for the sequencing of distillation columns. AIChE Journal 1985, 31, 686-689.

[5] P. B. Gadkari; R. Govind. Analytical Screening Criterion for Sequencing of Distillation Columns. Computer & Chemical Engineering 1988, 12, 1199-1213.

[6] M. Bezzina; L. Pibouleau; S. Domenech. A Statistical Approach to the Synthesis of Separation Sequences. Computer & Chemical Engineering 1989, 13, 651-660.

[7] Floquet, P.; L. Pibouleau.; S. Domenech. Separation Sequence Synthesis: How to Use Simulated Annealing Procedure? Computer. Chem. Eng 1994, 18, 1141-1148.

[8] An, W-Z.; Yuan X-G. A Simulated Annealing-Based Approach to the Optimal Synthesis of Heat-Integrated Distillation Sequences. Computers and Chemical Engineering 2009, 33, 199-212.

[9] F. Wang; Y. Luo; X. Yuan; A Formulation Methodology for Multicomponent Distillation Sequences Based on Stochastic Optimization. Chinese Journal of Chemical Engineering 2016, 24, 1229-1235.

[10] Boozarjomehry R. B.; A. Pourahmadi Laleh; W. Y. Svrcek. Automatic Design of Conventional Distillation Column Sequence by Genetic Algorithm. The Canadian Journal of Chemical Engineering 2009, 87, 477-492.

[11] Boozarjomehry R. B.; A. Pourahmadi Laleh; W. Y. Svrcek. Evolutionary Design of Optimum Distillation Column Sequence. The Canadian Journal of Chemical Engineering 2012, 90, 956-972.

[12] Andrecovich, M. J.; A. W. Westerberg. A Simple Synthesis Method Based on Utility Bounding for Heat-Integrated Distillation Sequences. AIChE Journal. 1985(b); 31: 363-375.

[13] Andrecovich, M. J.; A. W. Westerberg. An MILP Formulation for Heat-Integrated Distillation Sequence Synthesis. AIChE Journal 1985(a), 31, 1461-1474.

[14] Andrecovich, M. J.; A. W. Westerberg. Utility Bounds for Nonconstant QΔT for Heat-Integrated Distillation Sequence Synthesis. AIChE Journal 1985(c), 31, 1475-1479.

[15] C. A. Floudas; G. E. Paules IV. A Mixed-Integer Nonlinear Programming Formulation for the Synthesis of Heat-Integrated Distillation Sequences. Computers & Chemical Engineering 1988, 12, 531-546.

[16] Jaroslav, J.; Radim Ptacnlk. Synthesis of heat integrated rectification systems. Comput. Chem. Engng 1988, 12, 427-432.

[17] Douglas, J. M. Conceptual design of chemical processes. New York, McGraw-Hill, 1988.

541877 Research Challenges and Opportunities in Pathway Toward Sustainable Society

Abstract:

Maintaining adequate future supplies of energy and water requires the creation of a pathway to sustainable energy and water that is both multi-faceted and complex. An effective pathway aims to improve the reliability, security, and affordability of energy and water by: utilizing a least-cost strategy to reduce the negative impact of energy consumption on climate change and water availability; emphasizing the gradual decarbonization of the global energy system; increasing energy efficiency and conservation; and preserving natural resources and the environment.

A successful pathway to sustainability must include short-term, transitional, and long-term plans that must be launched simultaneously. This presentation will discuss the critical contribution of Chemical Engineering research to the successful implementation of these plans.

In this presentation the role of Chemical Engineering research in carbon capture processes, particle pulverization as a recycling tool for polymeric and elastomeric materials, computational fluid dynamics (CFD) research in the production of natural gas from unconsolidated hydrate reservoirs, production of biofuel and renewable energy research including heat transfer medium for concentrated solar power (CSP) systems will be presented.

541881 Incorporation of Safety and Sustainability in Conceptual Design Via a Return on Investment Metric

Abstract:

Traditionally, process design and selection of alternatives has mainly relied on techno-economic analyses. A general practice is that once a design has been selected, the assessment of other aspects such as environmental impact and process safety implications is carried out at a later stage, at which time limited modifications can be done to the design in order to enhance such performances. Therefore, there is a need for a tool to simultaneously consider economic, environmental and safety objectives during the conceptual design phase of a project. In this work, a new approach for the simultaneous incorporation of process profitability, environmental implications and safety considerations is presented. This approach is based on extending the conventional return on investment (ROI) metric to include positive or negative impacts of potential design changes on the three aforementioned aspects. In particular, a safety and sustainability weighted return on investment metric (SASWROIM) is introduced to systematically integrate those multiple objectives and provide valuable information at the conceptual design stage. To illustrate the usefulness of the proposed approach, two case studies for the production of butadiene and methanol are considered. It is shown how the new metric can be used by decision makers to reliably combine the economic viability of a project with the assorted impact on the environment and safety characteristics of the process.

529438 Condensed Phase Reactions of Polysaccharides during Fast Pyrolysis

Abstract:

The goal of this study is to investigate the early stages of polysaccharide deconstruction during fast pyrolysis. Fast pyrolysis of polysaccharides is typically studied by analyzing volatilized products, which precludes direct measurement of non-volatile intermediate products, namely anhydro-oligosaccharides. Focusing only on the volatile products biases theories against the role of large, intermediate products. We aim to overcome this analytic weakness.

We have employed two unique reactors that rapidly heat biomass and quench products, allowing us to truncate pyrolysis before its completion and recover solid products for chemical analysis. The smaller reactor, termed the Controlled Pyrolysis Duration (CPD)-Quench reactor, can control reactor residence time on the order of a few seconds. The larger reactor, a modified free fall reactor, has a fixed residence time of about one second but has the advantage of creating larger quantities of intermediate products for analysis. These methodologies allow us to examine condensed phase reactions during the early stages of polysaccharide pyrolysis.

We performed truncated pyrolysis on cellulose, the most abundant biomass polysaccharide, at three fast pyrolysis temperatures: 400, 450, and 500 °C. The CPD-Quench reactor, combined with gel filtration and high-performance liquid chromatography, shows that small anhydro-oligosaccharides form before substantial levoglucosan production – an indication that cellulose rapidly depolymerizes via cracking reactions before unzipping reactions produce significant levoglucosan. The signal from a flame ionization detector (FID) directly connected to a conventional micropyrolyzer correlated well with the time evolution of levoglucosan from the CPD-Quench reactor, suggesting that it could be used as a proxy for determining the rate of levoglucosan production in future cellulose depolymerization studies.

The larger quantities of intermediate products created in the free fall reactor make possible more detailed studies. The non-volatile products of truncated pyrolysis of cellulose, which are large, water-insoluble anhydro-oligosaccharides, were measured by size-exclusion chromatography and multi-angle light scattering (SEC-MALS)—the first known analysis of its kind—along with other analytical techniques.

The results from these experiments agree well with a simple population balance model of cellulose depolymerization. Using recently published rate parameters for cracking and unzipping of cellulose, the model predicts rapid depolymerization to small anhydro-oligosaccharides before significant levoglucosan is formed. The ultimate yield of levoglucosan from cellulose is strongly dependent upon both the yield of levoglucosan from the dimer cellobiosan as well as the relative rates of cracking and unzipping of anhydro-oligosaccharides. Deciding which of these mechanisms is most important in determining levoglucosan yields from cellulose requires further experimentation. Furthermore, there is evidence that the absolute rate of unzipping needs to be much higher than suggested in the literature to explain the experimentally observed evolution of levoglucosan during fast pyrolysis.

531034 Origins of Enhanced Levoglucosan Yields during the Co-Pyrolysis of Cellulose and High-Density Polyethylene

Abstract:

The widely available forest, crop, and wood resources in the United States have the capability to produce at least one billion dry tons of renewable biomass per year. Cellulose is the largest fraction of biomass and the most abundant biopolymer in the world. Converting cellulose in a biorefinery into value added chemicals or fuels presents a unique opportunity to create cost-effective and sustainable chemical production processes in regions where biomass is widely available. Fast pyrolysis is one of the simplest methods for biomass conversion, but producing value added chemicals via fast pyrolysis is still challenging due to its low selectivity. As a result, traditional application of cellulose pyrolysis is to produce biofuels via a series of small oxygenated molecules followed by upgrading. Levoglucosan (LG) is the primary pyrolysis product of cellulose, with yields as high as 70%. Although the objective of conventional biofuel production has been to minimize its yields, LG has gained recent interest as a value added chemical due to its potential applications in polymer, pharmaceutical, and consumer product industries. In this work, we present a strategy to enhance LG yields during cellulose pyrolysis by incorporating high density polyethylene (HDPE). Our experiments have shown that during cellulose/HDPE co-pyrolysis under low-pressure conditions, LG yields can be enhanced by as much as 26%. Co-pyrolysis experiments with different mixing patterns indicate that the origins of this enhancement could be from the reduction of thermohydrolysis reaction at low pressures and the inhibition of the mass transfer processes, such as evaporation, escape, and ejection, of the small sugar molecules by the addition of HDPE. Our work suggests that the interplay between chemical kinetics and mass transfer can be manipulated to alter the yields of key products during cellulose pyrolysis.

531190 Understanding Fast Pyrolysis of Microalgae Model Components Via Py–FTIR and Py–GCMS

Abstract:

Microalgae, fast growing aquatic plants, are considered as prospective source of renewable energy to reduce the dependency on fossil fuels. No competition with food crops is foreseen as they grow in water unlike lignocellulosic biomass. Thermochemical conversion of microalgal biomass is a potential technique to convert them into valuable fine chemicals and fuel molecules. Fast pyrolysis of microalgae proves to be a promising conversion technique to produce bio-oil. Owing to the complexity in structure of microalgae, a wide variety of compounds containing oxygen and nitrogen such as aldehydes, ketones, carboxylic acids, nitriles, amides, pyrroles and indoles are present in the bio-oil from microalgae. Microalgae consist of three key components namely proteins, lipids and carbohydrates. Tweaking the culture conditions can significantly change their composition. There is limited literature available on the interaction among the constituents under fast pyrolysis conditions. The effect of interactions between proteins-lipids, proteins-carbohydrates and lipids-starch on the formation of various products should be investigated in detail for a better understanding and implementation of large scale bio-oil production from fast pyrolysis of microalgae.

In this study, bovine serum albumin (BSA), sunflower oil (SO) and potato starch (PS) were chosen as microalgae components representing proteins, lipids and carbohydrates, respectively. Prior to fast pyrolysis experiments, BSA, SO and PS were characterized to determine their thermal stability, elemental composition, high heating value and functional groups. Fast pyrolysis of individual components and mixtures of BSA, SO and PS emulating microalgae behavior were investigated using a single shot micropyrolyzer (Frontier Laboratories) and CDS Pyroprobe® reactor at 500 °C. Pyrolysates evolved were analyzed using a Shimadzu QP 2010 plus gas chromatograph–mass spectrometer (GCMS) for the organic composition and Agilent Cary 660 Fourier transform infrared (FTIR) spectrometer for the time evolution of major functional groups. The effect of BSA-to-SO, BSA-to-PS and SO-to-PS ratio (1:0, 2:1, 1:1, 1:2, 0:1 wt./wt.), and BSA-to-SO-to-PS (1:1:1 wt./wt./wt.) on the pyrolysate composition and temporal evolution of C-H (methyl as well as aromatic), C=O, C-C, N-H and carbondioxide functional groups were evaluated. The presence of interactions among the various model components was assessed by comparing the actual selectivity of a functional group with predicted selectivity, determined using the additivity rule. Significant positive (increase in production) and negative (decrease in production) interactions were observed in the selectivity of various functional groups. Interaction of SO with BSA and PS resulted in enhanced selectivity towards long chain carbon compounds (>C15) owing to repolymerization reactions. For co-pyrolysis of BSA with SO, high selectivity (30–40%) was observed for esters as a result of possible esterification reactions between carboxylic acids and alcohol intermediates. Due to the interactions during co-pyrolysis of BSA and PS, selectivity of carboxylic acids (10–25%) increased. Interactions between BSA and PS also decreased the selectivity towards aromatic compounds by increasing the selectivity towards N- containing cyclic compounds. Mixtures of SO and PS is shown to inhibit decarboxylation reactions resulting in very high selectivity of carboxylic acids (45–48%). Selectivity of hydrocarbons was highest (30%) when three component mixture was pyrolyzed. The presence of SO and PS altered the pyrolysis mechanism of BSA as they inhibited the formation of aromatic hydrocarbons and N-containing cyclic compounds. The maximum vapor evolution time for BSA, SO and PS varied between 9–50 s. Importantly, the evolution of all the functional groups started simultaneously as soon as pyrolysis started. BSA and PS accelerated the pyrolysis of SO as maximum vapor evolution time was brought down from 50s to below 10 s for the major functional groups. This study provides a better understanding of the interactions between model compounds representing the algae components. This can in turn be useful in tailoring the composition of microalgae species to improve the yield of specific organics from algae. More interesting results will be addressed during the presentation.

536635 Renewable Materials from Catalytic Fast Pyrolysis

Abstract:

Polymers and advanced carbon can be prepared from the bio-oil produced from catalytic fast pyrolysis (CFP), and this approach offers sustainable and cost-effective pathways to drop-in, functional replacement and performance advantaged materials. Furthermore, diverting some of the materials from biofuels production in an integrated biorefinery could help improve the overall economics. We will discuss research being conducted to separate oxygenated chemical products from the oil and aqueous streams from CFP as part of an integrated biorefinery, where a majority of the carbon is being directed towards hydrocarbon biofuels production and smaller streams are diverted to high value co-products. Removing oxygenated molecules from the oil has the added benefit of reducing the demand for hydrotreating to produce hydrocarbons, and reducing the carbon in aqueous streams decreases the cost for waste water treatment. We are also investigating a strategy whereby the heavy residues that remain after distillation of bio-oil are converted into high value advanced graphite that can be used electrical storage devices such as lithium ion batteries (LiBs). These heavy residues are often very difficult to hydrotreat and diverting them to graphite production could dramatically improve the biorefinery economics. The presentation will discuss recent advances in polymers and carbon production and their potential impact on biofuels cost.

537787 The Effects of Pretreatments in Improving the Quality of Bio-Oil Products from Non-Catalytic and Catalytic Pyrolysis of Lignocellulosic Biomass

Abstract:

Lignocellulosic biomass is a promising feedstock for the production of chemicals and fuels since it is available in vast amounts. One promising approach that has gained interest for converting lignocellulosic biomass in the past few years is fast pyrolysis. Fast pyrolysis is a thermochemical conversion process in which biomass is decomposed to produce a liquid (bio-oil), a solid (biochar), and a non-condensable gas mixture (syngas) by exposing biomass to high temperatures, typically between 400 and 600°C, in the absence of oxygen. The liquid mixture of bio-oil is composed of wide variety of chemical compounds having different physicochemical properties. These chemical compounds include aromatics, carbohydrates, phenolic compounds and oxygenated compounds, such as aldehydes, ketones and organic acids. To be used as a feedstock for the production of liquid transportation fuels, bio-oil ideally must contain high amounts carbohydrates, aromatics and phenolic compounds and low presence of oxygenated compounds. High presence of oxygenated components in bio-oil is undesirable since it causes bio-oil to be chemically unstable, highly polar, acidic and low in heating value. The high polarity of bio-oil makes it immiscible with crude oil, making it difficult for using bio-oil as a co-feedstock in petroleum refineries. Furthermore, while many chemical constituents in bio-oil are valuable, their contents are low, making their recovery technically difficult and costly.

Bio-oil can be upgraded to make it more suitable for use as a liquid transportation fuel by using zeolites as a solid acid catalyst. Catalytic upgrading of biomass using H-ZSM-5 has been shown to produce a product mixture with lower oxygen content and higher amounts of single ring aromatics. Unfortunately, the increased aromatic content from catalytic pyrolysis lowers the overall liquid bio-oil yield due to higher production of non-condensable gases (primarily CO_2).

We have found one way to improve bio-oil quality while still increasing the yield of aromatics is to carry out acid hydrolysis pretreatment of the biomass. In the experiments, acid hydrolysis pretreatment steps were conducted by using phosphoric acid at concentrations of 0, 2, 4, 6.5, and 9 wt% in boiling water under reflux for one hour. Non-treated and some of the acid-treated biomass feedstocks were subjected to torrefaction process, in which biomass was heated under inert atmosphere at 220°C for 30 minutes. All fast pyrolysis experiments were conducted by using a micropyrolyzer system which was connected to a GC/MS for analyzing the chemical composition of bio-oil products and a TCD gas analyzer for analyzing the composition of non-condensable gas products. To evaluate the effects of the type of feedstocks, pinewood and switchgrass were used to represent woody and grassy type lignocellulosic biomass, respectively.

The experimental studies indicated that acid hydrolysis pretreatment on both biomass feedstocks have a significant effect in increasing selectivity to aromatics and sugars when pyrolyzed. 2 wt% phosphoric acid was the most effective concentration for pretreating both biomass feedstocks. Surprisingly, the pretreatment improved aromatic yields using a standard non-catalytic pyrolysis of pinewood without the addition of the HZSM-5. Although still a minor component of the bio-oil, in the absence of HZSM-5, there was over five times the aromatics produced. A similar improvement was found for the switchgrass. The selectivity towards undesirable low molecular weight oxygenated compounds was decreased up to 60% compared to that of bio-oil produced from non-treated biomass. For catalytic pyrolysis, there was about a 3-fold increase in selectivity towards the desired aromatics and a 60% decrease in undesirable oxygenated compounds. Since catalytic pyrolysis with HZSM-5 is known to produce significant yields of single ring aromatics, the 3-fold improvement is noteworthy.

The increase of product selectivity towards sugars and aromatics on non-catalytic and catalytic pyrolysis of acid-treated biomass may be attributed to the breakdown of the lignin-hemicellulose bonds surrounding the cellulose structure making the cellulose more available for further conversions to sugars and aromatics. Acid hydrolysis also reduces the amount of inorganic compounds in the solid biomass, which lowers the degree of inorganic metal-catalyzed secondary reaction giving undesired low molecular weight oxygenated compounds.

Torrefaction treatment on acid-hydrolyzed biomass did not significantly improve the selectivity for aromatics or sugars. The treatment can be used as a biomass drying method which improves the overall process for producing bio-oil in an actual plant. Torrefaction on the acid hydrolyzed biomass produces a dry feedstock which can be easily stored and feed continuously to a process without the complications arising from having high water content in the feed or resulting bio-oil product. Overall, the combined pretreatment/torrefaction treatment improves both the processing and desired aromatics production from a plant carrying out catalytic pyrolysis.

*Contact: Dr. Justinus Satrio (justinus.satrio@villanova.edu)

527417 Non-Oxidative Conversion of Methane into Light Hydrocarbons Using Single-Atom Platinum Catalysts

Abstract:

Nanoceria-supported atomic Pt catalysts (denoted as $Pt_1@CeO_2$) have been synthesized and demonstrated as advanced catalysts for non-oxidative, direct conversion of methane. These catalysts were synthesized by calcination of Pt-impregnated porous ceria nanoparticles at high temperature (ca. 1,000 °C), with the atomic dispersion of Pt characterized by combining aberration-corrected high-angle annular dark-field scanning transmission electron microscopy (HAADF-STEM), X-ray photoelectron spectroscopy (XPS), X-ray absorption spectroscopy (XAS) and diffuse reflectance infrared Fourier transform spectroscopy (DRIFTS) analyses. The $Pt_1@CeO_2$ catalysts exhibited much superior catalytic performance to its nanoparticulated counterpart, achieving 14.4% of methane conversion at 975 °C and 74.6% selectivity toward C_2 products (ethane, ethylene and acetylene). Comparative studies of the $Pt_1@CeO_2$ catalysts with different loadings as well as the nanoparticulated counterpart reveal the single-atom Pt to be the active sites for selective conversion of methane into C_2 hydrocarbons.

527863 Highly Selective Nonoxidative Coupling of Methane (NOCM) over Pt-Bi Bimetallic Catalysts

Abstract:

There is widespread interest in converting methane, primary component of natural gas and shale gas, into valuable chemicals. As an important direct methane transformation technique, despite extensive research conducted for decades, oxidative coupling of methane (OCM) remains industrially uneconomic owing to low selectivity towards valuable target products (C2 species, ethane/ethylene). In the present work, we describe that ZSM-5 zeolite supported bimetallic Pt-Bi catalysts stably and selectively convert methane to C2 species with high carbon selectivity (> 90%) at relatively moderate temperatures (600-700 C). Based on experimental observations, it is proposed that the surface Pt in catalysts functions as the active site for methane activation, while Bi addition as promoter plays an important role for C2 species formation and catalyst stability. This is a promising start toward technologies suitable for exploitation on the industrial scale

528802 Oxidative Coupling of Methane: The Role of the Tungstate Promoter in Mn-Na_2WO_4

Abstract:

The recent increase in recoverable natural gas resources has renewed interest in using this resource beyond combustion via conversion to higher value chemicals. One reaction route of particular interest is the conversion of methane to ethylene via oxidative coupling (OCM). This reaction has been studied for many decades but is still missing an efficient catalyst to-date. In the present work, we hence revisit the mechanism of oxidative coupling of methane over Mn-Na_2WO_4, one of the most promising catalysts studied for this reaction, with specific focus on elucidating the role of the tungstate promoter in this catalyst system.

Through careful fixed bed reactor studies using both Mn-Na_2WO_4 mixed oxide catalysts as well as individual manganese oxide and tungstate catalysts, we found—in agreement with previous studies—that lattice oxygen of unpromoted Mn-oxide catalysts is effective for methane activation, but results in poor C_2 yields since the undesired total oxidation dominates the surface reaction. The addition of the well-known promoter Na_2WO_4 improved reaction selectivity strongly by both suppressing CO_2 formation and increasing C_2H_4 yield. Interestingly, however, we found the same improvement in C_2 selectivity when using a simple physical mixture of separate Mn-oxide and Na_2WO_4 catalysts: Product distribution with the single, mixed oxide catalyst and the mixture of two pure metal oxides was experimentally indistinguishable. This contradicts the prevailing mechanism in the published literature that concludes that the promotion by tungstate is based on an exchange of lattice oxygen between the two metal phases.

A series of further kinetic tests and catalyst characterization performed to elucidate the role of Na_2WO_4 revealed that methane activation occurs exclusively on Mn-oxide, while Na_2WO_4 is essentially inactive towards all hydrocarbon species in the system but can be easily reduced by H_2. The promotion of Mn-oxides by Na_2WO_4 hence does not rely on a lattice oxygen sharing mechanism but instead seems to be based on a concerted reaction mechanism in which the tungstate selectively removes H_2 during the C_2H_6 dehydrogenation step in OCM. Thus, the combination of Mn-oxide and tungstate catalysts—either in a single mixed oxide or a via physical mixture of two individual catalysts—synergistically combines the methane activation over the Mn oxide, resulting in the formation of ethane, and the oxidative dehydrogenation of ethane to ethylene over the tungstate phase. Our results hence show that the currently prevailing understanding of the reaction mechanism over this catalyst is incomplete and opens a new direction for rational catalyst design for oxidative coupling of methane.

531145 Microkinetic Modeling of Direct, Non-Oxidative Conversion of Methane to Value-Added Chemicals over Iron/Silica Catalyst

Abstract:

The abundance of shale gas has generated a lot of interest for production of chemicals. Currently, methane is converted into fuels and chemicals through syngas, an energy intensive and large capital investment process. Direct methane conversion routes have received considerable interest for their potential to provide more efficient conversion routes. Recently, it has been reported that isolated iron atoms embedded in a silica matrix (Fe@SiO_2) can catalyze methane conversion under non-oxidative conditions to produce olefins, aromatics and hydrogen at high temperatures (1223 â€“ 1363 K). It has been proposed that methane conversion is initiated by catalytic generation of methyl radicals followed by subsequent gas-phase reactions.[1] Although the concept was demonstrated experimentally, fundamental understanding of the reactions is missing. Herein, we model this process by combining detailed catalytic and gas-phase reaction mechanisms. The latter includes up to 600 species and 10,000 reactions. For the former, we investigate the C-H activation of methane and C-C coupling reactions using density functional theory (DFT) calculations. The combined catalytic and gas-phase reaction mechanisms allow not only to capture the trends in terms of conversion and selectivity but also to assess the predictive capability of the model. The interplay of catalytic and gas-phase reactions is delineated. Simulations show that ethylene is a primary product. Under differential conditions, an increase in selectivity towards acetylene and higher molecular weight cyclic species (benzene, toluene and naphthalene) at the expense of ethylene and ethane with increasing temperature is observed.

1. Guo X, Fang G, Li G, et al. Direct, nonoxidative conversion of methane to ethylene, aromatics, and hydrogen. *Science.* May 09 2014;344(6184):616-619.

529756 Co-Aromatization of Methane with Propane over Zn/HZSM-5: The Methane Reaction Pathway and the Effect of Zn Distribution

Abstract:

The co-aromatization of methane and propane over the Zn/HZSM-5 catalyst under a condition similar to the desired reaction condition for industrial applications is investigated. The methane participation demonstrates different reaction pathways and incorporation sites at different stages of the reaction, which might be related to the Zn catalytic site redistribution on the catalyst surface as well as the product molecules including benzene and toluene. Initially, methane favors the benzylic sites of the product molecules. Then methane mainly contributes to the phenyl ring formation as the reaction continues. As the reaction proceeds further, the methane incorporation site in the co-aromatization product molecules is on the substitution groups of the formed aromatics. These phenomena are witnessed by GC-MS analysis of the products, and confirmed by a ^{13}C isotope labeling study and SSNMR spectroscopy. The reaction intermediates formed on the catalyst surface are studied by DRIFT, SIMS and theoretical calculations. The reaction between methane and propane in the co-aromatization reaction over Zn/HZSM-5 might be closely related to the interaction between methane and the unsaturated species upon propane activation, particularly the reaction between the C_1 and C_3 pieces produced from methane and propane activation. Zn sites bonding to the zeolite framework are critical in the co-aromatization process by stabilizing the intermediates. XAS, XPS, XANES, and ^{67}Zn SSNMR spectroscopies provide further support to the SIMS results on the catalytic zeolite and the importance of Zn. Theoretical calculations predict a lower energy for Zn in the inner pores compared with the external surface in the pristine catalyst, resulting in a higher Zn concentration in the inner pores, which is confirmed by STXM images. The XANES spectra of the pristine catalyst and spent catalysts after the co-aromatization reaction reveal that the Zn concentration on the external surface is increased after the reaction. This phenomenon is further confirmed by EFTEM and XPS analyses, indicating that the Zn species is transferred to the external surface in the reaction and exhibits an interaction with the oxygen sites exposed on the external surface of the zeolite framework, where the space on the external surface is sufficient to hold intermediates and product molecules with large molecular sizes. The migration of Zn is attributed to the decreased energy gap between inner pore Zn sites and those on the external surface. The methane participation pathway is consequently altered to the incorporation to alkyl substitution groups of the formed aromatics at the later stage of the reaction.

534706 Catalytic Aromatization of Methane: Strategies for Improving Active Chemistry and Stability of the Catalysts

Abstract:

Intensifying the use of CH_4 as a chemical feedstock for its conversion to liquid fuels and other added value chemicals is of increasing interest since CH_4 is the main constituent of natural gas, of which vast natural reserves are being exploited. Paradoxically, a significant amount of natural gas is wasted in flaring, due to the expensive handling and transportation of natural gas. Among the different processes that exist to convert CH_4 to added-value liquids, there has been an emerging interest in the heterogeneous catalytic conversion of methane, CH_4, to a mixture of benzene and hydrogen, by the non-oxidative methane aromatization reaction (6 CH_4(g) â†' C_6H_6(g) + $9H_2$(g)), which happens in one step. Our group is carrying out a fundamental and systematic study of this catalytic process with the goal to answer some fundamental questions, which can mitigate the technological challenges of this process.

Zeolite-supported molybdenum catalysts have so far been the most widely studied catalysts for the CH_4 aromatization reaction. Rapid catalyst deactivation and low conversions are disadvantages that impede commercialization of the methane aromatization process. It is agreed that Mo carbide species, originated by exposure of the Mo oxide species to the reactant methane, are responsible for methane activation, however, the nature and amount of carbide phases formed in the induction period of the reaction are not controlled.

We will present out results related to the exploration of the pretreatment conditions that are necessary to purposefully prepare zeolite-supported metal carbide phases with the best catalytic properties and relate the reaction and deactivation pathways to the structure of a specific carbide phase.

536966 Performance and Phase Stability Studies of Gadolinium-Doped Barium Cerate in Oxidative Coupling of Methane and the Impact of Zr Doping

Abstract:

Oxidative coupling of methane (OCM) is a promising technology for the direct conversion of methane to ethylene and ethane (C_2). This process is yet to be commercialized mainly due to the occurrence of undesired reactions that lead to deep oxidation of methane to CO and CO_2. Oxygen ion (O^{2-}) conducting membrane reactors are currently under investigation as an alternative to conventional packed bed reactors (PBRs) used in OCM. These membrane reactors operate at lower oxygen partial pressures by distributing the oxygen feed, which promotes C_2 product selectivity. A practical design for these membrane reactors would include combining a selective catalyst, preferably O^{2-} conducting, with an O^{2-} conducting membrane. In this contribution, we study an O^{2-} conducting material, gadolinium-doped barium cerate ($BaCe_{0.8}Gd_{0.2}O_{3-\delta}$ or BCG), to evaluate its potential applicability as a catalyst and membrane in OCM membrane reactors. From PBR tests, we found that this material was active for OCM, and achieved a maximum C_{2+} yield of 14% at 1023 K. Furthermore, a C_{2+} selectivity of 90% was obtained at low oxygen partial pressures expected to occur in membrane reactors. Although the C_{2+} yield from this material was stable over 48 h on stream at high methane conversions, X-ray diffraction data showed that the BCG perovskite phase, which is required for its O^{2-} conductive properties, decomposes into carbonate and oxide phases. We show that this phase instability can be suppressed by doping BCG with Zr without significantly affecting the OCM performance in a PBR.[1]

(1) Igenegbai, V. O.; Meyer, R. J.; Linic, S.. *Appl. Catal. B Environ.* **2018**, *230*, 29–35.

538521 Effects *of Controlled Crystalline Surface of Hydroxyapatite on Methane Oxidation Reactions*

Abstract:

Hydroxyapatite (HAP, $Ca_{10}(PO_4)_6(OH)_2$) has a hexagonal prismatic structure that exposes two crystalline surfaces: prism-faceted a- and basal-faceted c-surfaces. In this work, the predominant exposure of c-surface was controlled and its influences in methane oxidation reactions (combustion and oxidative coupling over HAP and lead-substituted HAP (Pb-HAP), respectively) were studied. The c-surface exposure was realized by crystal orientation in HAP-based catalyst film, which was created by an electrochemical deposition of HAP seeds on a titanium substrate, followed by hydrothermal crystallization and peeling off of the crystalline films from the substrate. In comparison to a-surface that is prevalently exposed in unoriented HAP-based catalysts, the c-surface (i.e., (002) crystalline plane) of HAP-based catalysts exhibited up to 47-fold enhancement in areal rate in both reactions. The distinct catalytic activity between these two crystalline surfaces is attributed to the preferential formation of oxide ions and vacancies on c-surfaces. The oxide ions and vacancies in turns function as actives sites for promoting methane activation and complete oxidation into CO_2. Density functional theory calculations confirmed the close relationship between different catalytic activities in c-surface of oriented and a-surface of unoriented HAP through the tendency of vacancy formation. Without the presence of vacancies, the methyl or methylene group after methane activation forms ethane or ethylene via coupling. The present study explored the effects of HAP crystal orientation in methane oxidation reactions, which revealed distinct catalytic behaviors of crystal surfaces in HAP-based materials.

541296 Application of Simplified LOPA and Development of Risk Matrix for a University or Small Operational Company

Abstract:

A disquieting number of recent worldwide accidents in laboratories and small operating companies demonstrates that there is a need to improve their hazard identification and risk assessment practices. These entities seem to face common challenges including a lack of process safety knowledge and a lack of human resources that can be used to properly recognize hazards and evaluate the risks. Moreover, there is a lack of distinction between the concepts of hazard and risk among the people outside of the large process industry. It seems, they often do not understand that the hazard corresponds to the potential for harm and is usually independent of scale, e.g. a small quantity of hazardous chemical is the same hazard as a big quantity of the same. It seems that at the same time, they do not understand that risk is related to the combination of the likelihood of the accident scenario occurring and the severity of the potential consequence, which may lead to have a large risk even if the hazards look to be small. This risk has to be minimized by introducing more layers of protection in place and the better quality of the layers to lower the likelihood, and thus lower the risk.

This work presents the development of the method which can help the entities with low human resources, like laboratories and small operating companies, to evaluate the risk level and the need for layers of protection. A simplified Layer of Protection Analysis (LOPA) and Risk Matrix were developed and proposed to be implemented into Preliminary Hazard Analysis (PreHA) methodology as an easier tool than quantitative or semi-quantitative approach, and yet effective for laboratories or small companies with low human resources. In this method, the layers of protection are identified based on their quality and no reliability numbers are required. Examples are provided and discussion is given.

541303 Role of Academia in Educating Students and Industry Professionals in Process Safety: Experience from MKOPSC-Qatar

Abstract:

Building competency in process safety is critical to the success of process safety management programs in the process industry. This involves teaching the fundamentals of process safety to young engineers which can be challenging given the importance and the complexity of the topic. Academia has a central role to play in this area by incorporating process safety in curriculums for engineering students and by supporting the industry in the training young engineers through continuing education programs.

This presentation will summarize the experiences gained from teaching process safety under the umbrella of the Qatar branch of the Mary Kay Oâ€™Connor Process Safety Center young professionals in the Middle East. It will discuss how academic and research activities can be used as opportunities to practice the principles of Process Safety Management in support of conventional process safety courses. Finally, this presentation will provide two-yearâ€™s experience on how academia can team up with industry to provide higher quality and strong continuing education programs in process safety beyond the short courses.

538852 Waste Management Enhancement Strategies Learned from Hurricane Harvey

Abstract:

Abstract: Hurricane is the most frequent and severe natural disaster along the Gulf Coast. Even though most of the refineries and chemical plants made a quick recovery after hurricane Harvey, hazardous chemicals and toxic contaminants were released into the environment, ozone pollution and surface water pollution was observed. The scale of Harvey far exceeded the capacity of plantsâ€™ and regionsâ€™ common waste management practices. This paper will present the environmental challenges associated with hurricanes in a region heavily concentrated with refining and chemical industries, and discuss strategies to enhance regional resilience from waste management perspective.

522018 The Future Is Here: Robotic Catalyst Removal - Field Results 2018

Abstract:

WorleyParsons has been developing new technology using vacuum catalyst removal from refinery and petrochemical vessels via a remote-operated, screw-propelled vehicle. The development of CAROL^{TM} (Catalyst Amphirol) responds to increasingly stringent requirements to reduce human risk in inert confined-space entry. It is the culmination of a three-year development effort from conception and design to prototyping and testing and represents a potential game-changer in the industry.

The number of fixed-bed catalytic vessels in the global refining and petrochemical industryis estimated to exceed 58,000. Based on current technology a rough estimate of the number of worker days of risk exposure due to confined-space entry during catalyst unloading exceeds 10,000 days per year. Current practices for catalyst removal from each of these vessels pose a risk to safety and/or the environment. This remains true regardless of whether dumping, vacuuming, or water flooding is employed. CAROL^{TM} provides an alternative option that minimizes risk to workers.

CAROL^{TM} is a simple, one-of-a kind machine that has been demonstrated to achieve vacuum catalyst removal without the placement of workers inside the vessel. A review of operating characteristics in a single bed test vessel is provided. Additionally, two case studies from plant demonstrations are examined:

- Catalyst removal from a narrow naphtha hydrotreater at a refinery in the United States.
- Adsorbent removal from a wide dehydration vessel at an LNG plant in Australia. This test is underway at the current time.

This paper showcases the challenges associated with current catalyst removal techniques, and the inherent safety advantages that CAROLTM has shown. The presentation includes video of the test vehicle operating characteristics and discussion on industry acceptance progress and preliminary test results.

531006 Back from the Future: What Nanotechnology Risk Analysis Can Teach Us about Process Safety Today

Abstract:

Nanotechnology. Even today, nearly two decades since the onset of a rapid acceleration in nanotechnology research and development, it still has a futuristic aura about it. Nanotechnology may evoke sweet dreams of tiny machines that vanquish disease-causing invaders in our bloodstream, or horrific nightmares of nanobots that disassemble our flesh to feed their voracious self-replication. In reality, and more prosaically, nanotechnology is understood to be the intentional manipulation of matter to exploit novel phenomena that emerge when one or more dimensions of the object are between about 1-100 nanometers.

At first blush, safety professionals employed in mature industries like the oil and gas and petrochemical sectors would seem to have little in common with those working in emerging technology areas such as nanotechnology. The former are dealing with conventional equipment, materials, and processes whose hazards are generally well understood, if not always perfectly managed. This is not necessarily the case for nanotechnology production, where many fundamental questions about risk remain unanswered. And the points of intersection may grow as nanotechnology continues to penetrate the materials marketplace in many sectors. Nanotechnology risk researchers have had to both learn from past research linking particulate matter to public health outcomes and anticipate a future in which conventional materials and their means of production are radically different. This presentation makes the case that the societal approach to understanding and managing potential risks from nanotechnology can be instructive for those working in process safety. Lessons learned from the US Chemical Safety Board's first 20 years will be interwoven throughout.

515793 Measuring Safety Decision-Making Effectiveness with an Engineering Process Safety Research Instrument (EPSRI)

Abstract:

A common approach toward improving process safety decision-making of a student or employee is increasing their knowledge of design techniques such as proper vent sizing and inherently safer design or operational measures such as HAZOP. However, another key part of process safety is in-the-moment decision making of individuals. In these moments individuals are often forced to make a decision in the context of a dilemma, where competing incentives and disincentives (such as cost savings at the risk of decreased safety) make decisions more difficult.

This work presents an Engineering Process Safety Research Instrument (EPSRI) which can be used to judge in-the-moment process safety decision making of individuals in a Neo-Kohlbergian ethical context. The EPSRI measures how much an individual values selfish motivators in the context of decision making during a dilemma as opposed to value-based systems. The goal of the EPSRI is to represent a validated assessment method of the effectiveness of process safety training, coursework or other interventions on decision making. Details on development, content validation and testing of the EPSRI in real life scenarios will be presented and discussed.

536268 Integration of CO_2Â Biomass Gasification with SOFC as a Viable Pathway for Carbon Capture and Sequestration

Abstract:

Global climate change highlights a pressing need for carbon capture and sequestration (CCS). CCS technology can act as a crutch to limit anthropogenic emissions until the energy sector shifts in favor of supplying energy from low-carbon or carbon-neutral renewables. Herein, we explore CO_2 gasification of biochar (Reverse Boudouard reaction), a method of carbon monoxide-rich feedstock production for solid oxide fuel cells. Through gasification, combustion of biochar is bypassed; instead, the carbonaceous fuel is converted primarily to carbon monoxide. Though considered a poison for some fuel cells, carbon monoxide is a viable feedstock component for SOFCs. Its oxidation in the fuel cell during electricity generation produces a carbon dioxide-rich exhaust stream suitable for sequestration. Theoretically, exhaust processing for CCS for this fuel cell approach will require less energy, making it a promising alternative pathway that can be integrated more seamlessly with carbon storage processes.

Studies of the Reverse Boudouard reaction have been performed on various biomass waste feedstocks (acorns, oak, walnut shell) using thermogravimetric analysis with effluent analysis by gas chromatography. Experimental mass loss data show that the reaction is favored at temperatures near 850ºC, consistent with thermodynamics and with typical operating conditions of SOFCs. The addition of an iron catalyst to the biochar markedly increases the rate of gasification. The raw biomass was found to be more reactive than its biochar counterpart.

535084 Energetic and Environmental Assessment of Biomass to Ethanol Processing in the Sugar-Cane Industry

Abstract:

Bio-fuels have come to be recognized as environmentally friendly alternatives to fossil fuels. This work studies the Bio-ethanol production process complementing a sugar mill.

The process flowsheet examined includes three major processing stages typical of distillery installations for fuel-grade production in sugar-cane processing facilities: (1) fermentation, (2) separation, and (3) dehydration stages.

Using plant data from a bio-ethanol pilot plant in the Northwestern of Argentina (NOA) a kinetic model is formulated for a fed-batch molasses fermentation process. The model for this unit, along with a user-defined separation module, is then integrated into a commercial process simulator. The distillation step is optimized by performing rigorous energy analysis and heat integration on the process utilizing the resources made available by the sugar mill, while pervaporation is used in the dehydration step. Different schemes meeting equivalent performance standards in terms of purity and recovery are examined in this paper.

Particular attention is given to energetic integration by performing a pinch-analysis to each of the fundamental stages to find a global optimum. Optimization is achieved by minimizing a given objective function that accounts for bio-ethanol recovery, separation efficiency, and minimization of effluents. Life Cycle Analysis is used to complete the assessment of operating parameters/conditions on environmental effects.

536999 Effects of Overliming and Activated Carbon on Carbonyl Inhibitors Removal and Butanol Fermentation in Biomass Hydrolysates

Abstract:

Lignocellulosic biomass as the most abundant sustainable resource can be converted into biofuels by enzymatic hydrolysis and subsequent microbial fermentation. However, considerable amount of inhibitory compounds were generated from the dilute acid pretreatment process. The carbonyl aldehydes and ketones have been suggested to the most potent inhibitors in the biomass hydrolysate, which significantly limit the efficient utilization hemicellulose hydrolysates. This study was aimed at the identification of these carbonyl inhibitors and their removal by overliming and activated carbon. Forty-six carbonyl compounds (including furans, aliphatic derivatives, aromatic monomers and aromatic dimers) were identified as potential inhibitors by gas chromatography-mass spectrometry (GC-MS). It was observed that overliming and activated carbon treatment could remove 73.9% and 82.2% of carbonyl inhibitors, respectively, but the treated hydrolysates were still not fermentable. The sequential overliming and activated carbon removed 92.6% of carbonyl inhibitors and resulted in remarkable fermentability and high butanol yield (0.22 g g$^{â€“1}$ sugar). The organic acid in the prehydrolysates were also analyzed by a liquid chromatography connected to a quadrupole time-of-flight (LC-Q-TOF) mass spectrometer. Activated carbon removed 92.9% phenolic acids, while overliming removed 40.2% of them. This study revealed the distinctive effects of overliming and activated carbon treatments on the prehydrolysates detoxification. Overliming removed more dialdehydes and diketones than activated carbon treatment, while activated carbon removed more phenolic acids than overliming.

538669 Green and Sustainable Nanomaterials

Abstract:

Following the American Chemistry Society's 2017 Green Chemistry and Engineering Conference, thought leaders in green and sustainable nanomaterials published as part of a special issue, individual research efforts. This poster reviews outcomes that include the use of biological materials to synthesize nanomaterials, current approaches to the simulation of nanomaterial synthesis, the details of low impact synthesis methods, scaling relationships, current trends in flow through synthesis methods, and approaches or methods to synergize green chemistry and sustainability metrics. The authors also propose that traditional chemistry-process engineering is a first generation design approach, that life cycle methodologies can be contrast with many "design for X" approaches, and that achieving the Brundtland commission's goals requires more broad systems considerations including the upcycle of wastes and circular economy.

527808 Methods for Tracking the Evolution of Refractory REE Mineral Decomposition in Strong Acid Media

Abstract:

Rare earth elements (REEs) are considered critical materials for the development of advanced technologies in applications such as clean energy production. In many ore deposits, the REE-bearing minerals are very complex and refractory in nature, requiring the use harsh processes such as acid baking for their treatment. Acid baking is a hydrometallurgical process that employs elevated temperature and high-strength acid to break down and transform minerals in an ore concentrate in order to recover the REEs into water-soluble sulfates. Aggressive conditions used such as concentrated (98 wt%) sulfuric acid and temperatures in the range of 155-230°C [1] render characterization of the resulting products very difficult, and as such the reaction mechanisms and associated kinetic limitations are not well understood. To this end, the focus of this research was on the development of innovative approaches for the characterization of the solid residue and precipitates throughout the reactive process, especially in the presence of residual sulfuric acid. This may provide an opportunity to optimize the REE recovery, and lead to cost reductions in the final extraction flowsheet.

Laboratory-scale experiments were performed using a time series at baking temperatures ranging from 80°C to 250°C, with a constant acid: ore ratio of 1:1.4 (g/g), with the goal of identifying progressive stages of the sulfation process. To effectively stop the reaction and prevent hydration of any newly-produced phases and the extremely hygroscopic residual sulfuric acid, rapid cooling down to ~8°C was achieved under desiccant. In order to obtain information on the phase transformations, a portion of the resulting paste-like material was retained to prepare high-quality polished cross-sections amenable to detailed characterization. To maintain the pristine nature of this hygroscopic and reactive acid-bearing product, vacuum drying techniques that minimize sample modification were explored. Another portion was subjected to water-rinsing, to identify the soluble phases that had formed up to that stage of the reaction. This was done using cold water (~10°C), to reduce the contribution of exothermic precipitation of alkali-bearing rare earth double sulphate salts[2]. The $<$0.45 µm filtered rinsate was analyzed by inductively coupled plasma spectroscopy and polished sections of the stable dried residue could be easily prepared using conventional preparation techniques. Textural, structural and chemical properties of the stabilized paste and the rinsed sample were characterized by X-ray diffraction, field-emission scanning electron microscopy, wavelength-dispersive X-ray microanalysis, and the vibrational spectroscopy (Raman, Fourier-Transform Infrared).

[1] Sadri, F., ***et al.*,** "A review on the cracking, baking and leaching processes of rare earth element concentrates," ***J. Rare Earths*, 35** (8), pp. 739-752 (2017). **[2]** Lim, H., ***et al.*,** "Leaching of rare earths from fine-grained zirconosilicate ore," ***J. Rare Earths*, 34** (9), pp. 908-916 (2016).

533040 Optimization of Dye Degradation Process By Oxidative Technology

Abstract:

Optimization of dye degradation process by oxidative technology

Mayur M. Yenkie*, Pratibha S. Agrawal, M.G.Bhotmange, M.K.N. Yenkie

Department of Chemical Technology, LIT, RTM Nagpur University, Nagpur-444033

Abstract

Background

Increasing population affects the food demand and with the modernization, various food additives are competing in market with each other. Different food flavors and dyes are now a days added for its better presentation. Excess use and disposal of such additives is dangerous for aquatic life. Various conventional methods are used for removal of such dyes which are transferring the contaminant from one phase to another phase only. Wastewater comprised of dyes are hard to treat, since the dyes are recalcitrant organic molecules that are resistant to aerobic digestion and are stable to light, heat and oxidizing agent. The Fenton reaction has been used in industrial wastewater purification [1]. Degradation of different dyes like acidic, reactive, direct, cationic, disperse and vat dyes by the Fenton process was reported [2].

In the present work, degradation of Sunset Yellow under solar, UV and dark Fenton was carried out using different oxidants. The rate of photocatalytic degradation was measured spectroscopically by measuring dye concentration at regular time interval. Optimization of experimental parameters were evaluated for the effective degradation of dye.

Methodology

All chemical were of analytical grade. All experiments were carried out in batch mode in a 1.0 liter jacketed thermostatic photo reactor (diameter 81mm and height 320 mm) equipped with low pressure mercury lamp (8 W, UV-C, manufacture by Philips, Holland) placed axially at its center. The UV lamp was encased in a quartz tube to protect it from direct contact with an aqueous solution flowing through the annulus between the inner surface of the vessel and the outer surface of the quartz tube, located at the axis of the vessel. The reactor was placed over a magnetic stirrer and a teflon coated magnetic needle was used for stirring the solution at the constant rpm using a dimmerstat. All the reactions were performed at atmospheric pressure. Constant reactor temperature was maintained by circulating water from a cryostat bath (Fourtech Systems, Mumbai, India). A gas tight syringe is used to collect the sample at desired intervals from the sample-port of the reactor.

All the stock solutions, standards and pure compounds are stored in dark, below room temperature. The adjustment of pH is made with 0.1 N /1.0N solution of H_2SO_4 or 0.1N/1.0N NaOH. Experiments were carried out in batch mode. For studying the photodegradation of sunset Yellow, the photoreactor was charged with 800 ml of 3x10^{-4}mol/L solution of the pollutant in deionised water and the required concentrations of iron salt $FeCl_3$ as a source of Fe(III) were added. A 30% aqueous hydrogen peroxide solution was injected into the reactor at required concentrations and the UV-light was switched on at the same time. The change in pollutant concentration was measured with UV-visible double beam spectrophotometer (Spectrascan UV 2600, Chemito, India). The experimental data is used to study the reaction kinetics of degradation of pollutants.

Results

Comparison among various degradation processes

Solar, fenton and fenton like processes were compared for decreasing the cost of degradation.It has been observed that, the degradation rate is minimum for dark fenton like reaction and solar reaction. In other words, solar light can induce degradation in absence of FeCl3 also. For the same set, rate has been found to be increased to 4X 10 -7, in case of UV assisted degradation. Further addition of Fenton reagent in such a reaction vessel increased rate to ten times. Still highest rate is observed for solar assisted Fenton like reaction.

Conclusion

In this work, the decolorization of SY in aqueous solution by Fenton and Fenton like oxidation process has been studied at different experimental conditions, including pH value of solution, initial concentration of H2O2 and Fe2+ dose, as well as the initial dye concentration. The examined Fenton's reaction was found to be more effective at pH 3.5. The best removal efficiency was achieved for solar assisted Fenton like reaction wheredecolorization efficiency was achieved within 60 min of reaction for initial SY solution of 2x 10-4 Molar. The Kinetic study indicated that the degradation kinetics of SY followed the second-order kinetics

References

[1] N. Klamerth, S. Malato, A. Agüera, and A. Fernández-Alba, "Photo-Fenton and modified photo-Fenton at neutral pH for the treatment of emerging contaminants in wastewater treatment plant effluents: A comparison," *Water Res.*, vol. 47, no. 2, pp. 833–840, 2013.

[2] X. R. Xu, H. Bin Li, W. H. Wang, and J. D. Gu, "Degradation of dyes in aqueous solutions by the Fenton process," *Chemosphere*, vol. 57, no. 7, pp. 595–600, 2004.

SR.NO	System Sunset Yellow	Values & Equation
1	Sunset Yellow	Lamda max = 440nm
2	Caliberation equation	K = 3913.1 & R^2 = 0.9925 Y = 3913.1 X
3	3ml H_2O_2 + 2ml $FeSO_4$ + UV	K = -2 x 10^{-6} & R^2 = 0.9816 Y = -2 x 10^{-6} X + 0.000

4	3ml H_2O_2 + Solar	$K = -2 \times 10^{-7}$ & $R^2 = 0.984$ $Y = -2 \times 10^{-7} X + 0.000$
5	UV + 3ml H_2O_2	$K = -4 \times 10^{-7}$ & $R^2 = 0.951$ $Y = -4 \times 10^{-7} X + 0.000$
6	$FeCl_3$ + 3ml H_2O_2 + Solar	$K = -2 \times 10^{-5}$ & $R^2 = 0.835$ $Y = -2 \times 10^{-5} X + 0.000$
7	$FeCl_3$ + 3ml H_2O_2 + Dark	$K = -2 \times 10^{-7}$ & $R^2 = 0.955$ $Y = -2 \times 10^{-7} X + 0.000$

535314 Assessing the Effect of Substrate and Catalyst on Catalytic Waste Gasification

Abstract:

This research is primarily aimed to extend proof-of-concept gasification experiments as a suitable alternative for waste management and sustainable technology. While initially intended to make more efficient use of resources in space exploration missions by producing high value reaction products from normally unused or wasted materials, the same principles can be applied to municipal waste as well.

The experiments performed examine gasification dynamics for different substrates promoted by different catalysts. It utilizes a high pressure batch scale reactor which allows examining controlled multi-phase catalytic isothermal oxidation of solid and melting substrates in the presence high purity air. The reaction time was varied between 2 and 24 hours and a preliminary overall kinetic model was formulated based on conversion, selectivity, and yield determinations. The studied substrates, Polyethylene and Cellulose, are both long chain organic polymers, and make up a substantial portion of both space and municipal waste composition. Although similar in nature, these substrates exhibit marked differences as it pertains to gasification, and were therefore selected as model substrates. Namely, cellulose contains oxygen as part of its molecular structure, while polyethylene does not. Since the technology investigated is aimed at working to relatively low temperatures (below 350 ^{o}C), cellulose remains as a solid particulate at the investigated reaction temperature while polyethylene melts.

Following the reaction phase, both the liquid and solid reaction residues, along with the gaseous products, are collected for characterization and analysis. Gas quantification is performed using a Gas Chromatograph. While thermal characterization (DSC) and Scanning Electron Microscopy (SEM) are used to characterize solid and liquid residuals.

Preliminary results suggest that catalytic gasification provides potential control for the liquid-phase oxidations. Under appropriate conditions, oxidation was found to follow the incomplete oxidation pathway which resulted in significant yields of hydrogen and methane, which are potentially high value products as they can be used for energy recovery.

Comparisons between the gasification rates for the two substrates are drawn. Similarly, comparative studies of different catalysts are also presented. The results are used in conjunction with a kinetic model for the gas-phase to formulate a methodological model for the gasification of solid waste. The results will be presented along with a preliminary extension for a continuous gasification process.

534018 Site-Specific Cross-Linking and Immobilization of Agarase Enzymes for Conversion of *Gelidium Amansii* into Biologically-Active Sugars

Abstract:

Red macroalgae *Gelidium amansii* is currently considered as a prospective renewable resource due to its ubiquity, high carbohydrate and low lignin contents. Its major cell wall carbohydrate is agarose, and sugars derived from agarose have various industrial potentials, i.e. D-galactose can be fermented to ethanol whereas neoagaro-oligosaccharides, neoagarobiose, and 3,6-anhydro-L-galactose can be utilized as biofunctional materials owing to their anticariogenic, anti-inflammation, probiotic, skin whitening and moisturizing properties. Selective production of these sugars without degradation is needed to preserve their bio-functionality. In this context, *G. amansii* can be effectively pretreated by using ionic liquids (ILs) as non-derivatizing solvents combined with enzymatic hydrolysis. However, its large scale feasibility is still challenged due to the high cost of enzymes (i.e. agarases), which have low thermostability and IL tolerance. Herein, tyrosinase-catalyzed site-specific cross-linking and covalent immobilization of agarase (i.e. Aga2 and AhgI from *C. omnivescoria* sp. nov. W5C and Aga50D from *S. degradans* 2-40) is proposed to tackle these challenges. Tyrosine residues (Y-tag) were introduced to the enzymes via genetic engineering, and mushroom tyrosinase was utilized to catalyze the crosslinking and immobilization reactions. Site-specific crosslinking and immobilization of agarases improve their activities, thermal stabilities, IL tolerance and recyclability for repeated production of agarose-derived sugars. This work was supported by the Basic Science Research Program and Korea Research Fellowship Program through the National Research Foundation of Korea (NRF) funded by the Ministry of Science and ICT (No. 2015H1D3A1062172 and 2016R1C1B1013252), and the Ministry of Education (No. 22A20130012051(BK21Plus)).

535718 Identifying and Addressing Potential Barriers Towards Commercialization of Novel, Thermocatalytic Non-Food Sugar to Acrylonitrile Process

Abstract:

Southern Research (SR) has developed a novel, thermo-catalytic biomass derived non-food sugar to acrylonitrile (ACN) process under a cooperative agreement with the Department of Energy. Acrylonitrile (ACN), a precursor monomer of high quality, automotive grade carbon fiber, is currently being produced commercially via ammoxidation of propylene, a high volume commodity chemical subject to the volatility of the petrochemical market. Propylene accounts for approximately 70% of the total ACN production cost. As an alternative, the novel, SR developed process replaces propylene with renewable, more ubiquitous and low cost biomass derived feedstock which makes the process cost effective with significantly reduced environmental footprint. The project has successfully completed the Phase I laboratory scale tests and is currently undergoing pilot scale phase II operation. A techno-economic and life cycle assessment (TE/LCA) conducted based on phase 1 results reveals that the process is capable of producing acrylonitrile at \$0.7-\$0.8/lb with co-product credit and reduce GHG emission by ~ 37% in the process.

As SR is scaling up the process from lab to pilot scale, several potential barriers towards commercialization are being addressed. Some of these barriers include, impacts of commercial sugar impurities (inorganic and organic) on catalysis and final product (ACN), ACN purity, catalyst scale-up, cost-competitive separation of intermediate products to obtain desired purity levels, collection and monetization of co-products and potential site and scale of ACN plant. Commercial sugar hydrolyzates consisting of ~70 different inorganic and organic impurities have been tested. Separately, sugar feeds spiked with ~15 of these impurities (one at a time) were tested to understand the fate and subsequent impact of each impurity. Purity of final ACN product has been tested and compared against commercial ACN samples to obtain drop-in quality and to provide feedback on sugar impurity levels. Each catalyst has been synthesized in large batch and tested in the final industrial form. ASPEN Plus simulation for intermediate separation has been conducted using industrially relevant units (distillation, flash drums). Experimental data are being collected on these separations using lab scale distillation column. Collection, purity and monetization potential of several co-products are being examined. This poster will highlight some of these addressed barriers and the resulting outcomes.

538513 Electronic Waste to Nanoparticles: Influence of Precursor Purity on Nanoparticle Synthesis

Abstract:

The synthesis of gold nanoparticles typically uses a 99.9% pure gold chloride precursor. However, natural resources of precious metals, such as gold, are in a deficit and estimates project depletion in the next 50 years. With high concentrations of precious metals contained in electronic waste, recycled e-waste can serve as an alternate source. To minimize the waste generated during the refining of metals stripped from e-waste, we explored the potential to synthesize gold nanoparticles using a mixed metal precursor solution where gold was not the dominate species. The mixtures tested included binary mixtures of gold and copper or gold and nickel, as well as a ternary mixture of gold, copper, and nickel. Varying molar ratios (1:1 to 1:25) of gold to copper and/or nickel were prepared for gold nanoparticle synthesis by two widely used methods, the Turkevich method and sodium borohydride chemical reduction method. It was determined that gold nanoparticles formed without aggregation at ratios up to 1:2 gold to copper/nickel for both reducing agents. This also translated to the ternary mixture with the formation of gold nanoparticles up to a ratio of 1:1:1 gold to copper to nickel. Calculating the E-factor and other green chemistry metrics suggest that the Turkevich method is less waste generating compared to the sodium borohydride chemical reduction method. The results of this study demonstrate that lower purity gold precursors can still produce gold nanoparticles suggesting that use of alternate sources of gold and/or less separations steps are viable routes to reducing the environmental impact of gold nanoparticle synthesis.

536699 Techno-Economic Analysis of Biofuels Production Via Localized Fast Pyrolysis and Electrocatalytic Upgrading

Abstract:

Based on the DOEâ€™s projections made in their 2016 Billion-ton Report, the annual harvestable biomass in the U.S. (that can be produced at $60/ton or less) by the year 2040 is not sufficient to meet the annual energy demands of the U.S. transportation sector, in 2017. Therefore, for biofuels (derived from biomass) to be considered as a realistic alternative to fossil fuels, their energy content must be upgraded to match that of liquid hydrocarbons like gasoline and diesel. Biomass fast pyrolysis with electrocatalytic upgrading in regional depots, using renewable electricity, followed by further upgrading in central refineries, offers a pathway for the production of fuels with high energy density. To investigate the merits of this novel biofuel production strategy, a mass, energy, and carbon flow analysis was performed for the proposed system. These were compared to the traditional approach of cellulosic fermentation to ethanol, as described in NRELâ€™s 2011 report. The comparative study revealed the advantage of the proposed process in terms of carbon and energy yields of the produced fuel. However, to implement the proposed bioenergy system on a large scale, it is important to thoroughly quantify the commercial potential of the process. Therefore, a techno-economic analysis (TEA) was performed for the proposed system, with the objective of determining the cost per gallon of the fuel produced. Sensitivity analyses were also performed to establish the key parameters that contribute to the overall cost of the fuel, and hence identify an economically optimized scenario for the proposed process.

530619 The Valorization of Sprayed Lignins with Ozone

Abstract:

Advanced biorefineries will be designed with the capability to process a diverse array of feedstocks. In addition, products should be easy to isolate and meet quality standards. Lignin biomass, the most abundant source of renewable aromatics, is as varied as its many plant sources. Constructed of globular units, alkene linkers within lignin differ between plant type. Oxidative scission of these lignin alkenes can partially depolymerize the material to afford valuable aromatic aldehyde products such as vanillin and deconstructed lignin. Here we discuss the ozonation of acid-solubilized lignins in an ambient temperature and pressure continuous spray reactor. Reaction products include vanillin, and 4-hydroxybenzaldehyde (>4.5 wt. % lignin) with an absence of overoxidized acid byproducts. Scanning electron microscopy reveals spherical lignin structures with a diameter of 10 nm which correlate with ~100k Dalton polymer species detected with gel phase chromatography. Spectroscopy of samples before and after reaction indicate that ozonized lignins is effectively treated for downstream globule depolymerization. This versatile process does depend on the lignin feedstock, but may be used to treat both grass and hardwood lignins.

538771 DES Pretreatment Leading to Highly Concentrated Sugar Hydrolysate and Valorizable Lignin

Abstract:

This study investigated high-solid loading deep eutectic solvent (DES) pretreatment and resultant valorizable lignin and concentrated sugar hydrolysate. Switchgrass was pretreated using acidified choline chloride:ethylene glycol (ChCl:EG) at high solid loadings of 20-27%. Highly concentrated sugar hydrolysate was obtained from the pretreated switchgrass via 20% solid loading enzymatic hydrolysis after only 48 h, which contained 164 g/L glucose with a glucose yield of 85%. ChCl:EG was highly recyclable, showing good performance when it was reused for new pretreatment cycles. 2D HSQC NMR revealed that ether bonds in switchgrass lignin were cleaved and guaiacyl lignin degraded during the pretreatment. The degree of bond cleavage and lignin subunit degradation depended on the solid loading for pretreatment. The lignin derived from a higher solid loading pretreatment preserved the original structure of natural lignin better, thus producing increased amounts of phenolic monomers upon pyrolysis. This study demonstrated that highly efficient high-solid loading DES pretreatment can not only solubilize lignin and hemicellulose but also enable the production of valorizable lignin and sugar hydrolysate with high concentration.

527961 Supply Chain Analysis and Process Evaluation for Advancing Sustainable Material Recovery from Post-Consumer Waste

Abstract:

To protect human health and the environment, communities should have sustainable infrastructures to manage waste. Critically, organic post-consumer waste (e.g., livestock) may cause some air quality degradation, soil quality detriment, and water pollution issues. For instance, nutrient pollution from organic waste contributes to tropospheric ozone and acid rain, and can origin negative impacts on aquatic ecosystems, terrestrial biodiversity, and drinking source waters. Better nutrient management initiatives are essential to addressing the wide ecological and economic impacts caused by nutrient pollution.

Despite the significant environmental benefits by performing management initiatives (e.g., nutrient recovery) of post-consumer materials, the deployment of recovery technologies for obtaining value-added products and renewable energy is not a current practice due to the high investment and production costs and product market value uncertainty. Furthermore, such technological alternatives vary in their technical performances and cause different levels of economic and environmental impacts throughout their life cycles. Therefore, understanding the tradeoffs in processing post-consumer materials and their supply chain is an important part of informing decision-makers who are considering recovery and recycle of these materials.

This work presents a supply chain design framework to conduct simultaneous economic and environmental analysis of post-livestock organic material to value-added products under different government incentive scenarios[1]. In addition, this contribution describes a process evaluation from a sustainability perspective by applying some GREENSCOPE indicators[2] to the recovery technologies as a gate-to-gate sustainable objective function in which the four GREENSCOPE target areas (*i.e.*, efficiency, energy, economics, and environment) are evaluated. The material management and process evaluation frameworks are applied to a case study for dairy farms in the State of Wisconsin (U.S.). The analysis reveals that the best strategy to manage waste is to synergize the deployment of technologies that conduct simultaneous recovery of liquefied biomethane and nutrients.

The selection of a suitable supply chain configuration for advancing the sustainable management of post-consumer waste is a complex decision-making process that is highly dependent on the spatial locations of the waste and technology placement, management technology, the destination of products, geographical prioritization, and on local economic (market) and policy conditions. In addition, providing a process performance evaluation is an essential first step towards a holistic sustainability analysis, with social, economic, and environmental implications.

[1] Sampat, A. M.; Gerardo J. Ruiz-Mercado, and Victor Zavala. "Economic and Environmental Analysis for Advancing Sustainable Management of Livestock Waste: A Wisconsin Case Study." ACS Sustainable Chem. Eng. (2018). DOI: 10.1021/acssuschemeng.7b04657.

[2] Ruiz-Mercado, G. J.; Smith, R. L.; Gonzalez, M. A. Sustainability Indicators for Chemical Processes: I. Taxonomy Ind. Eng. Chem. Res. 2012, 51, 2309â€“ 2328 DOI: 10.1021/ie102116e.

544482 Optimization Based Improved Water Recycle Strategies for Modern Oil Refineries

Abstract:

Due to the present global scenario of rising fresh water demands on scarce water resources, increasingly stressed by water pollution and climate change, opportunities for improving industrial water management have been gaining interest. Towards this end, refineries being one of the major water consuming industry, it is important to improve the disposal, management and treatment of industrial waste water in achieving sustainable development goal (SDG) in a refinery. To explore substantial recycling and reuse with zero liquid discharge practice in an oil refinery complex, here we propose a mathematical optimization framework. Synergistic integration of process knowledge with mathematical models has been used to develop an integrated superstructure of water network in refinery plants in this work. It is shown that the formulation of the optimization problem with both environmental and economic objectives and a systematic utilization of process based decision making helped to converge the formulated NLP problem of a large refinery complex to global optimal solution with significant minimization of computational efforts. The novel features of proposed optimization model include total dissolved solids (TDS) based segregation of waste water streams for distributed treatment plant (Low TDS and High TDS units), phenolics based segregation of sour water, and optimal cycles of concentration based operation of cooling tower for improvement in water foot print and reduction of waste water treatment load in a refinery complex. The application of proposed model for analysis of economic benefits and optimal water routing in the refinery water system has been shown on a representative case study of a large sized refinery complex. A new efficient routing scheme of treated waste water to boiler and cooling tower in optimal ratios has been identified in the resulted global solution of the optimization problem.

Keywords: Refinery water system, Sustainable water management, Integrated superstructure, simultaneous optimization, Distributed effluent treatment plant

533846 Life Cycle Carbon Footprint of Renewable Electricity Generation from Aspen Forest Harvest in Wisconsin, USA

Abstract:

Life Cycle Carbon Footprint of Renewable Electricity Generation from Aspen Forest Harvest in Wisconsin, USA

Olumide Winjobi[1, 2], Michelle Cisz[3], Colin Phifer[3], Rodney Chimner[3], Sigrid Resh[3], David R. Shonnard[1, 2]

1 Sustainable Futures Institute, Michigan Technological University, MI USA 49931

2 Chemical Engineering, Michigan Technological University, MI USA 49931

3 School of Forest Resources and Environmental Science, Michigan Technological University, MI USA 49931

The Renewable Portfolio Standard mandates increased production of electricity from renewable sources. There are large quantities of timber net growth in excess of removals in the Great Lakes region, for example greater than 2:1 in Michigan. The excess timber wood can be a promising feedstock to address the forest products industry while also providing renewable feedstock for energy production. In this study, we aim to evaluate the life cycle greenhouse gas (GHG) emission of renewable electricity produced from cogeneration of aspen biomass in Wisconsin. The typical harvest age for aspen is 40 years. We propose that one way to provide biomass for energy is to intensively harvest aspen (by shortening rotation cycles) and by collection of logging residues for electricity generation. A life cycle assessment model is used to quantify the cradle-to-grave GHG emission of renewable electricity, including soil and landscape C change, wood harvest, and combustion in fluidized bed boiler. The soil and landscape C change in this LCA model is simulated using the DAYCENT ecosystem model. The life cycle GHG emissions of renewable electricity will be compared with that of Wisconsin grid mix.

524309 A Life Cycle Cost Analysis of a Cattle-Based Anaerobic Digester Operation in Iowa

Abstract:

A Life Cycle Cost Analysis of a Cattle-based Anaerobic Digester Operation in Iowa.

Aui A.[b] and Wright M. M.[ab]

[a]Bioeconomy Institute

[b]Department of Mechanical Engineering

Iowa State University

Anaerobic digestion (AD) is an attractive and beneficial process for the conversion of agricultural, industrial and commercial waste into clean and useful renewable natural gas. Anaerobic digestion is a promising approach to achieving the economic and environmental goals set forth in the Iowa Energy Plan. This project aims to provide a life cycle cost assessment (LCCA) for Iowa anaerobic digesters and to identify opportunities for their profitable operation. Recent technological and policy developments have created opportunities to develop anaerobic digestion by providing an array of options to producers, farmers and businesses.

This study evaluates the different costs that affects the conversion of manure into biogas for heat, power, and renewable natural gas markets. It describes the capital and operating costs involved in the industrial operations of anaerobic digesters, while also evaluating the role of federal and state incentive programs in reducing commercialization risks. Finally, this project creates a business plan for stakeholders to evaluate the different opportunities and feedstocks such a cornhusk, rye and wheat, available for the development of anaerobic digesters in Iowa.

Current results indicate that an anaerobic digester attached to a 2400 head of cattle operation, that is co-digested with glycerin and cornhusk has 950 kW of generation capacity. The discounted cash flow rate of return (DCROR) analysis suggest that at a capital cost of $4.78 million, this project could achieve an internal rate of return of more than 5.50% at electricity prices of 12.0 ¢/kWh. By replacing cornhusk with rye and wheat, the internal rate of return is still in the upper range of 5%. Additionally, wheat has a higher net present value average in compared to the other feedstocks. The main contributors to the cost include capital, labor, and operating capacity. Solid digestate credit is an important source of revenue based on its C:N content. The role of tipping fees largely depends on the energy content provided by the feed. In particular, glycerin has been shown to enhance the biogas potential of animal manure. Future work will include investigating the aspects related to upgrading biogas, its environmental impacts, and exploring other major policies or incentives that influence an AD system.

531459 Environmental Impact Assessment for High Conversion Synthesis of <10 Nm Silver Nanoparticles Using Microwave Assisted Heating By Life Cycle Techniques

Abstract:

Environmental impact assessment for high conversion synthesis of <10 nm silver nanoparticles using microwave assisted heating by life cycle techniques

Adarsh Bafana[1], Shishir Kumar[1], Prasad Pawar[1], Sila Temizel-Sekeryan[2], Si Amar Dahoumane[3], Liv Haselbach[2], Clayton Jeffryes[1,*]

[1]Nanobiomaterials and Bioprocessing Laboratory (NABLAB), Dan F. Smith Department of Chemical Engineering, Lamar University, Beaumont, TX 77710

[2]Department of Civil and Environmental Engineering, Lamar University, PO Box 10024, Beaumont, TX 77710, USA

[3]School of Biological Sciences & Engineering, Yachay Tech University, Hacienda San José s/n, San Miguel de Urcuquí 100119, Ecuador.

*Corresponding Author: cjeffryes@lamar.edu

Abstract:

A microwave reaction to convert 99 ± 1% of Ag^+ to silver nanoparticles (AgNPs) of size <10 nm within 4.5 min with a specific production rate and energy input of 5.75 mg AgNP L^{-1} min^{-1} and 5.45 W mL^{-1} reaction volume was developed. The glucose reduced, and food grade starch stabilized particles remained colloidally stable with less than a 4% change in the surface plasmon resonance band at 425–430 nm at t > 300 days. TEM determined the size of AgNPs, while TEM-EDS and XRD verified elemental composition. The conversion was determined by inductively coupled plasma atomic emission spectroscopy (ICP-AES) and thermal gravimetric analysis (TGA). Additionally, the required silver to starch input mass ratio, 1.0:1.3, to produce colloidally stabilized AgNPs is significantly reduced compared to previous studies. The antibacterial activity of freshly prepared AgNPs and AgNPs aged >300 days was demonstrated against *E. coli* as determined by agar diffusion assays. This result, corroborated by spectrophotometric and TEM measurements, indicates long-term colloidal stability of the product. We attempted novel synthesis routes that reduce resource requirements and use benign chemicals, while maintaining control over their unique properties. Thus, we evaluated the potential environmental impacts of this method using Life Cycle Assessment (LCA) techniques which are used to assess the environmental impacts of a product's life through all the stages from raw material extraction to disposal/ recycling. GaBi 6.0 software was used to carry out the life cycle impact assessment on a declared unit of 1 kg of 3.0 ± 1.2 nm diameter AgNPs. The results indicate that the impacts are predominantly on acidification (AP), human health particulate air (HHAP) and human toxicity non-cancer (HTNCP) potentials. These impacts are mainly from the production of silver metal and electricity used. The starch and glucose used to produce AgNPs of 3.0 ± 1.2 nm appear to be environmentally benign as they have negligible environmental impacts.

535289 Anhydrous Bio-Ethanol Production: Life Cycle Analysis of Distillation and Dehydration Steps

Abstract:

Bio-ethanol production is a rapidly growing market due to the possible uses of the anhydrous product in energy generation. This production however, is weighed down by the energetic demands and the environmental effects of the separation and dehydration steps needed to deliver fuel-grade bio-ethanol.

This paper examines the distillation and dehydration steps of a bio-ethanol production plant complementing a sugar mill. The distillation step is optimized by performing rigorous energy analysis and heat integration on the process utilizing the resources made available by the sugar mill, while pervaporation is used in the dehydration step. The user-defined module is integrated in commercial process simulators to study the environmental and energetic impact of different process configurations. Different schemes meeting equivalent performance standards in terms of purity and recovery are examined in this paper.

Particular attention is given to energetic demands as comparison metrics. Life Cycle Analysis of the optimum process configuration is further used to identify critical sources of environmental impact.

527181 Planning, Design and Operation of Sustainable and Efficient Multi-Product Rice Value Chains Using Multi-Objective Spatio-Temporal Optimisation

Abstract:

The rice value chain, especially in the Philippines, has a lot of potential for improvement in terms of sustainability and efficiency, as well as additional value creation by effectively utilising the by-products from rice production. Rice straw and rice hull that pose disposal problems can be used as low-carbon sources of energy. Rice bran and rice hull ash, which are currently considered, as wastes are finding uses in the animal feed and cement industries, respectively. Generating multiple products will lead to a complex decision-making process in supply chains especially in the case of rice crop which is vital in the Philippinesâ€™ food security. Thus, systematic planning techniques are needed to address important decisions along the stages in rice value chains.

In this study, a mixed integer linear programming (MILP) model was developed for planning, design and operation of multi-product rice value chains. A value chain considers all activities from planting, cultivating and harvesting of the crop to distributing the products to customers. In order to capture the spatial-dependencies of the problem, such as the candidate locations for rice plantations and processing facilities, and location of demands, the Philippines is represented as a grid of 50 km squares. The model considers a long planning horizon, out to 2050, in order to model the staged investment in and retirement of technologies. It also captures seasonal variations in rice yield/productivity and changes in demands for rice and the by-products from rice production. The model determines interdependent decisions such as where to locate the plantations and processing facilities, when to invest in them and what size/capacity; what products to produce, how to transport and store resources, centralised or distributed production, among others. Different objectives are considered such as maximisation of profit and minimisation of GHG emissions. Pareto sets are generated in order to determine optimal solutions that represent a balance between economic gain and environmental protection. Case studies using the Philippine rice sectors are used to illustrate the capability of the model for providing useful insights on planning, design and operation of multi-product rice value chains.

References:

[1] B.A. Excode (2016). Rice production and consumption trends in the Philippines from 2000 to 2015. Yuchengco Center, De La University; Philippines.

[2] C.T. Briones (Ed.) (2016). The PhilRice strategic plan 2017-2022. Philippine Rice Research Institute; Philippines.

[3] A.B. Mataia, A.B. Mataia, R.G. Manalili, J.C. Beltran, B.M. Catudan, N.M. Francisco and A.C. Flores (2016). Analysis of the rice value chain in the Philippines. Philippine Rice Research Institute; Philippines

[4] R. Sims, A. Flammini, M. Puri, S. Bracco (2015). Opportunities for Agri-Food Chains to become energy-smart. Food and Agriculture Organization of the United Nations and the United States Agency for International Development; United States of America.

[5] S. Samsatli, N.J. Samsatli (2018). A multi-objective MILP model for the design and operation of future integrated multi-vector energy networks capturing detailed spatio-temporal dependencies. Applied Energy. DOI: 10.1016/j.apenergy.2017.09.055.

[6] S. Samsatli, N.J. Samsatli, N. Shah (2015). BVCM: a comprehensive and flexible toolkit for whole-system biomass value chain analysis and optimisation - mathematical formulation. Applied Energy, 147, pp. 131-160.

527334 Kinetic Parameter Estimation for Electrocatalytic Hydrogenation of Model Compounds Derived from Fast Pyrolysis of Biomass

Abstract:

According to the DOEâ€™s Billion-ton report published in 2016, there is not enough biomass (that can be produced at less than $60/ton) in the U.S. to meet the annual energy demands for the U.S. transportation sector. Biofuel made using one billion tons of biomass, available by 2040, would not result in the complete displacement of petroleum-derived transportation fuels consumed in 2017. In this context, carbon and energy efficiency are paramount for maximizing the scale of displacement. Cellulosic fermentations to produce ethanol are inherently carbon and energy inefficient, as biomassâ€™s lignin is not converted into liquid fuels and one-third of the carbohydrate carbon is lost as carbon dioxide during fermentation.

An alternative biofuel production strategy is proposed that combines fast pyrolysis with localized electrocatalytic hydrogenation (ECH) and centralized petroleum refining to produce a gasoline-like fuel. In this regard, a comparative study of the energy, mass, and carbon flux through these two processes was performed. Pyrolysis-electrocatalysis processes potentially overcome the carbon and energy inefficiencies that limit cellulosic ethanol. ECH of bio-oil, the liquid product of fast pyrolysis, at a localized depot is a key component of the proposed biofuel production scheme. ECH utilizes renewable electricity from wind turbines or solar photovoltaics to split water and make *in situ* hydrogen and subsequently catalyze the reduction of reactive aromatic compounds. As bio-oil is a complex mixture of hundreds of organic compounds, ECH has been primarily studied by analyzing individual representative compounds present in bio-oil, such as phenol, syringol, and guaiacol. An unwanted side reaction associated with ECH is the hydrogen evolution reaction, which adversely impacts the Faradaic efficiency. Faradaic and voltage efficiency impacts the number of membrane-electrode stacks required to fully saturate bio-oil, therefore a mechanism for determining the overall reaction rate is needed. In this regard, a kinetic model has been developed to estimate the kinetic rate constants for the electrochemical and catalytic surface reactions for the ECH of phenol as a model bio-oil compound. Experiments were performed in a two compartment cell with the cathodic and anodic chambers separated by a proton exchange membrane. Ru/ACC was employed as the catalytic cathode while a Pt wire was used as the anode. Bulk species concentration data were collected at regular intervals of time to estimate the reaction rate constants by means of non-linear sequential parameter estimation in Matlab. The developed model includes equations to characterize the diffusion of bulk species to the catalyst surface as well as equations to describe the adsorption, desorption, and surface reactions occurring on the cathode. The next step is to perform the same analysis for other representative bio-oil compounds to better characterize the electrocatalytic hydrogenation of biomass-derived bio-oil. This leads to a better understanding of the process and serves as a crucial step towards scale-up of ECH as a keystone of novel pyrolysis-based bioenergy systems.

538411 The Search for Sustainability in an Integrated Economic-Ecologic-Social Model through the Use of Feedback Loops

Abstract:

The Search for Sustainability in an Integrated Economic-Ecologic-Social Model Through the Use of Feedback Loops

Pablo T. Rodriguez-Gonzalez[a], *Vicente Rico-Ramirez*[a*], *Ramiro Rico-Martinez*[a] *and Urmila M. Diwekar*[b]

[a]Tecnologico Nacional de Mexico en Celaya, Departamento de Ingenieria Quimica, Av. Tecnologico y Garcia Cubas S/N, Celaya, Guanajuato, Mexico 38010

[b]Vishwamitra Research Institute, 368 56th Street, Clarendon Hills, IL, 60514, USA

pbl_tet@hotmail.com, vicente@iqcelaya.itc.mx, ramiro@iqcelaya.itc.mx,

urmila@vri-custom.org

Abstract

It has been recognized that the study of human-ecosystems sustainability must consider an integrated point of view. Current developments on sustainability theory intend to integrate social and economic responses to environmental problems, including the simultaneous analysis of various human systems (finance, politics, production, energy, transportation, etc.). The goal is to support the understanding of the interactions between human activities and natural resources as well as the effect of economic and social issues on the sustainability of such systems. In the end, it is expected that such basic understanding will allow the development of systematic approaches that can be used to provide sustainable policy guidelines for complex integrated systems.

Literature reports considerable modeling efforts to represent complex integrated ecosystems. As a recent contribution, we presented a sustainability assessment of an enhanced compartmental ecological-economic-social model that proposes the incorporation of social inequity as a key concept in social sustainability. Such concept was introduced by including two human compartments in order to represent the differences in income, demand and consumption levels of the population; in addition, external stochastic influences were incorporated through Ito processes (time dependent uncertainties) on demographic parameters.

This work represents an extension of such work. In particular, we try to analyze the effect of incorporating feedback loops as a mechanism to support the resilience of a system that is being perturbed by external-internal stochastic factors which may drive the system to undesirable states (a system collapse). The feedback loop under study is related to the demographic parameters, since we expect that such parameters can indeed be modified by human response. Our results indicate a favorable impact of such loop in the resilience and sustainability of the system.

Keywords: *Sustainability, modeling, feedback loops*

543980 Water Pinch Points in Thermoelectric Power Generation

Abstract:

Sustainable and affordable energy and water resources require understanding regions where water availability is projected to limit current thermoelectric power generation technology. The National Energy Technology Laboratory is carrying out systems analyses to understand water pinch points in thermoelectric power generation in the U.S. The analyses utilize GIS data collected by the United States Geological Survey and results from a Sandia National Laboratory study. Maps have been developed to identify regions where water availability affect thermoelectric power generation technology option (e.g. need for dry or hybrid cooling or different minimum water use concepts). The results are illustrated by projected 2030 estimates in the Southwest. Four watersheds have planned thermoelectric power growth in HUC8 watersheds which are projected to have zero water availability in 2030. Agriculture and thermoelectric power accounted for 76% of freshwater withdrawals in the U.S. in 2015. While agriculture increased water withdrawals 2% from 2010 to 2015, thermoelectric power decreased withdrawals 18%. Locations where water resources are required substantially for both agriculture and thermoelectric power are include California, Colorado, Arkansas, and Mississippi.

523181 Process Design Optimization for *Saccharina japonica* Based Biorefinery: A Superstructure Based Approach

Abstract:

In this study, we address the problem of optimizing *Saccharina japonica* based biorefinery coproducing bioethanol and protein rich solids. The proposed superstructure consists of multiple design alternatives for feed pretreatment, solid processing, and enzymes manufacturing. The goal was to find optimal plant design by formulating robust techno-economic model i.e., net present value by considering mass and energy balances, capital and manufacturing constraints. The resulting superstructure optimization model was a mixed integer nonlinear programming (MINLP) problem and implemented in GAMS to investigate two separate optimization scenarios (maximize net present value and maximize yield). Separable programming was used to efficiently compute the solution by approximating the initial MINLP problem into a mixed integer linear programming (MILP) problem. The results indicated that the production of bioethanol and protein rich solids from *Saccharina japonica* is economically viable. The minimum ethanol selling price (MESP) of $1.97/gal whereas the net present value of $61.5 million was obtained for optimal design. Sensitivity analysis of MESP and yield was performed to identify potential economics improvement. The developed framework can be efficiently used to identify optimal processing pathway under different objective functions.

Acknowledgements

This research was respectfully supported by Engineering Development Research Center (EDRC) funded by the Ministry of Trade, Industry & Energy (MOTIE). (No. N0000990).

522618 Optimization of Macroalgae Based Biorefinery Producing Fuel and Chemicals with Zero Carbon Emissions Potential

Abstract:

Process flowsheet optimization of seaweed-based biorefinery was performed to find the optimal processing route to produce bioethanol, protein rich solids, succinic acid, and microalgae. Biochemical processes such as fermentation produce considerable amount of carbon dioxide as a byproduct. This immense amount of carbon dioxide is a threat for global warming. Products like succinic acid and microalgae use carbon dioxide as a precursor, which may reduce the greenhouse gas emissions from biorefineries. Most promising industrial design alternatives were selected to build a superstructure. The proposed superstructure contains thirty design alternatives (including different pretreatment steps, fermentation, carbon dioxide mitigation alternatives, and different separation technologies). Furthermore, four design alternatives with respect to products and chemicals were also considered. A techno-economic model was formulated by considering mass and energy balances, capital and manufacturing constraints. The resulting model was mixed integer nonlinear programming (MINLP) and solved in GAMS software. Global optimization was performed by using LINDO solver. The aim was to determine optimal biorefinery design which produces minimum amount of carbon dioxide as well as high economic potential. Numerical results were computed for three separate optimization objective functions namely maximize NPV, maximize yield, minimize carbon emissions. Sensitivity analysis was performed to identify the major cost drivers to process. The proposed framework can be utilized to investigate a range of important questions: what is the best process design for minimum carbon dioxide emissions? What is the best manufacturing strategy of biofuel and chemicals to maximize NPV?

Acknowledgements

This research was respectfully supported by Engineering Development Research Center (EDRC) funded by the Ministry of Trade, Industry & Energy (MOTIE). (No. N0000990).

533859 Two Different Synthetic Approaches to Obtain Levan Nanoparticles

Abstract:

Levan is a fructose biopolymer with some useful properties related to different biomedical applications. For instance, that polymer shows cell recognition as well as the cells overexpress GLUT5, which is a fructose transporter overexpressed in different carcinoma cell lines, such as colorectal or liver lines (1). In addition, levan forms nanoparticle when is put in contact with water, due to its amphiphilic properties. Therefore, those properties make possible to create nanoparticles of this polymer with specific cancer recognition (2).

Levan is obtained with the transfructosylation reaction in which an enzyme (levansucrose) breaks the sucrose bond to produce glucose and fructose, to later polimeryze the fructose monomers. This reaction can be carried out by microorganism, or by using directly the isolated enzyme in a cell-free system.

This work aims to compare levan nanoparticles that are obtained from the above explained methodologies. According to the results, cell-free systems provide levan nanoparticles with a smaller particle size distribution in comparison with the particles from microorganisms. Furthermore, nanoparticles produced by cell-free systems have more glucose residues, although its molecular weight is lower compared with the values from the polymer obtained from microorganisms. Both systems were characterized by TEM, TGA, FTIR and its cell internalization process was studied in different cell lines. Both nanoparticle systems were non- toxic. They are taken up by the cells by an endocytic mechanism.

Finally, levan from the cell-free system is obtained faster and with a higher yield than the polyfructose from microorganisms due to the reduction of steps in its downstreaming.

536814 Inhibitory Effect of Hemicellulose Hydrolysates on Glucose Transport across Cell Membrane in Yeast Fermentation

Abstract:

Biomass pretreatment generates a wide range of toxic compounds from lignin and sugars degradation. These toxic compounds significantly inhibit cell growth and production rate in microbial fermentation. Phenolic aldehydes and ketones have been suggested to the most toxic compounds to microbes. However, the mechanism of inhibition has not been elucidated. Glucose transport across the yeast cell membrane typically is mediated by specific membrane carriers. The effect of hemicellulose hydrolysates on glucose transport and cell-free fermentation with yeast was studied. Unexpectedly, it was observed that cell-free yeast extracts consumed all the glucose in the undetoxifed hydrolysates within 12 h and produced 9.5 g/l of ethanol, but the intact yeast could not ferment the undetoxifed hydrolysates. It was also noticed that no intracellular glucose was detected in yeast after incubation with the undetoxifed hydrolysates, while 46 mg/g intracellular glucose was presented inside of yeast after incubation with the detoxified hydrolysates. The volumetric ethanol productivity was correlated very well with the intracellular glucose concentration. This study indicated that the toxic compounds mainly inhibit the glucose transport step, but not the fermentation step in yeast or bacteria.

525719 Biochar Enhanced Alcohol Production from Syngas By *Clostridium Carboxidivorans*

Abstract:

Biological production of ethanol and butanol from syngas is an alternative route to traditional ABE (acetone-butanol-ethanol) fermentation. Ethanol and butanol can be mixed with gasoline. Butanol can be converted to jet fuel via a hydrogenation process. Biochar made by gasification and pyrolysis is a carbon-rich material with high surface area, pH buffer capacity and cation exchange capacity (CEC). Biochar use in biofuel production has not been extensively reported. This study examined the properties of biochar made from switchgrass (SGBC), forage sorghum (FSBC), red cedar (RCBC) and poultry litter (PLBC), and their effects as a medium additive on ethanol and butanol production by *Clostridium carboxidivorans*. Experiments were performed in 250 mL bottle reactors with a 50 mL working volume at 37 °C for 360 h. Syngas mixture containing $CO:H_2:CO_2$ (40:30:30, by volume) was fed every 24 h. Results showed that PLBC and SGBC enhanced ethanol production by 90% and 73%, respectively, and butanol production by fourfold compared to standard rich yeast extract medium without biochar (control). FSBC and RCBC also enhanced ethanol and butanol production but lower than PLBC and SGBC. Total CO and H_2 utilization was also enhanced in biochar media compared to control medium. PLBC had the highest concentrations of Na, K, Ca, Mg, P and Mn, pH buffering capacity and CEC compared to other biochars tested, which possibly contributed to its highest enhancement of ethanol and butanol production. This study showed the potential for biochar use as a medium substitute for expensive nutrients to enhance biofuel production via syngas fermentation, which increases the feasibility of the process.

530305 Butanol Production By Fermentation with New Symbiotic Strain TSH06

Abstract:

In spite of the increasingly serious problem of global energy shortage and environmental deterioration, butanol has been developed as a promising renewable energy because butanol has some characteristics better than that of ethanol. Traditional ABE fermentation use strict anaerobic strain to produce butanol. In our research, a new symbiotic strain TSH06 was isolated to produce butanol, which could produce butanol under micro-aerobic condition. The results indicated that ratio of acetone: butanol: ethanol reached 1:8:1, butnaol yield reached 0.31, and butanol concentration reached 19.2g/L, respectively in batch fermentation after optimized technology. Furthermore, butanol fermentation integrated with gas stripping technology was also investigated.

538651 Methane Conversion Using Catalytic Melts to Produce Separable Carbon and Hydrogen or Electrical Power

Abstract:

Methane conversion without the co-production of carbon dioxide remains an important challenge for the production of chemicals and electrical power. Partially oxidizing methane to carbon instead of carbon dioxide prevents greenhouse gas emission and allows for the carbon to be stored or used as a valuable solid product. Catalytic processes to convert methane to carbon and products have been challenged due to carbon deposition and fouling; however, molten catalysts have the potential to overcome this challenge because carbon can float and be continuously removed. Recently, we reported liquid metal catalysts to convert methane to carbon and hydrogen without deactivating[1]. In this system, active metals for methane pyrolysis are dissolved in inactive low-melting temperature metals to produce stable molten catalysts, and density function theory calculations indicated that the active metal was negatively charged and its charge correlated with experimentally observed activity. The carbon produced is soluble in the melt and precipitates as graphitic carbon flakes.

These concepts are also used for a catalyst for the partial combustion of methane, whereby methane and oxygen are converted to carbon, steam, and heat (reaction 1).

$CH_4 + O_2$ -> C + 2 H_2O DG° = 424 kJ/mol, DH° = 497 kJ/mol (1)

Electrical power can be produced from the heat and steam. A molten salt is used to catalyze the reaction and separate the carbon, and a steam cycle is proposed to co-generate power. Selectivity of 90% at 56% methane conversion in a 12 cm bubble column reactor is reported. The reaction mechanism is investigated by feeding intermediates, altering salt composition, and determining apparent activation energies. A kinetic model will be presented which is used to explain the product distribution when reaction intermediates are fed. Carbon is characterized by Raman, SEM, and EDX which suggest carbon black is formed and several mechanisms of removal are discussed.

1 Upham, D. C. *et al.* Catalytic molten metals for the direct conversion of methane to hydrogen and separable carbon. *Science* **358**, 917 (2017).

526296 NiCe@SiO$_2$ Multi-Yolk-Shell Nanotube Catalyst for Tri-Reforming of Methane

Abstract:

Greenhouse gas emissions have been continuously increasing over the decades, with carbon dioxide (CO_2) being the primary greenhouse gas making up about 80%, followed by methane (CH_4), composing of 10%. Technologies such as tri-reforming of methane (TRM) can mitigate CO_2 emissions with generating syngas for the production of valuable chemicals. The TRM combines dry reforming, steam reforming, and partial oxidation of methane, thereby can directly utilize flue gas from a combustion source without requiring any gas separation. Nickel-based catalysts are widely studied for TRM, as they can show high activity with equilibrium CO_2 and CH_4 conversions reaching up to 80% and 97% at 750 °C, respectively. Despite the high activity of Ni-based catalysts, different morphologies of the catalyst such as yolk-shell structure have been proposed, since they can maintain high resistance to coke deposition and sintering due to the confinement effect, and provide long term stability.

In this work, NiCe@SiO$_2$ multi-yolk-shell nanotube catalysts have been prepared at different yolk and tube sizes, and their TRM activity has been evaluated at different oxidizer ($CO_2 + H_2O + O_2$) to methane (O/M) feed ratios. At high O/M ratio, NiCe@SiO$_2$ consisting of small yolk particles lost its activity, possibly due to the re-oxidation of active Ni components. On the other hand, the catalyst with bigger yolk particles showed stable activity without deactivation by re-oxidation. At low O/M ratio condition, an opposite behavior was observed for the small and big yolk particles. The catalyst with small yolk particles maintained its initial activity with high resistance to carbon deposition whereas NiCe@SiO$_2$ with big yolk particles deactivated fast under the low O/M ratio. The stable activity observed on the small yolk particles could be explained by the fast oxidation ability of the small yolk particles, which can enhance oxidation of carbon deposits on the catalyst. A detailed analysis of the fresh and spent samples characterized through STEM, XRD, H_2-TPR, TGA and Raman will be discussed in the presentation. These results confirm that the morphology of the NiCe@SiO$_2$ yolk-shell nanotube catalyst can be engineered for maintaining a high TRM activity at different gas feed concentrations.

537441 Dynamics and Mechanism of Carbon Filament Formation during Methane Reforming on Supported Nickel Catalysts

Abstract:

Inactive carbon deposits can form during catalytic CH_4 reforming reactions on Ni-based catalysts, leading to commercial processes that must use CO_2/CH_4 or H_2O/CH_4 ratios above those required by stoichiometry in order to suppress coking. These carbon deposits hinder CH_4 reforming reactions by blocking pores and can lead to catalyst disintegration and elution as fines from reactors [1]. Carbon formation is generally considered to proceed via the following sequence: (a) dissociative chemisorption of CH_4 on the metal surface, (b) dissolution of carbon atoms into the metal, (c) bulk diffusion of carbon atoms through the metal, and (d) precipitation of carbon atoms at the metal-carbon interface to form a filament at the supra-equilibrium chemical potentials required for nucleation of the solid phase. Bulk diffusion of carbon through the metal is considered the rate limiting step in this scheme and is driven by a gradient in the thermodynamic activity of surface carbon between the side exposed to reactants $(a_{C^*})_{gas}$ and that in contact with the filament $(a_{C^*})_{fil}$ [2]. Carbon deposition rates and the morphology of the carbon deposits thus depend on $(a_{C^*})_{gas}$, which is set by the gas composition for equilibrated mixtures, but which depends on the kinetics of formation and removal of C-atoms during steady-state catalysis and thus on the structure and composition of the active surfaces. Most previous studies have investigated carbon formation using equilibrated binary gas mixtures, such as CO-CO_2 or CH_4-H_2, and developed relationships for the carbon activity under equilibrium conditions [3-4]; few studies have systematically examined the kinetics of carbon deposition for CH_4-CO_2-H_2O-CO-H_2 gas mixtures during steady-state catalysis [5]. The conditions that lead to carbon formation during CH_4 reforming before equilibration is achieved and the nature of the requisite driving forces have remained speculative and uncertain. This study describes rate data and their mechanistic interpretation for the formation of surface carbon during CH_4 reforming and develops the appropriate expression for $(a_{C^*})_{gas}$ over at a wide range of conditions.

Experiments were carried out on 7% wt. and 15% wt. Ni/MgO samples prepared by incipient wetness impregnation of MgO powders with aqueous $Ni(NO_3)_2 \cdot 6H_2O$ (Alfa, 99.9%), as reported previously [6]. Samples were treated in H_2 (Airgas, UHP, 50 cm^3 g^{-1} s^{-1}) at 1123 K (0.167 K s^{-1}) for 3 h before their use in reactions. The dynamics of carbon formation on these catalysts were measured during reforming reactions using a tapered element oscillating microbalance under controlled reactive environments at 793-973 K; changes in the oscillation frequency of the tapered quartz element are correlated to changes in the mass of the catalyst sample due to deposited carbon. The thermodynamic carbon activity $(a_{C^*})_{gas}$ was varied through systematic changes in the inlet CH_4, CO_2, CO, H_2, and H_2O concentrations and in the residence time and chemical conversion for a given inlet gas composition. The effluent was continuously monitored by on-line mass spectrometry to determine CH_4 reforming rates. High-resolution transmission electron microscopy (TEM) was used to characterize Ni particle sizes and the morphology of any carbon deposits formed.

The mechanistic interpretation of CH_4 reforming rates on Ni-based catalysts, provided by extensive isotopic and theoretical inquiries [6], lead to $(a_{C^*})_{gas}$ values that are proportional to either $P_{CO}P_{CH4}/P_{CO2}$ (χ) and $P_{H2}P_{CH4}/P_{H2O}$ (φ) at CH_4 reforming gas mixtures far from equilibrium. The equivalence of these two ratios reflects their proportional values through the equilibrium constant for water-gas shift, a reaction that is at equilibrium during CH_4-CO_2 and CH_4-H_2O reforming reactions. For each Ni/MgO sample and temperature, carbon deposition rates depended only on these concentration ratios at the conditions of our study and increased with $(a_{C^*})_{gas}$. Three carbon growth regimes are evident from TEM analysis: minimal carbon deposition (regime I), filamentous carbon growth (regime II), and metal particle encapsulation by carbon layers (regime III). These regimes also aligned with trends noted in measured carbon deposition rates. TEM detected only surface carbon patches on 5.4 nm Ni crystallites in regime I ($\chi < 1.1$ kPa); these patches minimally covered the Ni particles. CH_4 reforming reaction rates remained constant with time, and carbon formation rates were negligible for these low χ values. Hollow carbon nanotubes with diameters similar to those of the metal crystallites formed in regime II ($1.1 < \chi < 4.2$ kPa). CH_4 reforming and carbon formation rates also remained constant with time throughout this regime, even after the deposition of $C/Ni_{surface}$ ratios >800 and the extensive formation of filamentous carbon structures evident from TEM. The filaments do not hinder reaction rates in this regime because the rate of dissolution and diffusion through the Ni particle remains equal to the rate of C-addition to the filaments, thus continuously unblocking active Ni surfaces as chemisorbed carbon forms. Encapsulation of the metal particle in onion-like carbon shells was evident in micrographs for regime III ($\chi > 4.2$ kPa), during which reforming and carbon formation rates both decreased with time after an initial induction period. Such induction periods are characteristic of nucleation-growth processes that require supersaturation levels of carbon at metal surfaces to nucleate a new phase. These onion-like layers in regime III appear to reflect the simultaneous nucleation of a carbon phase at several points in Ni particles, thus breaking the symmetry and preferential diffusion direction present on particles with a single nucleation point, which contain a growing filament that continuously deplete the C-pool and maintain the carbon activity below supersaturation levels. Carbon formation rates were consistently higher on 11.1 nm than on 5.4 nm Ni particles (by ~ 0.3 C-atoms surface-Ni^{-1} s^{-1} ; 873 K) at each given value of χ (or φ), apparently because of the more facile nucleation of carbon in larger particles and lower carbon activity for filaments with larger diameters, which tend to form on larger particles [7]. Carbon deposition rates decreased with increasing temperature for a given value of χ (or φ), reflecting presumably lower concentrations of surface carbon and surface carbon-forming precursors at higher temperatures. Removal of carbon from the catalyst by the reverse of their formation was detected at $\chi <$ 0.07 kPa on samples with carbon deposits formed during deposition experiments.

These results are consistent with the diffusion-limited growth of carbon filaments, driven by a gradient in carbon activity between the gas-side Ni surface and the precipitated carbon filament. Carbon deposition does not occur until the carbon activity in the Ni particle exceeds that required for nucleation at the metal-filament interface, which depends on carbon activity of the filament and thus on its diameter and that of the Ni particle. The carbon activity was described by the partial pressure ratio χ or φ, and carbon formation rates were a single valued function of these parameters on each supported Ni catalyst and at each temperature. Carbon deposition rates increased with Ni crystallite size because of the lower barrier for nucleation and lower carbon activity in the filament in these particles. These results provide a framework for determining CH_4 reforming conditions at which $(a_{C^*})_{gas}$ can be kept below the threshold value that leads to carbon formation. It also provides predictive guidance for the design of CH_4 reforming catalysts that can locally control the values of χ and φ by imposing concentration gradients and by spatially preventing the formation of carbon filaments. Such properties can be conferred by metal nanoparticles that reside within void environments that impose such gradients, and which also prevent nanoparticle growth even at the high temperatures of CH_4 reforming catalysis [8]. Such small clusters would impose higher carbon activity thresholds for nucleation and growth as a result of their small diameters and their confined pore environment. Diffusional constraints for reactants and products in these confining materials are also expected to decrease $(a_{C^*})_{gas}$ in CH_4-H_2O reforming by increasing their intracrystalline H_2O/CH_4

concentration ratios because of the faster diffusion of H_2O than CH_4 within materials that have pore diameters similar in size to those of these molecules.

[1] Rostrup-Nielsen, J. R., Catalytic Steam Reforming. In *Catalysis: Science and Technology*, Anderson, J. R.; Boudart, M., Eds. Springer Berlin Heidelberg: 1984; pp 1-117.

[2] Holstein, W. L., *J Catal.* 152 (1995) 42-51.

[3] Audier, M., Coulon, M., *Carbon.* 23 (1985) 317-323.

[4] Debokx, P. K., Kock, A. J. H. M., Boellaard, E., Klop, W., Geus, J. W., *J Catal.* 96 (1985) 454-467.

[5] Jablonski, G. A., Geurts, F. W. A. H., Sacco, A., *Carbon.* 30 (1992) 99-106.

[6] Wei, J. M., Iglesia, E., *J Catal.* 224 (2004) 370-383.

[7] Rostrup-Nielsen, J. R., *Steam Reforming Catalysts.* Teknisk Forlag: 1975.

[8] Otto, T., Zones, S. I., Iglesia, E., *J Catal.* 339 (2016) 195-208.

Financial support from BP as part of the Methane Conversion Cooperative Research Program at the University of California at Berkeley and from MEXUS grant CN-02-76 are gratefully acknowledged.

534328 Carbonate-Catalyzed CO_2 Hydrogenation

Abstract:

CO_2 hydrogenation is a potentially valuable strategy for upgrading waste CO_2 to fuels and chemicals, but industrially practiced CO_2 hydrogenation is currently limited to the production of low-value C_1 products. To address this gap, new CO_2 hydrogenation chemistry must be developed which is capable of selective Câ€“C bond formation to yield higher value, multi-carbon products. We have recently reported a novel CO_2 hydrogenation process promoted and/or catalyzed by alkali carbonate (M_2CO_3) salts[1]. It was found that under elevated pressures and temperatures, and in the presence of small amounts of water, M_2CO_3 hydrates can promote CO_2 hydrogenation to formate, oxalate, and other C_{2+} carboxylate salts in the absence of solvents and transition metal catalysts. Notably, small amounts of M_2CO_3-catalyzed ethylene production are also observed.

Herein, we discuss recent developments in the carbonate-promoted CO_2 hydrogenation process. Emphasis is placed on maximizing catalytic production of ethylene, a high-volume precursor in polymer manufacturing. Impacts of both catalyst and reactor design are considered, and the resulting mechanistic insights are discussed. Dispersion of the M_2CO_3 catalyst onto high surface area metal oxide support improves catalyst performance and enables the use of more traditional packed-bed reactor configurations. Switching from a batch reactor to a packed-bed flow reactor design allows for the hydrogenation reaction to be studied under more practically relevant conditions and with finer process control and minimized background reactivity. Supported M_2CO_3 catalysts are characterized using electron microscopy, gas sorption, powder X-ray diffraction, and thermal analysis techniques, and the resulting insights into their physical and chemical properties are related back to observed CO_2 hydrogenation reactivity trends.

[1] Aanindeeta Banerjee and Matthew Kanan. Carbonate-promoted hydrogenation of carbon dioxide to multi-carbon carboxylates. *Submitted.*

528687 Bifunctional Catalysts for CO_2 Conversion to Plastics, Chemicals and Fuels

Abstract:

Recent advances in CO_2 hydrogenation have prompted investigations into engineering bifunctional catalysts that decouple CO_2 activation from hydrocarbon chain growth. By designing tandem catalysts with distinct active sites for CO_2 and CO hydrogenation, researchers can precisely manipulate the reaction mechanism, thus enabling high selectivity to light olefins with minimal undesirable side-products (CO and CH_4). Through down-stream processing, these light olefins can be oligomerized into plastics, chemicals and fuels, creating an economically attractive route to convert CO_2 into valuable products.

Transition metal carbides (TMCs) have been shown to outperform traditional precious bimetallic catalysts and are highly active and selective for the reverse water-gas shift (RWGS) reaction, the first step of CO_2 hydrogenation to light olefins.[1-2] By modifying TMCs with alkali metal promoters, the CO yield increases significantly and approaches the thermodynamic limit at 300 °C. These experimental results are supported by density functional theory (DFT) calculations, which show enhanced CO_2 adsorption and reduced CO_2 dissociation barriers on the alkali metal-promoted catalyst, compared to the pristine carbidic surfaces.[3]

For the second step of CO_2 hydrogenation, Fischer-Tropsch to olefins (FTO) experiments are conducted over metal oxide nanoparticles deposited within hollow zeolite nanoreactors to precisely control hydrocarbon selectivity.[4] Previous success for CO hydrogenation to olefins has been achieved over a similar $ZnCrO_x$/MSAPO catalyst, which activates CO and H_2 over the Zn-Cr mixed-oxide, while hydrocarbon chain growth is confined within the acidic MSAPO pores.[5] This type of tandem catalyst combining metal oxides with zeolites has been attempted for CO_2 hydrogenation through a methanol (MeOH) intermediate, but the high selectivity to CO (~50%) is generally unavoidable because of the instability of MeOH at the high temperatures required to activate CO_2.[6]

To better understand CO_2 hydrogenation tandem catalysts and improve selectivity to C_2-C_4 olefins, structure-property trends must be developed to relate zeolite properties (Si/Al ratio, pore size and topology) to CO and CO_2 hydrogenation activity and selectivity. Further research in this area will promote future breakthroughs into highly active, selective and stable catalysts for CO_2 hydrogenation to olefins.

References

[1] M. D. Porosoff, X. Yang, J. A. Boscoboinik, J. G. Chen, *Angewandte Chemie International Edition* **2014**, *53*, 6705-6709.

[2] S. Posada-Perez, F. Vines, P. J. Ramirez, A. B. Vidal, J. A. Rodriguez, F. Illas, *Physical Chemistry Chemical Physics* **2014**, *16*, 14912-14921.

[3] M. D. Porosoff, J. W. Baldwin, X. Peng, G. Mpourmpakis, H. D. Willauer, *ChemSusChem* **2017**, *10*, 2408-2415.

[4] S. Li, A. Tuel, D. Laprune, F. Meunier, D. Farrusseng, *Chemistry of Materials* **2015**, *27*, 276-282.

[5] F. Jiao, J. Li, X. Pan, J. Xiao, H. Li, H. Ma, M. Wei, Y. Pan, Z. Zhou, M. Li, S. Miao, J. Li, Y. Zhu, D. Xiao, T. He, J. Yang, F. Qi, Q. Fu, X. Bao, *Science* **2016**, *351*, 1065-1068.

[6] Z. Li, J. Wang, Y. Qu, H. Liu, C. Tang, S. Miao, Z. Feng, H. An, C. Li, *ACS Catalysis* **2017**, *7*, 8544-8548.

538949 Support and Promoter Effects on the Activity of Transition Metal Phosphide Catalysts for CO and CO_2 Hydrogenation

Abstract:

The development of new catalysts for converting synthesis gas ($CO/CO_2/H_2$) and CO_2/H_2 mixtures can unlock diverse feedstocks for the production of conventionally petroleum derived chemicals. Synthesis gas can be produced from biomass, natural gas and other sources and CO_2 can be collected from combustion and industrial processes to be hydrogenated to high value products. The use of CO_2 and renewable hydrogen or biomass-derived syngas, can generate fuels and chemicals which approach carbon neutrality. We have investigated transition metal phosphide catalysts for the hydrogenation of CO, CO_2 and their mixtures and have developed catalysts with high stability under changing feed conditions. Molybdenum phosphide catalysts were found to be active for the synthesis of a range of products including methane, hydrocarbons, methanol and higher alcohols. Promoter and support effects were significant in the control of the product distribution. The effects of support and promoters on both the MoP phase formation as well as catalytic activity were investigated using both ex-situ and operando characterization techniques such as XPS, XRD, and XAS.

537010 Tuning Ni-Catalyzed CO_2 Hydrogenation Pathways Via Ni-Ceria Support Interactions and Ni-Fe Bimetallic Formation

Abstract:

Conversion of CO_2 to CO for use as a Fischer-Tropsch feedstock provides an attractive option for offsetting CO_2 emissions.[1,2] Non-precious Ni catalysts are active for the reduction of CO_2 by H_2, but the catalysts are highly selective to CH_4 production.[1,3,4] We investigated the coordinated mechanistic contributions of the CeO_2 support and Ni admetal. We then modified Ni catalysts with Fe in order to tune the CO selectivity while maintaining high activity.

CO_2 hydrogenation over Fe-modified Ni/CeO_2 catalysts was investigated in a batch reactor using time-resolved *in situ* FTIR spectroscopy and in a pulse reactor using mass spectrometry. Low loading of Ni/CeO_2 was associated with high selectivity to CO over CH_4, while higher Ni loading improved CO_2 hydrogenation activity but reduced CO selectivity. *In situ* XANES and AP-XPS revealed Ni to be metallic for all catalysts including the CO-selective low loading 0.5% Ni catalyst, suggesting that the selectivity trend is due to structural rather than oxidation state effects. Isotope studies using $C^{18}O_2$ revealed that oxygen exchange with the CeO_2 support occurs beyond the surface layer. Subsurface oxygen participates in the reaction, and oxygen exchange is fast with respect to the reaction time. The results suggest that H_2 dissociation by metallic Ni promotes C-O bond breaking of CO_2 adsorbed on reduced ceria, and that oxygen vacancies are filled by oxygen from the adsorbed CO_2.[5]

Bimetallic Ni-Fe catalysts demonstrated improved selectivity towards CO without significantly compromising activity, coupling the high activity of Ni catalysts and the high CO selectivity of Fe. TPR results indicated bimetallic interactions between Ni and Fe, and XANES analysis showed that the majority of Fe in the bimetallic catalysts was oxidized.[6]

The trends in bimetallic modification and support effects should guide efforts to develop non-precious metal catalysts for the selective production of CO from CO_2 hydrogenation.

1. Porosoff, M. D., Yan, B. & Chen, J. G. Catalytic reduction of CO2 by H2 for synthesis of CO, methanol and hydrocarbons: challenges and opportunities. *Energy Environ. Sci.* **9,** 62–73 (2016).
2. Perathoner, S. & Centi, G. CO2 recycling: A key strategy to introduce green energy in the chemical production chain. *ChemSusChem* **7,** 1274–1282 (2014).
3. Sehested, J., Dahl, S., Jacobsen, J. & Rostrup-Nielsen, J. R. Methanation of CO over nickel: Mechanism and kinetics at high H2/CO ratios. *J. Phys. Chem. B* **109,** 2432–2438 (2005).
4. Tada, S., Shimizu, T., Kameyama, H., Haneda, T. & Kikuchi, R. Ni/CeO2 catalysts with high CO2 methanation activity and high CH4 selectivity at low temperatures. *Int. J. Hydrogen Energy* **37,** 5527–5531 (2012).
5. Winter, L. R., Chen, X., Chang, K. & Chen. J. G. Coordinated effects of metallic Ni and reduced CeO_2 for subsurface oxygen exchange during the reduction of CO_2 by hydrogen. *Chem. Eur. J.* (2018) In preparation.
6. Winter, L. R., Gomez, E., Yan, B., Yao, S. & Chen, J. G. Tuning Ni-catalyzed CO2 hydrogenation selectivity via Ni-ceria support interactions and Ni-Fe bimetallic formation. *Appl. Catal. B Environ.* **224,** 442–450 (2018).

548922 Which Factors Limit Renewable Fuel Production from Switchgrass?

Abstract:

Lignocellulosic biofuels generated from plant cell walls are a promising alternative to gasoline. There are a large number of potential lignocellulosic feedstocks including switchgrass (*Panicum virgatum* L.), a high-yielding, native U.S. grass species that is widespread throughout the eastern United States. Switchgrass is suitable for growth in a range of climates, and provides many positive ecosystem services. There are two distinct genetic forms of switchgrass and our previous work demonstrated that the same optimal thermochemical pretreatment conditions could be used for each. More recently we have investigated the effect of environmental and agronomic conditions experienced during switchgrass growth on fermentation of biomass-derived sugars. We observed that both the choice of harvest date and severe drought experienced by the plants can result in different degrees of fermentation inhibition, neither of which was predictable from the biomass or hydrolysate compositions. These results show the importance of running fermentation on potential materials, not simply relying on biomass and hydrolysate characterizations, in order to identify possible sources of inhibition.

537967 Photovoltaic/Thermal Energy Addition to Expeditionary Buildings

Abstract:

The US Air Force (USAF) has a goal of reducing field consumption of fossils fuels. One cost-effective solution is to retrofit existing field assets with the ability to harness renewable energy. This paper discusses the adaptation of USAF expeditionary shelters with photovoltaic thermal (PVT) hybrid panels. The resulting power is captured as both electrical energy and thermal energy.

An Air Force Civil Engineering team constructed steel buildings (known as K-Spans) at Tyndall Air Force base (Florida) as test structures for evaluation of the PVT hybrid technology, which is attached to the outside.

The capabilities, operating characteristics, and overall power gain of the configuration were examined, and results will be presented. Varying flowrates were used to investigate the trade-off between heated water and higher electrical efficiency. Similarly, control algorithms were used to control the flow and output water temperatures compared to a non-controlled baseline unit.

522844 An Integrated Approach to Water-Energy Nexus with Multiple Energy Sources

Abstract:

AN INTEGRATED APPROACH TO WATER-ENERGY NEXUS WITH MULTIPLE ENERGY SOURCES

Fadhil Y. Al-Aboosi and Mahmoud M. El-Halwagi

The Artie McFerrin Department of Chemical Engineering

Texas A&M University

College Station, Texas 77843-3122

ABSTRACT

Growing energy demand, water scarcity, and climate change have emerged as considerable challenges to sustainable development. Consequently, there is a necessity to create the proper management strategies for water resources (e.g., wastewater treatment) and to integrate conventional energy sources with renewables (e.g., solar energy, wind energy, biofuels, etc.). The objective of this study is to develop a design framework for integrating water and energy systems including multiple energy sources, cogeneration process, and desalination technologies. Solar energy is included to provide thermal power directly to a multi-effect distillation plant (MED) exclusively (to be more feasible economically) or indirect supply through a thermal energy storage system. Thus, MED is driven by direct or indirect solar energy, and excess or direct cogeneration process heat. The proposed thermal energy storage along with the fossil fuel boiler will allow for the dual-purpose system to operate at steady-state by managing the dynamic variability of solar energy. Additionally, electric production is considered to supply a reverse osmosis plant (RO) without connecting to the local electric grid. A multi-period mixed integer nonlinear program (MINLP) is developed and applied to discretize operation period to track the diurnal fluctuations of solar energy. The solution of the optimization program determines the optimal mix of solar energy, thermal storage, and fossil fuel to attain the maximum annual profit of the entire system. A case study is solved for water treatment and energy management for Eagle Ford Basin in Texas.

Keywords: Cogeneration, Process integration, Solar energy, Thermal storage, Desalination, Optimization

538748 The "Green Latrine": Deployable Latrine with Off-Grid Photovoltaic/Thermal Energy

Abstract:

The US Air Force has a goal of reducing field consumption of fossils fuels in order to improve energy efficiency and operational resiliency. One cost-effective solution is to provide field assets that can operate independent of the power grid by harnessing renewable energy. This paper details the design, construction, and assessment of a proof of concept deployable latrine powered exclusively with Photovoltaic/Thermal (PVT) energy technology that allows operation completely "off-grid". PVT arrays are traditionally used for parallel production of solar electrical power and water heating in residential recreational pools. The attachment of a heat exchanger to the back of a traditional PV panel allows for cooling of the panels, resulting in higher electrical output while simultaneously heating the water.

A standard military portable (containerized) latrine unit was retrofitted with PVT hybrid panels that generate hot water and electricity. Potable water is the only required input. The installed electrical production capacity totaled 2.475 kW with an associated battery bank (44 kWh). The PVT added 6.0 kW of additional potential thermal energy depending upon variation in the panel flow rates and PVT configuration.

The capabilities, operating characteristics, and overall power gain of the configuration was examined during studies at Tyndall Air Force Base, Florida, and results will be presented. The system's performance is examined under varying operating scenarios to assess electrical and thermal performance. Overall energy efficiency was determined in respect to electrical capacity, electrical storage, and operating utility.

539205 Optimization of Hybrid RO-PRO Membrane Processes at the Water-Energy Nexus

Abstract:

Water and energy are two of the most pressing problems faced by the world today. Membrane processes such as Reverse Osmosis (RO) and Pressure Retarded Osmosis (PRO) can help alleviate both challenges. This work focuses on systematic optimization of a two-stage hybrid RO-PRO process for energy-efficient seawater desalination which integrates RO for desalination and PRO for power generation into a single process. Such a process is of great significance at the water-energy nexus given that water production, pump energy consumption, and hydraulic/osmotic energy recovery are inherently intertwined. The optimization problem is formulated to minimize the Specific Energy Consumption (SEC) subject to specified total membrane area and total water recovery. By solving the optimization model the best design (e.g., allocation of membrane area among different units) and operating (e.g. applied pressure in each membrane unit) conditions are simultaneously determined. It is shown that the proposed two-stage RO-PRO is energy-efficient than other configurations currently available in literature for continuous membrane-based RO desalination. Challenges for commercial viability of RO-PRO are discussed.

Reference:

M. Li. Optimization of Hybrid RO-PRO Membrane Processes at the Water-Energy Nexus, under review, 2018.

520120 An Integrated and Distributed Anaerobic Digestion Waste-to-Energy System for Energy Recovery from Food Waste

Abstract:

High energy self-consumption and low energy production are two major reasons limiting the energy performance of waste-to-energy systems. As a major part of energy consumption, mixing in an anaerobic digester creates a homogeneous system enhancing mass transfer, and enables the solid wastes and microorganisms remain in suspension, but continuous mixing strategy is not cost-effective due to the demand of high electric energy. A high-energy efficient food waste-to-energy system was developed to combine anaerobic digestion and engine generator for energy recovery from food waste. Through investigating the effect of different mixing strategies on the performance of the anaerobic digestion process, mixing time was optimized to lower energy consumption and ensure that biogas production is economical and energy efficient. By testing an internal combustion engine generator, electricity generation potential and waste heat behavior were obtained when using different components of methane gas as fuel. Raising the methane content over 85% enhanced the performance of the engine generator, while the methane content at 60% was unable to drive the engine stably. In addition, the waste heatâ€™s utilization and electricity generation efficiency, as well as energy distribution were studied through the analysis of the whole system.

537732 Chemical Looping Based Hydrogen Production from Ammonia: System Analysis and Experiments of a Two Reactor and a Three Reactor System

Abstract:

Hydrogen (H_2) is a promising carbon neutral liquid fuel with a high specific energy of 39.8 kWh/kg (HHV). The market penetration of H_2 is limited by the high transportation costs associated with its low energy density of 0.8kWh/L (by comparison gasoline is 8.76 kWh/L) . A potential solution to reduce transportation costs for H_2 is to use ammonia (energy density: 4.25 kW/L). Ammonia (NH_3) is proposed to be an alternative H_2 carrier, due to its lower transportation costs and it is widely produced on an industrial scale using Haber-Bosch chemistry. From an end-user perspective, if ammonia is used as a H2 carrier and if H2 is still used as the primary energy source, there exists a technological need to convert NH3 back to H2 in an efficient way.

OSU has proposed a thermochemical solids looping based based NH3 to H2 (ATH) process that can operate at high thermal and H2 production efficiencies. The ATH system can be operated as a two reactor system or a three reactor system, wherein the first the reducer reactor utilizes an intrinsic O2 gradient driven by the reduction potential of NH3 and a metal oxide Fe2O3/Fe3O4) to efficiently crack NH3 to a mixture of N2, H2 and H2O. The reduced metal-oxide from the reducer reactor is re-oxidized in a second reactor oxidizer using H2O as the oxygen source.

This presentation will initially compare and contrast the two reactor ATH system to the three reactor ATH system. Specifics of the thermodynamic rationale for optimizing the H2 production from NH3 using these configurations, while minimizing the operating temperature will be presented. The study uses guidance from the thermodynamic simulations to design reaction-engineering experiments in a thermo-gravimetric analyzer and a fixed bed system for generating proof of concept data. A techno-economic analysis model based on reaction engineering parameters is constructed and analyzed using sensitivity studies that provide insights into further process development will also be presented.

540933 Process Intensification of Bio-Ethanol Dehydration Under Compressed Liquid Phase Conditions

Abstract:

Global ethylene demand is consistently growing as its uses as a precursor for plastics, detergents and lubricants expand. Although the majority of ethylene is produced from petrochemical sources, an existing green alternative is dehydration of ethanol derived from biorenewable feeds to produce ethylene. High feed costs and process inefficiencies limit bio-based ethanol as an economic alternative to crude oil for ethylene production. Specifically, the highly endothermic bioethanol dehydration reaction and poor heat transfer to the vapor phase necessitate operation through a series of adiabatic reactors and boilers to maintain high ethylene selectivity. In addition, agriculture and fermentation are highly localized, which necessitates costly ethanol transportation through trucking, waterways, and pipelines. Operation of ethanol dehydration in the compressed liquid phase has potential to intensify the dehydration process by improving heat transfer, product recovery, and energy efficiency. A process intensified reactor operating under liquid phase conditions has potential to reduce capital costs sufficiently to justify distributed ethylene production.

Catalytic activity in the liquid phase is a major knowledge gap that hampers development of process intensified, liquid phase ethanol dehydration reactors. To assess the effects of conditions on activity, we performed both liquid and vapor phase ethanol dehydration studies, studying both space velocity and the effect of the water content of the fed. Ethanol dehydration reactions were performed in a continuous phase packed bed reactor catalyzed using ZSM-5 zeolite. Ethylene activity was compared when operated under both liquid and vapor phase (0.1 or 24 MPa) and varying weight hourly space velocities (30 - 1200 $hr^{â€‘1}$) and water loadings (0 - 33 wt%). ZSM-5 is more active for ethanol dehydration reactions under vapor phase conditions compared to liquid phase conditions, however, differences in ethylene yield are negligible when operated at low space velocities and high water loadings. Characterization of ZSM-5 catalyst after reaction reveals only a minor loss in framework after both the vapor and liquid phase reaction. Ethanol dehydration throughput and heat transfer improvements using a compressed, liquid phase packed bed reactor compared to vapor phase reactors suggest a potential alternative to current industrial processes.

526555 Catalytic Etherification of Glycerol to Glycerol Oligomers in the Presence of Alumina Supported Ca/Sr Mixed Oxides

Abstract:

In this work, the solvent-less etherification of glycerol to oligomers using alumina supported Ca and Sr catalysts was studied. These catalysts were obtained with impregnation of strontium and calcium onto γ-alumina (γ-Al_2O_3) in $Sr(NO_3)_2$ and $Ca(NO_3)_2$ solutions. The evidence for synergistic effects between active Ca and Sr species was demonstrated by etherification tests over a series of Ca-Sr-Al catalysts with varied Ca-to-Sr molar ratios. Effect of calcination temperatures of these supported catalysts, from 600 °C to 900 °C, on the resultant reusability and conversion efficiency of glycerol to oligomers and the selectivity of linear and branched oligomer to cyclic ones were investigated. The catalyst properties were characterized by various instruments including BET, TPD, XRD and SEM, and the ICP-OES analysis was performed to confirm the leaching of the metal species in the liquid phase of the reaction mixture. With Ca and Sr supported on γ-alumina catalyst, the catalyzed etherification reactions were carried out at 250 °C with 2.5 wt% Ca/Sr/γ-alumina calcined at 900 °C in nitrogen atmosphere. Under optimal impregnation conditions, a conversion efficiency of glycerol 80 % glycerol could be obtained after 10-h reaction. Moreover, a selectivity near 70% of di- and tri-glycerol in linear and branched forms over the cyclic form could be achieved.

538721 The Role of Copper Stability in Selectively Condensation of Ethanol to Higher Alcohols

Abstract:

The recent emergence of a strong bioethanol industry has elevated interest towards its use as a sustainable platform molecule towards producing high valued chemicals; one such process is the selective conversion of ethanol into high value C_{4+} alcohols. This multistep reaction involving acid-base catalyzed condensation and hydrogen transfer tends to yield a complex mixture of side products such ethers, esters, and ketones that prevent efficient conversion. Our work has demonstrated 80% selectivity to higher alcohols with conversion above on MgAl hydrotalcite catalysts at low copper loadings, with stable catalyst lifetimes over 100 hours in a flow through-reactor setup. Lower concentrations of copper was shown to prevent the formation of ester and ketone side products without reducing the conversion of ethanol. Characterization of the catalyst by STEM and XAS EXAFS work reveal that this is largely due to high dispersion of copper at lower copper loadings with near absence of copper clustering, even under high temperatures and reductive conditions. Further *in operando* XANES and EXAFS study shows that copper only reduces partially to Cu

support matrix allow for the production of dispersed Cu^{+1} sites, which promotes the efficient catalytic reaction to higher alcohols while not contributing to side reaction pathways such as esterification and C-C scission. The dispersion and stability of these individual copper promoters is critical to the high efficiency of this reaction. This work will discuss the limiting copper movement within the support to lengthen the lifetime of catalyst activity as well as throughout multiple regeneration cycles.

538786 Effects of Alloying Pd and Cu on Tandem Dehydrogenation-Aldol Condensation Reactions

Abstract:

The successful reduction of carbon dioxide emissions as a result of the production and consumption of fuels and chemicals requires the use of biomass resources as a feedstock. As a result of this, the discovery of novel pathways to upgrade platform molecules to fuel precursors is of critical importance. More specifically, for the production of diesel substitutes, the formation of new C-C bonds is required, as platform molecules deriving from the acid-catalyzed hydrolysis of biomass or the fermentation thereof are in the C2-C6 carbon range.

In this work, we investigate the upgrade of fermentation-derived acetone-butanol-ethanol mixtures to renewable diesel precursors. To achieve this, we develop a bimetallic Pd-Cu catalyst. We show that alloying of Pd and Cu results in the suppression of undesirable decarbonylation reactions. We present X-ray absorption spectroscopy results showing the segregation of Cu to the catalyst nanoparticle surface. Kinetic measurements under flow conditions demonstrate that Pd-Cu surface ensembles are active for hydrogenation reactions, while Pd-Pd ensembles catalyze C-C bond scission reactions. Density functional theory and microkinetic modeling show that the PdCu ensembles are bind CO fragments weakly, and hence destabilize C-C bond scission transition states. Similarly, the weaker CO binding onto low-index planes compared to stepped surfaces results in higher decarbonylation rates over smaller nanoparticles. Based on these results, we generalize our conclusions for a range of reactions, under oxidizing and reducing conditions, and present the implications of the segregation of alloy components to the catalyst surface on reactivity and selectivity.

References:

1. *Nature* 491 (2013) 235â€“239
2. *Am. Chem. Soc.* 138 (2016) 6805-6812
3. *Catal. Sci. Technol.* 8 (2018) 314-327

530475 Lubricant Base Oils Production from Biomass

Abstract:

The gradual depletion of world fossil fuel resources and global environmental concerns have prompted the use of biomass feedstock for renewable products. Lubricant is one such product that is primarily obtained from petroleum and it is introduced to reduce friction between surfaces in mutual contact. Base oils are the main components of lubricant and the majority of them are petroleum-derived mineral oils and synthetic oils (e.g. poly-Î±-olefins). The oils are formulated with different additives into several grades to suit automotive and industrial applications. Production of lubricant base oils from biomass will be an attractive alternative to petroleum derived-based oils to improve sustainability. So far, a few efforts have been devoted to the synthesis of lubricant-based oil alicyclic hydrocarbons, from biomass furans.[1-2] In this work, we report a novel route for selective production of base oil hydrocarbons from biomass platform chemicals and natural oils. The overall yield of hydrocarbons could reach up to 80%. Importantly, the hydrocarbon structures can be tailored using different biomass derived compounds to tune properties for various applications.

1. Balakrishnan, E. R. Sacia, S. Sreekumar, G. Gunbas, A. A. Gokhale, C. D. Scown, F. D. Toste and A. T. Bell, Proc. Natl. Acad. Sci. USA 2015, 112, 7645-7649.
2. Balakrishnan, G. E. Arab, O. B. Kunbargi, A. A. Gokhale, A. M. Grippo, F. D. Toste and A. T. Bell, Green Chem. 2016, 18, 3577-3581.

530529 Selectivity Control during the One-Pot Conversion of Aliphatic Carboxylic Acids to Linear Olefins through Tandem Hydrogenation/Dehydration

Abstract:

Olefins are important building blocks for the synthesis of higher value chemicals in the petrochemical industry and they are typically produced from fossil fuels. As an alternative to fossil fuel-based chemicals, biomass-derived carboxylic acids could serve as a viable replacement feed for olefin production. However, the production of olefins from carboxylic acids in one step remains a challenge due to the stability of carboxylic acids and the tendency of olefins to over-hydrogenate during hydrogenation, leading to alkanes.

In the present study, we have studied the one-pot catalytic conversion of linear aliphatic carboxylic acids to linear olefins via tandem hydrogenation/dehydration reactions. Hexanoic acid was converted to a mixture of hexenes over Cu nanoparticles supported on commercial silica-alumina (Siral 70) in a continuous flow reactor. At 483 K and 5 bar H_2 gauge pressure, we obtained 98.5% conversion of hexanoic acid, with a molar product distribution of 88.9% hexenes and 11.1% hexane. Remarkably, we observed a brusque selectivity switch to 99.9% hexane at 100% conversion.

A product intermediates study shows that when switching from a feed of pure hexene to a mixture of hexene and hexanoic acid, overhydrogenation of hexene can be suppressed. In addition, hexene temperature-programmed desorption (TPD) and FT-IR spectroscopy reveals that adsorption of hexanoic acid prevents hexene from accessing the catalyst surface, thus avoiding overhydrogenation. Reactions using butyric acid as feed also exhibit similar phenomena. We have also performed reactions utilizing butyric acid derived from biomass fermentation, albeit with significant catalyst deactivation.

We propose that the presence of a small amount of carboxylic acid on the surface of the catalyst prevents overhydrogenation of the olefins. In light of these results, current efforts are targeted at applying this concept to explain similar phenomena in other reactions, particularly in biomass conversion. The presence of numerous carboxyl groups in native/pre-treated biomass might explain the difference in catalyst performance/selectivity when processing model feed versus real biomass. Furthermore, the addition of carboxylic acids to the reaction medium during biomass upgrading could afford more control over the selectivity of the reaction.

526827 Mechanism and Kinetics of Isobutene Production over Zirconia-Supported Zinc Oxides

Abstract:

Growing interest in utilizing biomass-derived platform chemicals to produce fuels and specialty chemicals has motivated the investigation of the reaction of ethanol or acetone to form isobutene, a valuable specialty chemical used in the production of fuel additives, polymers, and other high-value products. Reports in the literature have identified zinc-zirconium mixed oxides as effective and selective catalysts for the production of isobutene from ethanol at 723 K. In this investigation, zinc-zirconium mixed oxides prepared via incipient wetness impregnation were selective and highly active catalysts for the acetone to isobutene reaction, reaching near-theoretical isobutene selectivities of over 86%. The mechanism and kinetics of the acetone and ethanol to isobutene reactions were studied in order to elucidate the reaction pathway, the roles of active acidic and basic sites, and the role of water in promoting stability and selectivity. Zinc-zirconium mixed oxide catalysts with varying zinc loadings were synthesized and characterized with XRD, BET, CO2-TPD, and DRIFTS FTIR with pyridine adsorption. Kinetic studies suggested that the ethanol to isobutene reaction proceeds via dehydrogenation to acetaldehyde, followed by condensation to form acetic acid, aldol condensation to produce acetone, dimerization of acetone to form diacetone alcohol, dehydration to form mesityl oxide, then finally hydrolysis of mesityl oxide to produce isobutene and acetic acid, which undergoes further aldol condensation to produce acetone. Understanding the reaction pathway, mechanism, and roles of the acidic and basic sites can help to define the catalyst properties required to selectively produce isobutene from a variety of sustainably-derived platform molecules.

533068 Catalytic Upgrading of Sugar-Derived Polyols to Glycols in Absence of Externally Added Hydrogen

Abstract:

Due to the increasing demand for fossil fuels and their diminishing supply worldwide, it is of considerable importance to utilize renewable resources for sustainable development of human society. Replacing fossil-based feedstock with biomass-derived ones to produce renewable fuels and chemicals is one of the promising ways to address challenges facing the energy and chemical industries. Sugar polyols, such as sorbitol, xylitol and erythritol are attractive platform chemicals because they can be readily derived from cellulosic biomass. Hydrogenolysis of sugar polyols can be converted to a variety of industrially important chemical intermediates including ethylene glycol (EG) and 1,2-propanediol (1,2-PDO), which are widely used as antifreezes, solvents and pharmaceuticals.

Up to date, it is known most existing hydrogenolysis processes are carried out under high temperature (T > 220 ^oC) and high H_2 pressure conditions (P > 4 MPa), which cause several technological issues such as poor product selectivity due to significant side reactions, fast catalyst deactivation as well as high capital cost and energy consumption.

In this work, we proposed a novel one pot H_2 generation and hydrogenolysis process in absence of external H_2 under mild conditions (T: 160 ^oC-200 ^oC, P: 1.0 Mpa nitrogen atmosphere).We have designed a series of bimetallic Pd-M (M: Cu, Ni, Pt, *etc*) catalysts supported on active carbon and layered double hydroxides and confirmed their remarkable performances in catalytic conversion of xylitol, sorbitol and mannitol to 1,2-PDO, EG and lactic acid (LA). In particular, catalyst activity and selectivity will be studied for different catalysts. Structure-performance correlations for H_2 generation and hydrogenolysis will be established based on bulk and surface characterization using XRD, SEM, TEM and chemisorption. The bimetallic Pd-Ni/AC and Pd-Pt/AC catalysts showed superior hydrogenated activities as well as selectivities towards lactic acid and glycols, compared with other bimetallic catalysts under identical conditions. At nearly 60.0% sorbitol conversion, a selectivity of 36.5% to ethylene glycol, propanediol and other glycols was obtained on Pd-Ni/AC at 200 ^oC, 1.0Mpa nitrogen atmosphere in absence of external $H_{2.}$ Here we also researched reaction mechanism or pathways by studying metal active sites, H_2 spillover effect and the role of solid-base via regulating the growth of Pd-M and the ratio of acidity (basicity). It was concluded that H_2-spillover from Pd to M and solid-base acidity (basicity) may strongly influence the reaction mechanism to some extent, thereby influence the activity and selectivity of hydrogenolysis of biomass-derived to chemicals. Further details on reaction pathways and kinetic modeling will also be discussed in this presentation.

533584 Selective C-C Bond Scission of Primary Alcohols Using Cerium Oxide-Supported Palladium Catalyst

Abstract:

Biomass attracts significant attention as a carbon-neutral resource with the potential to suppress the CO_2 emissions. Recently, the biomass-derived chemicals such as glycerol, furfural, and levulinic acid are identified as the best suited alternatives for petroleum-derived chemicals. From the stand point of green sustainable chemistry, development of highly selective catalytic transformation of these compounds to valuable chemicals has been desired. To date, much effort has been devoted to selective C–O bond cleavage of high-oxygen containing biogenic polyols, such as glycerol, to produce valuable chemicals by hydrogenolysis and deoxydehydration.[1,2] On the other hand, C–C bond cleavage has not yet been widely researched despite its great potential to extend the utility of biomass-derived chemicals. The reported C–C bond cleavage methods often suffer from low selectivity toward the desired chemicals and limited substrate scope. In our group, we have reported the selective C–C bond scission of platform biomass-derived chemicals catalyzed by Ru/CeO_2 in the presence of water under 3 MPa of H_2.[3] In order to open new routes for industrially important chemicals from a wide range of biomass derivatives, development of highly selective and greener C–C bond scission catalysts is desired.

In this work, we found that cerium oxide-supported palladium (Pd/CeO_2) efficiently promoted the selective one-carbon scission of primary alcohols under air without H_2. The cleavage of C–C bonds occurs at positions adjacent to the hydroxymethyl groups (Scheme 1). The solid catalyst was reusable without appreciable loss of the activity and selectivity.

References

[1] T. Mizugaki, Y. Nagatsu, K. Togo, Z. Maeno, T. Mitsudome, K. Jitsukawa, K. Kaneda, *Green Chem.* **2015**, *17*, 5136–5139.

[2] T. Mizugaki, K. Togo, Z. Maeno, T. Mitsudome, K. Jitsukawa, K. Kaneda, *ACS Sustainable Chem. Eng.* **2016**, *4*, 682–685.

[3] T. Mizugaki, K. Togo, Z. Maeno, T. Mitsudome, K. Jitsukawa, K. Kaneda, *Sci. Rep.* **2017**, *7*, 14007.

535737 Renewable Energy Assessment and Its Integration in Angola

Abstract:

Generally, the predictable exhaustion of fossil fuels, the necessity of fighting against global warming, the awareness of the protection of the environment and the consideration of sustainable development in energy policies put the renewable energies in the heart of a strategic stake for the future of our planet. The challenge of transforming entire economies is enormous; even more so if a country is as fossil fuel based and emission intensive as in the case of Angola (which is the second major producer of Oil and Gas in sub-Saharan Africa). However, in an increasingly carbon constrained world and already now facing climate change impacts, Angola has to reduce greenhouse gas emissions intensity soon and decidedly. For the Republic of Angola, with decelerating annual GDP growth, it is almost vital to exploit its potential in renewable energies to ensure the continuous economic growth, to realize savings of currencies and to achieve the Millennium human development goals.

The Economic Community of Central African States (ECCAS), of which Angola is a member, has recently adopted the White Paper of CEEAC *Communauté Economique des États de l'Afrique Centrale* and CEMAC (Economic and Monetary Community of Central Africa) on a Regional policy for universal access to modern energy services, economic and social development (2014-2030), adopting specific targets regarding energy access, energy efficiency and renewable energies – in line with the Sustainable Energy for All initiative.

Unfortunately, the country knows the same energy situation of the countries of central Africa where the energy is plentiful but electricity is rare! Indeed, the current energy balance of the country is strongly overdrawn. 97% of the energy needs of the population are satisfied by oil productions and 95% of the Angolans households use the distillate fuel oil, diesel, motor gasoline and kerosene as domestic fuel. The cover rate for the electricity network is very low, about 43% (6.34 GW). Only 0.6% of the electricity production (with a total capacity installed of 40 MW) is made from a unique source of renewable energy (the biomass energy-sugarcane).

Nevertheless, the country has an important potential in renewable energies. At the level of the photovoltaic solar energy (PV), the technically exploitable solar potential is estimated to be between 1.355 and 2.068 kWh/m^2/year. At the level of the wind energy, the estimation of the currently exploitable potential is 3.7 GW and yet no form of wind energy (where the wind at a height of 80 meters above the ground reaches average speeds of more than 6 meters per second) is exploited in the country. At the level of the geothermal energy, the technically exploitable potential is not yet known (case to be studied) together with the tidal energy, also not yet estimated, potential exists and could be exploited in regions of the coast of Angola.

The energy plan for renewable energy integration that we propose, places a significant focus on the solar energy. Indeed, this one possesses advantages which the other forms of renewable energies do not have: it is a basic energy, present throughout the year in the entire country with ten months of sun, susceptible to be exploited for the production of the heat and electricity production. The overall electricity plan which we propose is the following one: on the horizon to 2025, we consider that the electricity demand will be situated near to 9.9 GW. Therefore, we recommend 45% of fossil fuels and natural gas (with a substitutable option in warm period by the electricity imported from Namibia to supply the south of Angola) and 25 GW of renewable energies (composed of 69.2% solar energy, 14.8% from biomass – sugar cane, 15.6% of wind energy and geothermal and tidal energies being the rest).

538660 Thermo-Economic Optimization of Mechanical Vapor Recompression System for Shale Gas Produced Water Treatment

Abstract:

The need for desalination and sustainable management of hydraulic fracturing produced water is increasing with the expansion of unconventional oil and gas extraction. Different desalination technologies could be potentially utilized to treat the produced water with varying energy consumption and different operational restrictions; Mechanical vapor recompression (MVR) is among the evaporative technologies that suit this purpose, given its ability to operate at high total dissolved solids (TDS), high salt rejection (99%), high recovery ratio, high energy recovery and its modular design. MVR is a work driven desalination process that requires considerable amount of electricity for operation, and relatively high capital cost for the equipment capable of working with high saline water. Therefore, systematic optimization of operating conditions, and capacity of the MVR technology is required to assess the economic feasibility of this technology for produced water treatment. Previous work on MVR includes designing and analysis of MVR systems for seawater desalination. However, very limited research exists on detailed analysis and optimization of MVR technology for high salinity produced water treatment.

The present work presents a new optimization model for MVR process design with the objective of minimizing the total cost of produced water treatment. The optimization process utilizes a modified version of particle swarm optimization (PSO) algorithm which allows for relatively fast computation, of the design criteria of the MVR process while simultaneously minimizing the unit cost of produced water treatment. The primary model is developed through mass and energy balances for MVR system elements (including horizontal tube evaporator, centrifugal compressor, desuperheater, feed/distillate preheater, pumps, and mixers), and capital and operating cost equations. The operating condition constraint, energy balance, and heat transfer equations are added to the objective functions as proper penalty functions. Finally, the annual unit cost of treatment and penalty functions are optimized using PSO algorithm with iterative decision variables consisting of top brine temperature, temperature difference in the evaporator/condenser, temperature of the feed leaving the preheaters, and geometrical characteristics of the heat exchangers regarding their heat transfer coefficient calculations.

The developed optimization model is run for a hypothetical desalination plant with a capacity of 0.5 million gallons per day (MGD), which concentrates the produced water with 10% (100,000 mg/Liter) TDS to 30% TDS corresponding to saturation condition to make the exiting brine system capable of integration with downstream crystallizers. Sensitivity analysis is performed to identify the impact of input parameters such as brine salinity, feed flowrate, and feed salinity, on the energy consumption and unit cost of desalinating produced water. Feed salinity and brine flowrate are analyzed further to analyze the impact of these input variables on initial investment and equipment sizes. The results show that the proposed optimization model has to potential to work as a powerful tool in MVR design. Additionally, the final calculated costs show promising results for the application of MVR in the field of produced water treatment with unit costs of \$2.2/m^3(feed). The implications of these findings for sustainable management of produced water will be discussed.

519029 Low Temperature Pyrolysis of Discarded Plastics

Abstract:

Plastics are petroleum based products which contain high value components. Most plastics are non-biodegradable, making their disposal difficult once they are no longer required for their intended purpose without impacting the environment.

With an increasing world population, demand for plastics and plastic products continues to rise, as does the amount of plastic waste. Between 1950 and 2015, an estimated 6300 million tonnes of plastic waste had been generated through various sources. Of this, only around 9% has been recycled and 12% incinerated. The vast majority of waste plastic is either accumulating in landfills or has found its way into the environment. If the current rate of plastic material disposal is carried forward, approximately 12000 Mt of plastic material will end up in landfills. Landfill sites have always been undesirable due to many reasons including, the emission of toxic gases, odour, land use and the unsustainable nature of this disposal method. Domestic and industrial sorting and transport of discarded resources to points where they are supposed to be disposed of are associated with a large carbon footprint. Other processes, such as recycling, have high costs and are labour intensive.

Having a sustainable method in place that is capable of converting the energy content within waste plastic into useful forms of energy would not only be highly advantageous to our environment, but could also be better economically. Low temperature (up to 300ºC) pyrolysis using heat pipe technology offers an attractive solution to converting discarded plastic materials into valuable energy vectors, such as syngas, bio-oils and biochar. Operating pyrolysis reactions at low temperatures ensures lower operating cost of the reactor in-comparison with current conventional, high temperature, pyrolysis reactors. Once the discarded plastic has been converted to other forms of energy (mostly biochar), they can be combusted to transfer the heat produced for use in, for example, process heating.

This paper will discuss the energy and composition of the products formed from pyrolysis of plastic materials, as well as the feasibility of low temperature pyrolysis for their conversion into energy products.

532395 Pelletization of Spent Coffee Grounds

Abstract:

Different types of raw materials as biomass can be densified to be used as biofuels. The produced pellets provide an alternative to solve the energy problem. With the densification, it is not only possible to improve the combustion process (increasing the calorific value). It also helps to solve the problems associated with loading, transporting and storing the product (thanks to the increased density).

In this project pellets were made of spent coffee grounds. Three kinds of these were characterized, from different sources. After this, the most appropriate biomass was selected to be used as fuel, according to three selection criteria: moisture content, ash content and superior calorific value. The pellets were made using not only the moisture content of the biomass, but also three other products as a binder: glycerin, crude palm oil and biodiesel.

The biomass with a moisture content of around 70% was dried in an oven at 80°C. Subsequently, pellets with different concentrations of moisture within the range between 4% and 20% were produced, obtaining a consistent product between 5% and 15% moisture. After defining the moisture range, the concentrations of each of the binders were varied between 2% and 10%. Pellets were formed between 2% and 8% concentration of binders, using moisture material between 4% and 8%.

Once the moisture content and concentration of binders were defined, a factorial design of experiments was carried out to obtain a response surface. The obtained pellets were characterized physically, chemically and mechanically, according to the German standard DIN 51731 for solid biofuels.

533005 Air Pollution Control: Pyrolysis Based on Waste to Energy Plant

Abstract:

Waste to energy (WtE) processes are processes that convert wastes to different forms of energy. WtE processes provide a source of energy and reduces the accumulation of wastes in landfills or other areas. Different WtE technologies are used worldwide, from fermentation to thermochemical conversions, these technologies turn different wastes to different products. The aim of this project is to design air pollution control units for a pyrolysis based waste to energy plant. The feedstock to the WtE plant was taken to be Municipal Solid Waste (MSW). The process produces pyrolysis oil, syngas, and char. The calculations for this project were done based on yields from fast pyrolysis. The process flow diagram was designed to remove the pollutants in the gas phase outlet streams. The pollutants include particulate matter of different aerodynamic diameters, and sulfur dioxide (SO2). The desulfurization unit uses limestone slurry as an absorbent, and produces gypsum. The particulate matter is removed by three main units, cyclone separators, baghouse filters and electrostatic precipitators (ESP). The cyclone removed relatively large particulate matters and the finer particulates were removed using the baghouse and ESP. The air pollution control units were sized in order to meet the outlet particulate matter and sulfur dioxide standards of the UAE government. The sizing of the equipment was performed using theoretical correlations and was later confirmed by simulation on SuperPro Designer. Based on the sizing that was calculated, the cost of each equipment was followed by the total APC units cost. HAZOP (Hazard and Operability study) was also considered and discussed in this project in order to study the environmental effects of the APC units as well in terms of potential failures. The project ended with future work which can be continued later in the future.

534670 Sustainable Ammonia Production through Process Synthesis and Optimization

Abstract:

Increasing the share of renewables is an important step in shifting toward a sustainable energy production and supply chain and reducing mankindâ€™s dependency on fossil fuels. While a transition to fully renewable energy production looks possible in the future, the intermittent and distributed nature of the renewable energy resources make the conversion and storage of renewables into useful and transportable energy vectors is a necessity and hence an outstanding challenge [1].While hydrogen is a promising energy vector to store renewables due to its high gravimetric energy storage density (33 kWh/kg), its low volumetric storage density (3 Wh/L under ambient conditions) makes both storage and transportation problematic [2]. Ammonia is a promising vector due to its proven production and transportation infrastructure.It has a high capacity for hydrogen storage, 17.6 wt.% based on its molecular structure and a hydrogen volume density of 105.0 kg/m^3, which is about 45 % higher than that of liquid hydrogen, which is reported as 71.2 kg/m^3[3].

Renewable resources are geographically distributed and an optimal solution at a location might not be feasible let alone optimal for another location. The distributed nature of the resources necessitates the concept of decentralized energy production. With this work, we want to contribute a study on the production of ammonia from various renewable sources, such as municipal solid waste, biomass, solar energy, and wind energy. In order to determine the economic and technical feasibility of novel ammonia production methods, a global optimization-based process synthesis approach is used to determine the optimal process topologies while minimizing the annualized cost of ammonia production. A superstructure of process alternatives is modelled as a large-scale mixed-integer nonlinear optimization (MINLP) model, with simultaneous heat, power, and water integration. The nonconvex MINLP model is solved to global optimality by using a branch-and-bound global optimization algorithm. Several case studies will be examined to investigate the types of available renewable resource and production scale on topological decisions.

1. WW Tso, CD Demirhan, JB Powell, EN Pistikopoulos, 2018, Toward Optimal Synthesis of Renewable Ammonia and Methanol Processes (RAMP), Proceedings of Process Systems Engineering, PSE 2018, Accepted for Publication.
2. D Teichmann, W Arlt, P Wasserscheid, 2012, Liquid Organic Hydrogen Carriers as an efficient vector for the transport and storage of renewable energy, Int. J. Hydrogen Energy 37, 18118- 18132.
3. G Thomas, G Parks, 2006, Potential Roles of Ammonia in a Hydrogen Economy, DOE.

534005 Modeling and Simulation of a Piston-Type Work Exchanger for Mechanical Energy Recovery

Abstract:

Energy efficiency is always a focal point in achieving process sustainability. Process integration, which was initiated for thermal energy recovery through heat integration, has been very successfully practiced in the process industry. Recovery of mechanical energy, however, has not received needed attention until recently, since a concept of work integration is introduced. It is conceived that a piston-type device could be a type of equipment to perform pressure-driven work exchange; this type of unit could be called direct work exchanger (DWE). An optimally designed network system using this kind of devices should be able to systematically recover mechanical energy. Note that work exchangers have been used widely for seawater reverse osmosis (RO) desalination, where liquid streams are pressurized or depressurized. This type of unit, however, cannot be directly used for mechanical energy involving streams in gas phase in chemical process systems.

In this paper, we introduce our investigation on developing a DWE that can be operated for mechanical energy recovery involving gas streams. We have developed a CFD based model and conducted various simulations to study the design of such a device, and its operational behavior under different operating conditions. Basic design and operational optimization has been also conducted to ensure cost effectiveness and energy recovery efficiency. Case studies will demonstrate application potentials for mechanical energy recovery involving gas streams in chemical process systems.

535018 Simulation-Based Computational Framework for Sustainability Assessment and Life Cycle Inventory Generation

Abstract:

In recent years, environmental consciousness has been growing and has become critical in the decision-making process in the chemical industry. This raised awareness has led to the incorporation of the concept of sustainability and sustainable development into process design and optimization, through pollution prevention and control technologies, releases minimization, sustainable process design, sustainability assessment (e.g., GREENSCOPE - Gauging Reaction Effectiveness for the ENvironmental Sustainability of Chemistries with a Multi-Objective Process Evaluator[1]) and life cycle assessment (LCA) tools (e.g., SimaPro[2], Open-LCA[3]). However, current methods in existing process design/optimization for sustainability studies are difficult to be customized for different applications[4]. One of the major obstacles for incorporating LCA and sustainability assessment tools into process design and optimization is the software integration among steps. Such steps are typically performed in separate software packages/interfaces, although they have overlapping data requirements[5]. For example, process design and optimization rely on conventional commercial tools, such as Aspen and CHEMCAD, while sustainability assessment tools (e.g., SustainPro[6] and GREENSCOPE) are typically developed employing Excel or other software that require life cycle inventory (LCI) information consisting of resource and energy consumption as well as environmental releases related to the process of interest.

To bridge this gap, an integrated framework with a process modeling/simulation environment, pollution control units (PCUs), LCI generation, and sustainability assessment is proposed. Specifically, a user-friendly automation interface is built via Microsoft Excel-VBA to enable the communication between CHEMCAD (a widely used commercial process simulation tool for chemical processes) and Excel tools, such as GREENSCOPE, PCUs and LCI generation. The employed Excel tools enable the augmentation of CHEMCAD's capability on waste treatment technologies and LCI estimation. The proposed framework is generic and can be implemented for other applications, with the intent of facilitating the communication between tools used by the LCA practitioner and process simulation packages used by process engineers.

The developed framework is applied to a case study of an Acetic Acid manufacturing process. The application results show process improvements in terms of sustainability performance represented by the GREENSCOPE indicators and LCI generation after the implementation of the appropriate PCUs for waste stream treatment. The case study thus demonstrates the effectiveness of the proposed framework for integrating CHEMCAD, GREENSCOPE, PCUs, and LCI generation for improving the sustainability performance of chemicals during their manufacturing life cycle stages.

References:

(1) Ruiz-Mercado, G. J.; Gonzalez, M. A.; Smith, R. L. Sustainability indicators for chemical processes: III. biodiesel case study. *Ind. Eng. Chem. Res.* **2013**, *52* (20), 6747–6760.

(2) SimaPro LAC software. https://simapro.com/about/ (accessed Apr 13, 2018).

(3) openLCA. http://www.openlca.org/ (accessed Apr 13, 2018).

(4) Bakshi, B. R. Methods and tools for sustainable process design. *Curr. Opin. Chem. Eng.* **2014**, *6*, 69–74.

(5) Li, S.; Feliachi, Y.; Agbleze, S.; Ruiz-Mercado, G. J.; Smith, R. L.; Meyer, D. E.; Gonzalez, M. A.; Lima, F. V. A process systems framework for rapid generation of life cycle inventories for pollution control and sustainability evaluation. *Clean Technol. Env. Policy.* **2018**. DOI: 10.1007/s10098-018-1530-6.

(6) Carvalho, A.; Matos, H. A.; Gani, R. SustainPro - A tool for systematic process analysis, generation and evaluation of sustainable design alternatives. *Comput. Chem. Eng.* **2013**, *50*, 8–27.

536601 Designing Manufacturing Sites Toward Local Sustainability By Understanding Spatial Variance of Industrial Air Pollution and Local Ecosystem Regulation

Abstract:

The consequences of a history of excluding ecosystem services from process engineering and decision -making are being revealed through many environmental impacts. Ecosystems are the basis of human survival as we rely on them for food, clean water, clean air, and many other services. Despite the many services ecosystems provide, traditional engineering has excluded these services as it has searched for solutions and technology to improve the lives of mankind, often impacting the ecosystems which we rely on for our well-being. Along with decreasing negative environmental impacts, including ecosystems in sustainable design can lead to innovative solutions which utilize the functions and co-benefits of nature as we search for sustainable solutions amid increasing population and demand of ecosystems.

In combining technological and ecological processes within one system, it is important to understand the scale and spatial variance of flows between the processes. Previous work in techno-ecological process design has lacked inclusion of the spatial heterogeneity of ecosystem service supply and the industrial demand of those services. This research aims to provide that understanding, specifically for air pollution and regulation by developing an approach for designing a manufacturing site as a network of technological and ecological systems. Previous research in the field of ecosystem service mapping yields methods for showing available ecosystem services across different spaces and scales, however, fails to connect these services to specific beneficiaries such as a manufacturing site. Connecting the service with the beneficiary requires understanding of ecosystem servicesheds, areas which provide specific ecosystem services to specific beneficiaries. Unfortunately, compared to ecosystem service mapping, research on servicesheds is lacking in application past concept. Applied to sustainable design, servicesheds can be understood as the spatial scale of a techno-ecological system, defining industrial pollution sources as beneficiaries. Focusing specifically on servicesheds which regulate air pollution from an industrial source, the ecosystems can be modeled as downstream unit operations of the technological process, using dry deposition to capture pollution from the air. Because methods for explicitly quantifying air quality regulating servicesheds are lacking in literature, defining the servicesheds mathematically was the first step towards bridging this ecological knowledge with engineering practice. Understanding the capacity of local ecosystems to regulate the chemicals released to the atmosphere from industrial processes can provide understanding of absolute sustainability, the condition where our demand on ecosystems is less than their available supply within the scale of interest.

Including ecosystems within sustainable process design enables eco-based solutions to increase the supply of ecosystem services, rather than solely focusing on decreasing the demand. This research advances the inclusion of ecosystems within sustainable design through exploring the spatial variation of both the ecological deposition and the dispersion of the air pollution from a given source. This knowledge can determine areas where land-use change can have the highest impact in meeting sustainability goals by implementing land restoration and management methods.

Acknowledging areas of high impact can enable sustainable land management solutions that minimize the area of land required along with restoration and management costs. The scale of these models, including the potential land which can be restored and managed, is determined through the quantification of the serviceshed.

To conduct this research, a case study was conducted on a biodiesel production facility in Cincinnati, OH. The case study focused on SO_2 stack emissions and dry deposition of local ecosystems. Using the EPA-approved atmospheric transport model, CALPUFF, the transport of the facilityâ€™s emissions was modeled along with the deposition fluxes of the surrounding land cover, calculated using ecological models within CALPUFF. The results yielded the information needed to determine the scale of the serviceshed of the specified service. Then, using land change scenarios within the serviceshed, optimization methods were used to find optimal inputs to the ecological model for the dry deposition equations, which depend upon the concentration of the pollutant. In addition to traditional engineering design decision variables, the ecological decision variable of this optimization includes the spatial arrangement of the various land cover classifications, while acknowledging the constraints of land use that could not be changed, such as bodies of water and urban development. This case study has led to the development of quantitative definitions of air quality regulation servicesheds and applications of optimization methods to understand the spatial heterogeneity of pollutant deposition and dispersion based on atmospheric chemical transport and land cover information. Understanding the spatial heterogeneity of the pollution and ecosystem services enables smarter industrial site design, including design of both the techno- and ecological processes.

538753 Sustainability Identification for N-Dimensional Systems

Abstract:

Sustainability Analysis of N-Dimensional Systems

Masih Jorat

Dept. of Chemical and Biomolecular Engineering, UCLA, Los Angeles, CA, 90095

Vasilios I. Manousiouthakis

Dept. of Chemical and Biomolecular Engineering, UCLA, Los Angeles, CA, 90095

The severity of degradation of the natural environment and look for a comprehensive solution in the form of flexible and multisided assessment tool analyzing the human impacts on its surrounding environment has led to the introduction of the concept of sustainability, to assist global, national and local policy making organizations, research institutes, and citizens at large, in assessing the current state of their environment and in developing solutions for its improvement. The development of sustainable strategies cannot be pursued in an efficient manner, without a comprehensive, rigorous, and unbiased definition of sustainability. Through utilizing the concept of positive invariant sets, the novel concept of Sustainability Over Sets (SOS) [1] provides a conceptual framework within which the time evolution of a system can be analyzed. SOS formalizes the incorporation of human input into the sustainability assessment process, first by defining a set in state-space which exactly quantifies human input regarding what is sustainable, and by then modifying the question "is a system sustainable?" to "is a system sustainable over a predefined set?" which is amenable to definitive (yes or no) answers. A system is sustainable over a set (SOS) in the system's state-space, if the system's state trajectories initiated within the set, remain for all time within the set, in other words the system's vector field is directing inwards at the set's boundaries. In this work we present computational algorithms that can identify sustainable regions in the state-space for n-dimensional systems which are not easily visualized due to their high (more than three) dimensions. This evolution of the SOS concept can respond to cases, in which sustainability restrictions are expressed in terms of multi variable functions. For example, Food webs are a central, if not the central, idea in ecology. A food web graphically represents the paths of nutrients and energy through the living components of an ecosystem and the context in which individuals exploit their prey and avoid their enemies.To demonstrate the practical applicability of the algorithm to be presented, it identifies the sustainable sets of a ten-dimensional food web, with nine different species and a nutrient pool [2].

[1] Manousiouthakis VI, Jorat M. Sustainability over sets. Environmental Progress & Sustainable Energy. 2017 Oct

[2] Fath BD, Cabezas H, Pawlowski CW. Regime changes in ecological systems: an information theory approach. Journal of theoretical biology. 2003 Jun 21;222(4):517-30.

530010 Sustainability Assessment and Targeting in Process Design: A Novel Method Based on Data Envelopment Analysis - Application to Liquid Fuels

Abstract:

Nowadays there is a growing interest on incorporating sustainability metrics in the chemical industry, particularly in the early stages of process design. Unfortunately quantifying and improving the level of sustainability related with chemical processes and products is a very challenging task since it requires balancing the inherent trade-offs between the goals of economic profit, environmental protection and social well-being support. To face this challenge and moving towards a more sustainable chemical engineering, alternative tools and approaches are needed in order to integrate sustainability concepts into the engineering design process. In this work we propose a novel framework that combines data envelopment analysis (DEA) with process system engineering tools which is tailored to help improving the sustainability of chemical processes. Considering a set of chemical processes alternatives characterized by transforming several inputs into several outputs, our framework allows to determine which processes are performing more efficiently in terms of sustainability. It systematically differentiates between efficient and inefficient processes, and for the latter, it provides quantitative targets for improvement that, if achieved, would make them more sustainable. Besides identifying those targets, our framework also determines the optimal design and operating conditions of the inefficient processes to reach the desired targets and therefore guaranteeing the feasibility of the alternative in practice. The capabilities of the proposed approach are illustrated through a real case study involving six transportation liquid fuels employed in the automotive fleet that are powered by internal combustion engines. The results provide the optimal operating conditions for the methanol process that would allow reducing its cost and global warming potential to a certain extent making it more sustainable. Our approach is intended to support decision makers towards a more sustainable chemical process design.

524736 Gas Phase Catalytic Oxidation of Lignin to Produce Phenolic Compounds over Vanadia Catalysts

Abstract:

The pretreatment of lignocellulosic biomass to create fermentable sugar streams *en route* to fuel/chemical production may also produce streams of insoluble lignin. These lignin stream can be burned to produce process heat and power. An alternative use of the lignin streams could be the production of valuable products such as phenolic compounds. This work explores the pyrolysis of lignin residue to produce gas phase products which can be passed of vanadia-based catalysts to oxidatively remove methoxy side chains to produce phenolic compounds, which could be used in the production of polymers and resins. Because lignin contains a complex mixture of linkages with methoxy groups, as well as both C-C and C-O bonds linking phenyl groups, there are number of bonds that must be broken in order to funnel lignin into the desired slate of phenolic compounds without methoxy groups (simple phenols). The subject of this presentation uses pyrolysis of lignin to create a mixture of gas phase species that can be passed over a catalyst to oxidatively cleave the methoxy groups to produce simple phenols. Because the goal is to partially oxidize the lignin pyrolysis vapors to produce phenols, as opposed to over-oxidation which would result in CO and CO_2 production, the oxidation activity of the catalysts must be controlled. Catalysts with varying vanadium-oxygen bond strength, which affects the oxidation activity, were designed by altering the catalytic support and synthesized using incipient wetness impregnation. Reaction experiments were conducted by passing lignin pyrolysis vapors over the catalysts at 500 °C and were shown to produce simple phenols from methoxy phenols. The effect of catalyst composition will be discussed along with relevant characterization techniques, coupled with computational modeling, to help understand the role of metal-oxygen bond strength in the catalytic activity.

526393 Mechanistic Study of the Hydrogenolysis of Diaryl Ethers Catalyzed By Heterogeneous Metal Catalysts

Abstract:

Manipulating reaction pathways of metal-catalyzed reactions is key to efficiently convert polar aromatic compounds. Diaryl ethers undergo C-O hydrogenolysis and hydrogenation with rates that depend on the specificity of the metal. We are studying the corresponding catalytic chemistry of Ni, Pd, Ru and Pt with reaction kinetics and experiments with isotopically labeled reactants. We have found a common reaction mechanisms for all studied metals but different isotopic effects, which points to the interaction of hydrogen with the metal as the factor determining selectivity. The details of the mechanisms, which are being tested by theoretical calculations, will be reported and selectivity rationalized as a function of the binding energy of reactants with the metal.

527899 Palladium-Iron Bimetallic Catalyst: High Activity and Stability for Aqueous Phase Hydrogenations

Abstract:

A novel palladium-iron bimetallic catalyst was synthesized that provided high hydrogenation activity and stability relative to palladium only under harsh hydrothermal conditions. To generate the catalyst, palladium was dispersed on carbon coated silica, which was doped with iron. The catalyst was applied to the hydrogenation of carbonyl groups in acetone, 2-pentanone, and propionaldehyde. The Pd-Fe catalyst gave more than a three-fold higher hydrogenation activity than the nanoparticle Pd only counterpart, while Fe nanoparticles were shown to be inert. The synthesis procedure was optimized for the bimetallic material, which demonstrated that the introduction of Fe contributed to a higher turnover frequency rather than simply increases the number of active sites. Interestingly, the same synthesis procedure was used to synthesize a Pd-Mo bimetallic catalyst that also increased activity.

536491 Hydrodeoxygenation of Guaiacol over Ni and Mo Nanoparticles Supported on SBA-15 and γ-Al_2O_3.

Abstract:

1. **Introduction**

The limitation of fossil energy resources and the increasing need for chemicals and liquid fuels require strong efforts in utilizing renewable energy sources. In this way, a technology such as biomass to liquid fuels (BTL) is being focus. The fast pyrolysis is a promising tool to transform various biomass feedstock's, lignocellulosic biomass, into an energy-dense and oily mixture. However, it possesses considerable amounts of oxygenates molecules, phenolic compounds and water in the material obtained by this method make it impossible to use the final product as motor fuel. For the use in chemical industry or fuel production an upgrading of the pyrolysis oil is necessary, which can be achieved via hydrodeoxygenation (HDO). This process consists in the treatment of bio-oil at temperatures between 200 and 450 °C with hydrogen at high pressure (10–300 bar) in the presence of a catalyst, so that the oxygen contained in organic molecules can be extracted with water formation[1]. Research efforts to study the chemistry of hydrodeoygenation of biomass-derived oil are gaining considerable importace over the last 20 years[1].The most frequently examined HDO catalysts commercially used for petroleum hydrotreatment products are NiMo/Al_2O_3 and CoMo/Al_2O_3.In the present study catalysts based on supported Ni and Mo are evaluated as catalysts for HDO reaction. The effect of metal phase and support properties, SBA-15 and γ-Al_2O_3 (reference support), as well as their interactions has been investigated. For the catalytic reactions, guaiacol was chosen as a model compound for lignin depolymerization fragments formed during bio-oil pyrolysis from lignocellulosic biomass.

2. **Methods**

The pure mesoporous silica SBA-15 was synthesized by hydrothermal method. The γ-Al_2O_3 was purchase from Degussa. The bimetallic nickel-molybdenum catalysts were prepared by wet impregnation. First, appropriate quantities of $(NH_4)_6Mo_7O_{24}.4H_2O$ and $Ni(NO_3)_2.6H_2O$ were dissolved in deionized water; the support was added and stirred for 2 h. Then, the suspension was freeze dried and the catalyst was calcined at 500 C for 2 h, under air flow. The metal contents were 15% of MoO_3 and 5% de NiO (weight percentage) with relation to support. The materials were characterized by nitrogen physisorption isotherms, transmission electron microscopy (TEM), X-ray powder diffraction (XRD), energy-dispersive spectroscopy (EDS) and with Raman and infrared spectroscopy. The total acidity of the catalysts was measured by Temperature programmed desorption of NH_3 (TPD-NH_3). The catalytic tests were carried out in a fixed bed flow reactor system under 15 kgf/cm^2 pressure. The catalysis, mass around 200 mg, was pre-treated in situ with 50 cm^3/min flow of pure H_2 at 500° C for 4 h and then cooled to the reaction temperature (200, 250 or 300° C). The feed mixture consisted of 2% w/w guaiacol (MERCK) in heptane P.A. (Aldrich) with 14.8

cm^3/h flow and 100 cm^3/min H_2 flow. The feed and the products were analyzed by gas chromatography Shimadzu CG17A with FID detector, capillary column CPsil 5CB and automatic injection.

3. Results and discussion

The fresh catalysts X-Ray diffractogram show that the Ni and Mo were successfully incorporated under the SBA-15 and Al_2O_3 supports as NiO and MoO_3 nanoparticles. After nickel and molybdenum incorporation the surface area of the SBA-15 decrease from 630 to 332 $m^2.g^{-1}$. The same behavior was observed for γ-Al_2O_3 support, the surface area chance from 279 to 233 $m^2.g^{-1}$. The temperature reduction of Ni and Mo oxide were determined by TPR-H_2, 500 °C under H_2 atmosphere for 4 hours. The catalysts total acidity for NiMo/SBA and NiMo/ALU were 470 and 548 $\mu mol.g^{-1}$, respectively. The results of catalytic activity for both materials are summarized in Table 1. For NiMo/SBA and NiMo/ALU the reaction temperature have a positive effect in guaiacol conversion, the higher results were observed at 300 °C 84.7% and 72.6% for NiMo/SBA and NiMo/ALU, respectively. However, the higher HDO values occurred for reaction temperature at 200°C. Thus, the selectivity for deoxygenated compounds were more pronounced in low temperatures for both catalysts. The principal products formed during the guaiacol HDO reaction were benzene, cyclohexane, cyclohexene, phenol and o-cresol. Compared to NiMo/ALU the NiMo/SBA showed the best performance in HDO, HDA and higher selectivity for benzene in all reaction temperature studied.

Table 1: The catalytic results of NiMo/SBA and NiMo/ALU in the hydrodeoxygenation of guaiacol.

	NiMo/SBA			NiMo/ALU		
Temperature (°C)	**200**	**250**	**300**	**200**	**250**	**300**
Conversion (%)	3.7	74.5	84.7	5.1	45.0	72.6
HDO (%)	92.1	74.3	66.5	37.9	31.5	30.8
HDA (%)	84.2	48.7	35.3	23.1	12.3	2.7
Selectivity (%)						
Benzene	16.7	8.6	1.3	3.3	1.1	0.3
Cyclohexane	29.1	13.9	11.0	6.8	3.3	0.4
Cyclohexene	38.5	26.2	19.5	12.9	7.9	1.9
Phenol	10.5	17.4	34.2	16.3	24.8	41.3
O-Cresol	0.0	18.4	8.0	7.7	8.5	11.3

With the catalysts thermogravimetric analysis were possible quantified the coke formation after HDO reactions. For NiMo/ALU and NiMo/SBA the coke formation were 1,65 and 1.14 $mg.g_{cat}^{-1}.h^{-1}$.

4. Conclusions

Compared with NiMo/ALU the NiMo/SBA show the best performance in guaicol conversion, selectivity for deoxigenated compound and in HDO percentage. The principal compounds formed during the HDO reaction with NiMo/SBA were cyclohexene, cyclohexane and bezene. For NiMo/ALU the principal compounds were phenol and o-cresol. Thus, the mesoporous silicon oxide (SBA-15) was the best support for nickel e molybdenum nanoparticles resulting in the catalyst for guaicol hydrodeoxygenation.

References

- Arun, R.V. Sharma, A.K. Dalai, Renew. Energy Rev., 48 (2015) 240-255.

Keywords

Hydrodeoxygenation, catalysis, nickel, molybdenum.

530385 Hydroprocessing of Biomass-Derived Oxygenates on Metal-Exchanged Zeolites Using Light Alkanes As the Source of Hydrogen

Abstract:

The broader use of natural gas and biomass for energy applications is limited because of the short hydrocarbon chains and/or high oxygen content of the molecules that comprise these resources. Current technologies for natural gas and biomass upgrading require high temperatures and suffer from poor selectivities. Furthermore, biomass deoxygenation requires external sources of H_2 molecules that are most commonly derived by sacrificing a portion of the feedstock in steam reforming reactions. In the absence of molecular hydrogen, a portion of the carbon feedstock is converted to arenes during hydrogen transfer reactions to satisfy the stoichiometry for oxygen removal as water. This work aims to improve selectivity and solve the hydrogen deficiency issue by developing new catalysts that can facilitate hydro-deoxygenation reactions using light alkanes as the hydrogen source. These hydrogen transfer steps have the potential to selectively terminate growing hydrocarbon chains via desorption as less reactive alkanes, thereby producing targeted classes of compounds, as opposed to a wide range of products over a broad carbon number distribution. Furthermore, this strategy has the potential to upgrade a portion of the light alkane feed because the hydrogen transfer steps generate alkenes that can participate in subsequent C-C bond formation reactions.

Reactions of isobutane and n-butanal (either as single reactants or co-feeds) on H-BEA zeolite and H-BEA samples partially exchanged with Zn cations were studied to probe the effects of alkanes as co-reactants during acid-catalyzed reaction of oxygenates. The Zn-exchanged BEA zeolite was synthesized via ion-exchange using aqueous solutions of zinc nitrate followed by treatment in flowing, dry air first at 373 K for 20 h then at 773 K for 4 h. This method has been previously shown to effectively exchange protons for Zn^{2+} ions within the cages of zeolites. n-Butanal reactions (2 kPa) on H-BEA predominantly form 2-ethyl-2-hexenal via aldol-condensation (60% selectivity to the aldol-condensation dimer) while also forming xylenes and cyclic ketones via cyclization/deoxygenation of 2-ethyl-2-hexenal (25% selectivity). n-Butene formation was also observed (selectivity of 15%) via hydro-deoxygenation of n-butanal. Addition of isobutane as a co-feed on H-BEA leads to minor changes in the product selectivity, indicating that the unsaturated species formed from n-butanal conversion are not strong enough hydride acceptors to activate isobutane via hydride transfer on Bronsted acid sites. Product selectivities for n-butanal conversion on Zn-H-BEA were similar to those on H-BEA, suggesting a minor effect of the Zn sites on n-butanal reaction paths. Co-reaction of isobutane (40 kPa) and n-butanal (2 kPa) on Zn-H-BEA exhibited a completely different product distribution. The predominant products were C_4 hydrocarbons (34% selectivity with 50% selectivity to n-butane within C_4) and C_8 alkanes (dimethyl-hexane and methyl-heptane isomers; 31% selectivity). n-Butane is the expected isomer formed from deoxygenation-hydrogenation of n-butanal, and the C_8 isomers that were produced were those that would be expected from deoxygenation of aldol-condensation (methyl-heptanes) and Prins-condensation (dimethyl-hexanes) products. Reaction of isobutane/n-butanal feeds on the Zn-H-BEA catalyst also exhibited lower selectivity to the aldol-condensation dimer (16% versus 70%) and to cyclic species (16% versus 25%) compared to isobutane/butanal reactions on H-BEA. Isopentane formation (5% selectivity) and C_7 alkanes (methyl-hexanes and n-heptanes; 9% selectivity) were also observed on Zn-H-BEA during conversion of isobutane/butanal feeds. These C_7 isomers are what would be expected from deoxygenation of C_7 ketones derived from Tishchenko esterification-ketonization routes. Reaction of isobutane (40 kPa) was also carried out on H-BEA and Zn-H-BEA to identify the extent of isobutane dehydro-oligomerization reaction paths. Isobutane conversion was only 0.1% on H-BEA and only 0.2% on Zn-H-BEA, indicating that dehydro-oligomerization of isobutane only occurs to a negligible extent in parallel with n-butanal conversion paths. This negligible conversion of isobutane is consistent with the absence of trimethyl-pentane isomers from the product distributions (indicating undetectable isobutane dehydro-oligomerization).

530346 Flowthrough Reductive Catalytic Fractionation of Biomass

Abstract:

As an abundant and renewable source of carbon, lignocellulosic biomass is an excellent feedstock for sustainable fuel and chemical production. Biomass primarily consists of carbohydrates (cellulose and hemicellulose) and lignin. Carbohydrates have been extensively studied and shown to be versatile feedstocks for the production of fuels and chemicals. Lignin, the largest source of natural aromatics, becomes highly recalcitrant during carbohydrate processing, hindering chemical production from lignin. Reductive catalytic fractionation (RCF) has emerged as a leading biomass fractionation and lignin valorization strategy that simultaneously extracts, depolymerizes and stabilizes lignin without disturbing the carbohydrate fractions. Here, flowthrough reactors were used to investigate RCF of poplar. Most RCF studies to date have been conducted in batch, but a flow-based process enables the acquisition of intrinsic kinetic and mechanistic data essential to accelerate the design, optimization, and scale-up of RCF processes. Time-resolved product distributions and yields obtained from experiments with different catalyst loadings were used to identify and deconvolute events during solvolysis and reductive stabilization. Multi-bed RCF experiments provided unique insights into catalyst deactivation, and initial rate studies enabled the determination of kinetic parameters intrinsic to the solvolytic step of RCF. Overall, this work demonstrates the concept of flowthrough RCF, which will be vital for realistic scale-up of this promising approach.

530547 Hydrodeoxygenation of Sorbitol to Monofunctional Fuel Precursors over Co/TiO_2

Abstract:

Over the next two decades, demands for heavier fuels such as jet and diesel are projected to increase while demands for lighter fuels such as gasoline are projected to decrease.[1] Production of these heavy C_8-C_{22} hydrocarbons remains a challenge when using biomass feedstocks in large part because the monomeric units which comprise cellulose and hemicellulose possess only five or six carbon atoms. One potential route for producing these fuels involves converting carbohydrates into C_{5+} monofunctional oxygenates (MFs) by aqueous-phase hydrodeoxygenation (APHDO) followed by performing C-C coupling reactions.[2] APHDO consists of feeding a highly-oxygenated species in water at moderate temperatures (~500 K) along with hydrogen at high pressures (~6 MPa) over a bifunctional catalyst usually utilizing a noble metal on an acidic support.[3] In this chemistry it is desired to control the cleavage of C-O bonds and minimize the cleavage of C-C bonds by retro-aldol condensation, decarbonylation and decarboxylation reactions.[4] However, over-dehydration to lower-demand alkanes is undesired. In addition, the noble metals commonly used are costly, but utilization of low-cost base metals is often limited in APHDO due to their propensity to leach into the aqueous phase. However, recent developments have shown that Co/TiO_2 catalysts can be stabilized for aqueous-phase hydrogenations utilizing high-temperature pretreatments which take advantage of the strong metal-support interaction (SMSI).[5] The purpose of this study is to investigate the potential for MF production in sorbitol APHDO using a SMSI-stabilized Co/TiO_2 catalyst.[6]

Sorbitol APHDO of Co/TiO_2 produced a 56% yield (on a carbon basis) to MFs at a moderate WHSV of 0.70 h^{-1}. This is the highest yield reported from sorbitol in the literature. We estimate the Co catalyst is 50 to 1,000 times cheaper than previously reported precious metal catalysts. Increasing space time leads to excessive deoxygenation and therefore selectivity toward alkanes and CO_x while decreasing space time leads to higher oxygenates, including diols, oxygenated heterocycles, unidentified O_{3+} species. The MFs produced are mostly primary alcohols (37% at 0.70 h^{-1}), secondary alcohols (32%), and heterocycles (23%). Ketones (7%) and aldehydes (<1%) are produced in much lower yields. 67% of the MFs are C_{5+} and would therefore be suitable for coupling. FTICR-MS shows that sorbitol/anhydrosorbitol species can oligomerize at lower contact times to form C_9-C_{18} species which appear to break down to some degree at longer contact times to eventually form MFs. Overall, the product distribution vs. WHSV suggests that about one third of the fed carbon reacts through the retro-aldol route and overall 20% of the carbon is lost to C_1 gases such as CO_2 and methane. The need to suppress retro-aldol condensation and decarbonylation in sorbitol APHDO is a clear limitation for the production of C_{5+} MFs. In the present study, Co/TiO_2 catalyst shows irreversible deactivation with TOS due to sintering and leaching of the Co. Overall, this work shows that Co/TiO_2 can be used to produce monofunctional fuel precursors from sorbitol at high yields, but suffers from irreversible leaching and sintering due to chelation from highly-oxygenated species.

References

1. ExxonMobil, 2017 Outlook for energy: A view to 2040. ExxonMobil: Irving, 2017.
2. Kunkes, E. L.; Simonetti, D. A.; West, R. M.; Serrano-Ruiz, J. C.; Gärtner, C. A.; Dumesic, J. A., Catalytic conversion of biomass to monofunctional hydrocarbons and targeted liquid-fuel classes. *Science* **2008,** *322* (5900), 417-421.
3. Vilcocq, L.; Cabiac, A.; Especel, C.; Guillon, E.; Duprez, D., Transformation of sorbitol to biofuels by heterogeneous catalysis: Chemical and industrial considerations. *Oil & Gas Science and Technology–Revue d'IFP Energies nouvelles* **2013,** *68* (5), 841-860.
4. Li, N.; Huber, G. W., Aqueous-phase hydrodeoxygenation of sorbitol with Pt/SiO_2–Al_2O_3: Identification of reaction intermediates. *Journal of Catalysis* **2010,** *270* (1), 48-59.
5. Lee, J.; Burt, S. P.; Carrero, C. A.; Alba-Rubio, A. C.; Ro, I.; O'Neill, B. J.; Kim, H. J.; Jackson, D. H.; Kuech, T. F.; Hermans, I.; Dumesic, J. A.; Huber, G. W., Stabilizing cobalt catalysts for aqueous-phase reactions by strong metal-support interaction. *Journal of Catalysis* **2015,** *330*, 19-27.
6. Eagan, N. M.; Chada, J. P.; Wittrig, A. M.; Buchanan, J. S.; Dumesic, J. A.; Huber, G. W., Hydrodeoxygenation of Sorbitol to Monofunctional Fuel Precursors over Co/TiO2. *Joule* **2017,** *1* (1), 178-199.

536942 Optimization of Sustainable Processes Incorporating Data Envelopment Analysis

Abstract:

Developing effective tools for process design is of paramount importance in the transition towards sustainable systems. In this work, we present a framework to incorporate sustainability principles in the design of chemical processes which combines a palette of tools, including life cycle assessment, surrogate modeling, objective reduction, multi-objective optimization, and multi-criteria decision analysis tools. In particular, we propose the use of Data Envelopment Analysis (DEA) to facilitate the post-optimal analysis of the Pareto front generated during the optimization. The use of DEA in the framework allows to narrow down the number of designs and rank them without the need to define weights in an explicit manner. Additionally, DEA also provides improvement targets for the sub-optimal alternatives that if attained would make them optimal. This characteristic of DEA can help to guide retrofit efforts when benchmark processes are compared against the best technologies available. The application of all the stages of the framework is demonstrated in the hydrogenation of CO_2 to produce methanol, where the process was initially modeled using a commercial process simulator to be further replaced by a surrogate model applying Artificial Neural Networks. The sustainability of the process was measured with the environmental categories included in the CML-2001 methodology and the total annualized cost. The objective-reduction stage narrowed down the number of total objectives to evaluate from 11 to five and the optimization was performed using a multi-objective Genetic Algorithm. The final application of DEA to analyze the solutions obtained from the optimization presented only 9 designs as the best options out of the 126 solutions that conformed the Pareto front. Finally, these 9 solutions were ranked and the best design identified. Overall, the methodology allowed an entire assessment of the process without compromising the performance of any sustainable indicator while reducing the computational resources used in the evaluation. The development of contributions such as the one presented in this work aim to facilitate the sustainable development of the chemical industry, not only by assessing new processes and technologies but also by identifying improvement targets for current sub-optimal ones.

535113 A Novel Model Predictive Control Scheme for Sustainability: Application to Biomass/Coal Co-Gasification System

Abstract:

Industrial, business and government communities have begun to shift from economic stand-alone focus to inclusion of sustainability in the decision-making process. This shift is due to the adverse environmental impact and unsustainable development caused by human activities, including chemical industry releases. As a result, process systems engineering (PSE) approaches have been developed, predominantly for incorporating sustainability into chemical process design and optimization [1]. However, the development of sustainability-oriented control schemes is still scarce, especially when compared to the number of steady-state design and optimization studies available [2]. Process control plays a critical role in realizing the efficient, sustainable and safe operation of the chemical and energy systems defined by the process design and optimization studies. The main barriers that have been prevented the integration of sustainability into process control are: 1) complexity of integrated large-scale chemical and energy processes; and 2) existence of multiple objectives for optimization involving social, economic and environmental issues.

To fill this gap, a novel model predictive control (MPC) scheme is proposed to drive the system to a sustainable operating point that is defined using a multi-objective optimization algorithm. In particular, the sustainable MPC is formulated based on dimensionless sustainability performance indicators from the U.S. EPA's GREENSCOPE (Gauging Reaction Effectiveness for the ENvironmental Sustainability of Chemistries with a Multi-Objective Process Evaluator) tool [3]. These established indicators are associated with the process state variables, releases, resource consumption, and products and can capture the sustainability information of the current process condition, including economic, environmental and social aspects. Such sustainability indicators are employed as hard/soft constraints in the implemented controller in order to maintain the process operation within a pre-defined sustainable zone, where the performance indicator values are higher than desired thresholds. To explicitly visualize the multidimensional sustainability indicators during transient, a multivariate plotting method is developed using dynamic radar diagrams.

The developed method is illustrated via a biomass/coal co-gasification process for syngas production with the end goal of methanol manufacturing. For this application, the whole process model is developed in Aspen Hysys based on existing literature information [4-5]. With the established models in Aspen and a link for data communication between Aspen and MATLAB, a multi-objective optimization problem is solved to maximize profit and optimize the process sustainability performance (e.g., environmental release and resource use minimization), by employing a genetic algorithm-based approach developed in MATLAB. In this presentation, the details on the application results of this novel framework for improving sustainability performance are discussed, focusing on the tradeoffs between using coal and biomass for the sustainable production of chemicals. The results show that the proposed sustainable MPC scheme can effectively drive the process to the optimal operating point, while maintaining the process within sustainable zones during transients.

References:

1. Daoutidis P, Zachar M, Jogwar SS. Sustainability and process control: A survey and perspective. J. Process Control. 2016;44:184–206.
2. Li S, Mirlekar G, Ruiz-Mercado GJ, Lima FV. Development of chemical process design and control for sustainability. Processes. 2016;4(3):23.
3. Ruiz-Mercado GJ, Gonzalez MA, Smith RL. Sustainability indicators for chemical processes: III. biodiesel case study. Ind. Eng. Chem. Res. 2013;52(20):6747–60.
4. Li S, Feliachi Y, Agbleze S, Ruiz-Mercado GJ, Smith RL, Meyer DE, Gonzalez MA, Lima FV. A process systems framework for rapid generation of life cycle inventories for pollution control and sustainability evaluation. Clean Technol. Environ. Policy. 2018. DOI: 10.1007/s10098-018-1530-6.
5. Robinson PJ, Luyben WL. Simple dynamic gasifier model that runs in Aspen Dynamics. Ind. Eng. Chem. Res. 2008;7784–92.

536285 Sustainability Dynamics and Control for Distributed Manufacturing of Renewable Energy

Abstract:

Prediction of a long-term sustainability performance of a large-scale system is always very challenging largely due to the existence of various types of uncertainties. Such a system usually consists of a number of interacted sub-systems, each of which could be structurally and operationally sophisticated. Distributed biodiesel manufacturing in a large geographical region could be a good example. A further challenge is the development of effective strategies, frequently for implementation in multiple stages, to improve regional sustainability performance in a time horizon of interest. Since sustainability is an act of balance among economic, environmental, and social domains, thus from system and control science point of view, sustainability assessment, performance prediction, and decision making is a non-conventional multi-objective, multi-input-multi-output control problem [1].

Recently, we developed a hierarchical control method to enhance manufacturing sustainability, which was successfully applied to a biodiesel manufacturing process [2]. In that framework, a two-layered hierarchical control scheme is introduced to implement strategic and tactical control for dynamic sustainability control. In this paper, we introduce an extended hierarchical control method, which is capable of conducting an integrated analysis of subsystems' dynamics and control strategies and developing strategies for coordinated control of the entire system to achieve a long-term sustainability goal under various uncertainties. The introduced methodology is tested through a case study on the sustainability performance prediction and improvement of a geographically distributed manufacturing of biodiesel using different feedstocks and different manufacturing technologies.

References:

[1] Moradi-Aliabadi, M. and Y. Huang, "Multistage Optimization for Chemical Process Sustainability Enhancement under Uncertainty," ACS Sustainable Chemistry and Engineering, 4(11), 6133-6143, 2016.

[2] Moradi-Aliabadi, M. and Y. Huang, "Decision Support for Enhancement of Manufacturing Sustainability: A Hierarchical Control Approach," ACS Sustainable Chem. Eng., 6(4), 4809–4820, 2018.

530944 An Analysis of Socioeconomic Impacts of Aviation Biofuel Development in Brazil

Abstract:

Aviation biofuel has been recognized by its potential to reduce greenhouse gas emissions of the aviation industry. Derived from renewable feedstocks, aviation biofuel is generally perceived sustainable. However, some of the sustainability issues such as land use change, food insecurity and biodiversity loss, are closely related to different phases of biofuel supply chains. Thus the overall impacts of biofuel production require a full investigation considering all dimensions of sustainability. While many studies have evaluated the environmental impacts or the techno-economic feasibility of aviation biofuel, very few have addressed the social or socioeconomic aspects in a systematic manner. The objective of this study is, therefore, to evaluate socioeconomic impacts of aviation biofuel production with a supply chain perspective.

A scenarios-based input-output (IO) analysis is used to evaluate socioeconomic effects including employment, GDP and trade balance (represented by imports) of aviation biofuel supply chains. This approach is exemplified for the local production of aviation biofuel in Brazil. Given that the commercialization of aviation biofuel is still at an early stage, available knowledge and data on the deployment of aviation biofuel are limited. A scenario analysis is useful to explore how possible futures of aviation biofuel would develop. The timeframe of the designed scenarios is set for 2050. The storyline of each scenario is elaborated on the market of aviation biofuel, conversion technologies, selection of feedstocks, and potential competition for biomass. Furthermore, the future demand of aviation biofuel under each scenario is estimated, which can subsequently be used in IO analysis to further determine the effects on employment, GDP and imports. In addition, we propose a stochastic simulation to address the uncertainty of IO analysis, by performing a Monte Carlo (MC) simulation for the technical coefficients in the IO matrix. The uncertainty analysis provides insights into the robustness and reliability of the assessment results.

Four scenarios are developed based on the diverging trends of two key driving forces: i) biofuel policies (conservative or proactive) and ii) technological advancement (gradual or breakthrough). Different demands of aviation biofuel are projected for each scenario, ranging from 3% to 15% of the total fuel demand. Three combinations of technologies and feedstocks are considered for producing aviation biofuel in Brazil: alcohol to jet (ATJ) with sugarcane, hydro-processed esters and fatty acids (HEFA) with macauba, and Fischer-Tropsch (FT) with eucalyptus.

Results of IO analysis show that the scales of socioeconomic effects on employment, GDP and imports are proportional to the estimated demand of aviation biofuel in each scenario. Specially, Scenario 3 (where proactive biofuel policies go hand in hand with advanced technology) leads to the highest employment, GDP and imports potential, whereas the lowest socioeconomic effects are expected in Scenario 1 (in which biofuel policies are conservative and biofuel technologies progress slowly). Furthermore, under the same scenario, the macauba-HEFA chain performs best in terms of employment creation, followed by the sugarcane-ATJ chain and at last the eucalyptus-FT chain. With regard to GDP, the macauba-HEFA chain leads to higher effects than the other two chains. On the other hand, the most significant effects on imports are estimated in the eucalyptus-FT chain, followed by sugarcane-ATJ chain, while the macauba-HEFA chain sees the lowest imports. Additionally, the uncertainty analysis suggests that socioeconomic effects calculated with deterministic method are generally higher than the stochastically simulated results. The variances range from 1% to 28%.

To conclude, it is clear that enabling policy environment, high level of technology readiness, as well as availability of sustainable biomass play a key role in maximizing socioeconomic benefits. The feedstock sectors are the key sectors responsible for employment and GDP generation. While this reveals substantial potential of job opportunities and social development, the cultivation of feedstocks is associated with many uncertainties such as climate, land availability, and agricultural practices. Only when the production of aviation biofuel are developed in a way without threatening food security or deteriorating land conditions, the positive socioeconomic impacts do not contradict sustainability. In addition, the proposed stochastic simulation approach has been useful to address uncertainty in IO analysis. This contributes to an informed decision-making process regarding aviation biofuel development, from the perspective of social sustainability. To the best of the authorsâ€™ knowledge, this is an innovative attempt to address uncertainty of IO analysis in the context of ex-ante socioeconomic assessment by using stochastic simulation to capture the uncertainty in the input-output matrix.

521791 A Coordinated Multi-Product Market for Organic Waste Management

Abstract:

One of the major challenges associated with organic waste management is determining an economic value for products. For instance, in current organic waste management systems, gate/tipping fees are normally required for processing, indicating that the waste supplier is willing to pay someone else for the waste disposal [1]. On the other hand, organic waste is a potential source for deriving value-added products such as biofuels, electricity, and fertilizers [2,3,4]. Thus, this suggests that waste suppliers can, in fact, earn income by letting others use their waste. Unfortunately, systematic waste exchange frameworks that can determine prices for waste streams are currently lacking. This is currently introducing significant levels of uncertainty on the market potential for derived products, technologies, and hindering investment.

We propose a coordinated market framework to exchange waste streams, processing, transportation services, and derived products. The framework is motivated and designed after coordinated electricity markets that currently exist in operation and that have led to high economic efficiency and technology innovation [5]. Also, the suppliers and consumers of waste and derived products, as well as technology and transportation providers bid into the market to offer their services. Prices for waste and derived products are obtained by solving a market clearing problem that balances products across a supply chain network that connects all the market participants. The market clearing problem provides prices for all products and services across the network.

We prove that the market clearing problem provides prices under which the profits of all market players are non-negative and provides a Pareto optimal solution (thus representing a true trade-off between the profits of all players). We provide illustrative examples to show the capability and properties of this coordinated market framework and demonstrate that the framework is flexible and can accommodate more sophisticated market players such as environmental remediation costs, waste storage, and policymakers and local/state/federal governments (which provide incentives/penalties to drive the market). The market can be cleared on a regular basis to adjust waste flows and prices due to seasonal effects (e.g., nutrient run-off during spring or poor air quality in particular areas). Moreover, the market naturally captures geographical effects (by providing space-dependent prices). This property can be used to redistribute waste streams across geographical regions (e.g., push nutrients away from endangered areas).

We use the proposed framework in a real case study of the Rock River basin in the State of Wisconsin. We consider (i) multiple organic waste streams including cow manure, wastewater sludge, and household food waste; (ii) multiple waste processing technologies that include nutrient recovery, energy production, and natural environmental degradation; (iii) regulatory drivers, such as RECs (Renewable Energy Certificates) and RINs (Renewable Identification Numbers) [6,7]; as well as (iv) environmental and social constraints on key factors such as TMDL (Total Maximum Daily Loads) [8].

References:

[1] R. Raven and K. Gregersen, "Biogas plants in Denmark: successes and setbacks," Renewable and sustainable energy reviews, vol. 11, no. 1, pp. 116–132, 2007.

[2] K. Krich, D. Augenstein, J. Batmale, J. Benemann, B. Rutledge, and D. Salour, "Biomethane from dairy waste," A Sourcebook for the Production and Use of Renewable Natural Gas in California, pp. 147–162, 2005.

[3] L. Young and C. C. Pian, "High-temperature, air-blown gasification of dairy-farm wastes for energy production," Energy, vol. 28, no. 7, pp. 655–672, 2003.

[4] Z. Ye, Y. Shen, X. Ye, Z. Zhang, S. Chen, and J. Shi, "Phosphorus recovery from wastewater by struvite crystallization: Property of aggregates," Journal of Environmental Sciences, vol. 26, no. 5, pp. 991–1000, 2014.

[5] V. Zavala, K. Kim, M. Anitescu, and J. Birge, "A stochastic electricity market clearing formulation with consistent pricing properties," Operations Research, 2017, 65(3): 557-576.

[6] "Renewable Portfolio Standard," Database of State Incentives for Renewables & Efficiency (DSIRE) available at http://programs.dsireusa.org/system/program/detail/564. [Online; accessed 06-December-2017]

[7] L. McPhail, P. Westcott, and H. Lutman, "The renewable identification number system and us biofuel mandates," ERS Report BIO-03, USDA, November 2011.

[8] D. Columbia, F L. Dodge, G L. Green, "Total Maximum Daily Loads for Total Phosphorus and Total Suspended Solids in the Rock River Basin," 2011.

536446 Vulnerability of United States Industrial Sectors Dependent on Insect-Mediated Pollination Service

Abstract:

Of the many critically important ecosystem goods and services that sustain human and industrial activity, one crucial service with high economic value is pollination by insects. The agricultural sector of the U.S. economy as well as many other industrial sectors (fiber, drugs, and fuel) are directly and indirectly dependent on insect-mediated pollination by both commercially-managed and wild species of bees and other insects. In forty-five years, production of pollination-dependent crops has increased more rapidly (50-62%) than the global population of managed bees (45%). Simultaneously, wild insects in all regions have declined up to and exceeding 75 percent over only the last twenty-seven years, both in biomass and with decreasing biodiversity. This disparity creates a concern to humans nutritionally and economically as many stress factors continue to influence the fitness of pollinators around the world. Most existing methods focused on environmental sustainability assessments across engineering including traditional life cycle assessment (LCA) ignore the critical role of ecosystem goods and services in product life cycles.

The present work utilizes field study data available in an extensive literature review to quantify the economic dependence of the United States agricultural and related sectors on pollination service mediated by insects, updating existing pollination dependence coefficients when possible and bounding uncertainty of the estimates with Monte Carlo or bootstrapping methods. Integration of InVEST model data and public price and production information available through the United States Department of Agriculture (USDA) National Agricultural Statistic Survey (NASS) and Census of Agriculture compiled at a national, state, and county level considers the spatial relationship between crop-specific economic dependence of pollination and landscape-specific pollinator forage suitability. Specifically, our results highlight regions of the U.S. most susceptible to pollinator loss or decline, having the highest demand for insect pollination services while simultaneously lacking appropriate pollinator habitat. Downstream, cascading dependencies of non-agricultural industrial sectors are assessed using input-output analysis framework. The implications of these findings include expanding understanding of pollination services and pollinator health and identifying dependence and vulnerabilities of agricultural and at risk industrial sectors. These findings also direct conservation and revitalization efforts, advise future policy development, and support the incorporation of valuable ecosystem services as a component of LCA through the creation of an impact category for insect-mediated pollination service.

526647 Low-CO_2 Integrated Networks for Heat and Electricity Based on Hydrogen: A Comprehensive Spatio-Temporal MILP Model for Planning, Design and Operation of Future Value Chains

Abstract:

Demands for space and water heating constitute a significant proportion of societyâ€™s overall demand for energy and can be several times those for electricity. As heat demands are primarily satisfied using natural gas boilers, the heat sector produces a significant amount of CO_2 emissions. An alternative low carbon solution is to combine heat pumps with renewable electricity but the cycling of the heat pumps can put a strain on existing electricity networks and renewable generation does not always match heat demands, particularly solar energy, which is highest in the summer months when heat demands are at their lowest. Further, large-scale seasonal electricity storage may not be possible for the foreseeable future. Therefore, another low-carbon alternative that can overcome these limitations is required and hydrogen is one of the possibilities. Low-carbon hydrogen can be produced using electrolysers powered by renewable electricity or from natural gas via steam methane reforming coupled with CCS. Clean hydrogen produced in this way has the following advantages when used as a carrier for heat: (1) it can be transported with little loss of energy, which is an advantage over electricity networks, which have losses of about 9% of production, and certainly much lower losses than transporting heat through a district heating network (which are limited in scale due to the cost and inefficiencies of transporting heat long distances); and (2) hydrogen can be stored with little or no loss. Long term direct thermal storage is generally not possible due to the difficulty in insulating the storage devices.

A mixed integer programming model was developed that can optimise different scenarios for generation, storage and transportation of renewable hydrogen to satisfy heat demands. The model considers the spatial distribution of both the heat demands and the availability of primary resources in order to make a comparison between centralised and distributed generation, as well as to determine the location of hydrogen plants and storage facilities and the transmission and distribution networks required. The temporal representation simultaneously captures the short-term operational issues, such as intermittency of renewable sources, and long-term planning decisions up to 2050 to examine different pathways from the present time to various potential optimal solutions. The model optimises the design and operational decisions to determine the most cost effective/environmentally-friendly transition to the future heat network while also determining what that network should be.

References:

[1] S. Samsatli, N.J. Samsatli (2018). A multi-objective MILP model for the design and operation of future integrated multi-vector energy networks capturing detailed spatio-temporal dependencies. Applied Energy. DOI: 10.1016/j.apenergy.2017.09.055.

[2] S. Samsatli, I. Staffell, N.J. Samsatli (2016). Optimal design and operation of integrated wind-hydrogen-electricity networks for decarbonising the domestic transport sector in Great Britain. Int. J. of Hydrogen Energy, 41, 447-475.

[3] S. Samsatli, N.J. Samsatli (2015). A general spatio-temporal model of energy systems with a detailed account of transport and storage. Computers and Chemical Engineering, 80, 155-176.

[4] S.M. Jarvis, S. Samsatli (2018). Technologies and infrastructures underpinning future CO2 value chains: a comprehensive review and comparative analysis. Renewable & Sustainable Energy Reviews, 85, 46-68.

529952 Membrane-Free Water Splitting for Hydrogen Generation

Abstract:

Power-to-Gas (P2G) technologies that convert renewable energies such as solar and wind to chemical fuels have been the focus of many scientific endeavors over the past decades. Specifically, hydrogen production by water electrolysis ($2H_2O$ â†' $2H_2 + O_2$) is a promising P2G technology. Photoelectrochemical (PEC) water splitting is a promising path to solar hydrogen production. However, separation of the H_2 and O_2 gas products as well as hydrogen collection and transport in large solar fields becomes an overwhelming technical challenge. State of the art water splitting technologies make use single cell units, separated into anode and cathode compartments by membranes. This implies that millions of PEC cell units in the solar field would have to be hermetically sealed and fitted with membranes, gas tubes and tube adaptors in order to separate and collect the hydrogen gas, resulting in a very complicated and expensive construction. The hydrogen would then have to be transported to the end user, either by pipelines or by high-pressure H_2 vessels. These obstacles, in addition to efficiency and stability challenges, render PEC hydrogen production economically questionable.

In this work, which was published recently in Nature Materials [1], we aim to solve these problems by totally separating the H_2-generation electrochemical cell from the O_2-generation PEC solar cell. This is achieved by introducing an additional set of electrodes, called the auxiliary electrodes. These are $Ni(OH)_2$/NiOOH electrodes, commonly used in rechargeable alkaline batteries, which can be cycled many times with minimal energy loss. By placing a charged (NiOOH) auxiliary electrode in the oxygen cell, and electrically connecting it to a discharged ($Ni(OH)_2$) auxiliary electrode in the hydrogen cell, electrolysis can be performed in two separate cells. During electrolysis, one auxiliary electrode charges while the other discharges. Thereafter, the process can be repeated by cycling the auxiliary electrodes between the charged/discharged states. Using suitable photoanodes, the PEC cell can generate O_2, which can then be discharged to the atmosphere, alleviating the need for sealing and piping. Since the separate cells are connected to each other by metal wires only, the H_2 can be generated at any location, for example, directly at a Hydrogen Refueling Station (HRS).

535417 Replicating Indian Point's Energy Generation Using Renewable Energy Pathways: A Technical and Economic Stochastic Analysis

Abstract:

Currently, the Indian Point Nuclear Center supplies approximately 25% of New York City and its surrounding areas with electricity. This is the closest nuclear facility to New York City. For this reason, it's imperative to keep a constant baseload power supply while keeping the surrounding area safe. The Indian Point Nuclear Center is scheduled to shutdown in 2020 and 2021. This study assesses the following five renewable energy pathways' ability to replace Indian Point's generation by assessing their renewable electricity productivity in terms of kWh/hectare: solar PV, willow biomass, onshore wind, onshore wind + solar PV, and onshore wind + willow biomass. More specifically, this study examines how to maximize the electricity productivity of the land while simultaneously minimizing land-use in Southeastern New York State and replicating Indian Point's generation profile. Stochastic analysis is performed in order to achieve the most realistic financial and technical outputs. This study is novel in its stochastic analysis of renewable energy pathways to replace Indian Point's generation.

533632 Ultra-High Thermal Effusivity Materials for Resonant, Ambient Thermal Energy Harvesting

Abstract:

Materials science has made progress in maximizing or minimizing the thermal conductivity of materials, however, the thermal effusivity â€“ related to the product of conductivity and capacity â€“ has received limited attention, despite its importance in the coupling of thermal energy to the environment. We design materials that maximize the thermal effusivity by impregnating high surface area copper (Cu) and nickel (Ni) foams with conformal chemical vapor deposited graphene (G) and octadecane (OD) as a phase change material, achieving values exceeding J m^{-2} $(s\ K)^{-1/2}$ â €“ the highest value in the literature, to date, for isotropic, ambient phase change materials. The graphene is shown to increase thermal conductivity up to 20% by enhancing interfacial thermal conductance and providing transport bridges over grain boundaries of the metal foams, whereas the foam porosity houses the OD for enhanced capacity in the form of latent heat. These composite materials are ideal for ambient energy harvesting in the form of what we call broadband thermal resonators to generate persistent electrical power from ambient thermal fluctuations over large ranges of frequencies. Theory and experiment demonstrate that the harvestable power for these devices is directly proportional to the thermal effusivity of the dominant thermal mass. To illustrate, we experimentally measure persistent energy harvesting from diurnal frequencies at an outdoor location in Cambridge, Massachusetts, extracting as high as 350 mV and 1.3 mW from approximately 10 ^{o}C diurnal temperature differences continually over 16 days using high thermal effusivity materials. Thermal resonance devices of this type may provide renewable and persistent energy sources over extended periods.

528133 Comparative Evaluation of Lead Emissions and Toxicity Potential in the Cradle-to-Gate Life Cycle of Lead Halide Perovskite Photovoltaics

Abstract:

Lead halide perovskites are an emerging class of photovoltaic (PV) materials that have drawn intense interest due to their power conversion efficiencies above 22% and their potential for low-cost fabrication. However, the toxicity of lead causes considerable concern about its use in perovskite solar cells at large scales. Here, we quantified the lead utilization and the toxicity potential of lead-based perovskite PV in potential commercial production. Lead intensity and toxic emissions in the lead halide perovskite PV life cycle can be 5-100 times lower than those in the coal fuel life cycle and those in a representative electricity mix for the United States, for PV operational lifetimes of 20 years. We introduce the metric â€œtoxicity payback timeâ€, which is <2 years for perovskite PVs produced by and displacing the U.S. electricity mix. Toxicity potential associated with the energy of manufacturing a PV system dominates that associated with release of embodied lead. Therefore, the use of lead should not alone preclude the commercialization of perovskite PVs; instead, effort should be focused on development of low-energy manufacturing processes and long service lifetimes. Additionally, detailed investigations are needed to quantify the full life cycle of commercial production of perovskites and to minimize potential emissions.

Preparation of Lithium Ion Battery Cathode Composites Using Leonardite-Derived Humic Acid

Justin Baker[1], Shuai Xu[1], Justin Mann[1], Andrew Dockter[1], Yong Hou[2], Michael Mann[1] and Xiaodong Hou[1]
(1) Institute for Energy Studies, University of North Dakota, Grand Forks, ND
(2) Clean Republic LLC, Grand Forks, ND

Abstract

Graphene can be used to greatly improve the performance of lithium ion batteries by increasing conductivity, improving cycling stability and low-temperature performance. High molecular weight humic acid has properties very similar to graphene oxide and has been shown to be capable of yielding graphene upon reduction. North Dakota leonardite is an oxidation product of lignite which has been found to have high levels of humic acid. This study focuses on using leonardite-derived humic acid to prepare a graphene-modified cathode material for lithium ion batteries, and the effect the modified cathode has on the electrochemical performance of the cell.

Introduction

Over the recent years, the demand for clean, sustainable energy has increased due to increased concern over the negative impact that fossil fuels have on the environment. Electrical energy storage technology is highly desired for sustainable energy systems that do not provide on-demand power. These systems require an energy storage medium for them to be viable, stand-alone systems. An ideal electrical storage medium should have the following properties: low cost, high energy density, good rate capabilities, excellent safety, and environmentally benign. Lithium ion batteries (LIBs) are currently most likely to meet all of these requirements [1].

Lithium ion batteries prepared with lithium iron phosphate (LFP) cathodes are of particular interest due to their intrinsic stability, long lifespan, abundant raw materials resources and low environmental impact [1]. One major drawback of LFP-based LIBs is their low specific capacity due to the low conductivity of the LFP active material. This can be overcome by adding carbon in the form of graphite [2], but this lowers the mass density of the active material, thus barring the material from having a high energy density. One solution to this is to use graphene-like material as a conductor in the active cathode material.

Graphene's high surface area and excellent ion mobility have been shown to improve several properties of LFP cathodes such as energy density, cycling stability, and rate capability [3]. For example, Zhou et al [4] mixed reduced graphene oxide (rGO) with LFP material resulting in graphene-wrapped, nano-sized LFP particles. These particles exhibited excellent capacity at high discharge rates, and low rate capacity decay even between 10 C and 20 C discharge rates. Mo et al [5] synthesized a graphene-modified LFP composite by mixing GO in an oil-water emulsion with LFP precursor solutions. The mixture was then dried and sintered under an Ar:H_2

atmosphere to reduce the GO. Cells prepared with this composite were able to achieve a capacity of 158 mAh/g at 0.1 C. While these methods are successful in producing high-performance graphene-modified LFP cathode materials, the high price of graphene oxide (>$100/g) [6] prevents their practical use.

A very promising alternative approach to prepare graphene is using coal as the precursor. Humic acids are the organic fraction of coal which is soluble in basic conditions. There is no defined molecular structure for humic acids, instead they exist as a wide range of molecular weights with various combinations of carboxyl and hydroxyl groups [7]. Among all the sources of humic acid, coal, and in particular low-rank coal like lignite or naturally oxidized lignite (leonardite), is the most abundant and commercialized source. Humic acids are commonly used as a fertilizer or soil supplement in agriculture and less commonly as a human nutritional supplement. In the past, studied have concentrated on extracting humic acid from soils, coal, and other organic matter [8]. The most common methods of extracting humic acid involve the use of aqueous solutions containing strong hydroxide bases [9] [10], though other organic solvents have been proven to yield similar results [7]. Humic acid produced by a single extraction step lacks the purity required for technical applications, so further purification is often required, such as reducing ash content by acid wash [11], and removing metallic impurities by chelating agents [12]. Recently several studies have shown that humic acid is capable of producing graphene upon thermal treatment between 400 and 1100 °C by carbo-thermal reduction (CTR) as well as other means [13] [14].

In this work we study the feasibility of using Leonardite-derived humic acid (HA) as a graphene precursor to prepare low-cost and high-performance graphene-modified LFP (LFP/G) cathode materials. Leonardite was chosen due to its high humic acid content, and its relatively low cost at $220/ton [15]. This work will focus on the optimization of humic acid extraction and purification procedure, and the effect the modified cathode has on the electrochemical performance of the lithium ion battery.

Experimental

Synthesis

Leonardite obtained from Leonardite Products, LLC in Willison, ND was used as the raw material. The humic acid was extracted using a proprietary procedure and used to prepare lithium iron phosphate (LFP) cathodes for electrochemical performance testing. The LFP was synthesized via sol-gel method using lithium acetate (Acros, 199842500), phosphoric acid (Fisher, A242-212), and iron nitrate nonahydrate (Aldrich, 216828) as precursors. The methods used to prepare the LFP and incorporate the humic acid are left out as they are currently being considered for patentability. A reference sample was prepared in which humic acid was replaced with dextrose. The samples were then sintered at 700°C for 12 hours under a nitrogen atmosphere.

Cathodes were prepared by applying onto an aluminum current collector a slurry comprised of 80% active LFP material, 10% conductive carbon black, and 10% PVdF binder suspended in N-methylpyrrolidone.

CR-2032 coin cells were prepared in an argon atmosphere (less than 1 ppm O_2 and H_2O) using cathodes prepared from the synthesized active material and metallic lithium chips as the counter electrode. Electrolyte used was 1 M $LiPF_6$ in EC/DMC 50/50 (v/v) solution.

Physical Characterization

Elemental compositions of leonardite and humic acids were determined by X-ray fluorescence (XRF) on a Rigaku Supermini 200. A Shimadzu TOC-V Total Organic Carbon Analyzer with a SSM-5000A Solid Sample Module was used to determine the carbon content of LFP samples. Crystal purity of the LFP samples was measured by X-ray diffraction (XRD) using a Rigaku Smartlab. Particle morphology was determined using a FEI Quanta 650 Field Emission Scanning Electron Microscope. The presence of graphene was determined using a Horiba Jobin Yvon ARAMIS Raman imaging system. Ash content was determined using ASTM D3174-12, Standard Test Method for Ash in the Analysis Sample of Coal and Coke from Coal.

Electrochemical Testing

Electrochemical testing was performed on a Neware BTS-4008 Battery Analyzer.

Results and Discussion

Extracted Humic Acid

The extraction and purification of humic acid resulted in a 40% yield from leonardite on a moisture-free basis. **Table 1** lists the elemental compositions of the leonardite starting material as well as the extracted humic acid used in the LFP preparation. It should also be noted that the organic component of leonardite includes humic acids as well as other organic matter. Many elements were significantly reduced, some down below the detection limit of XRF, as a result of the extraction and purification steps. Iron is perhaps the most significant impurity, as it has severe detrimental effects on the LFP material and electrochemical cells prepared with it. Ash is another component that negatively impacts the cathode materials prepared using humic acid. The leonardite raw material had an ash content of 20.9%, which was reduced to 0.72% in the purified humic acid. These results show that the purity of the obtained humic acid can satisfy the requirements to be used as a cathode material precursor.

Table 1. Elemental Compositions of Leonardite and Humic Acid

	Element	Na	Mg	Al	Si	P	S	Cl	K	Ca	Ti	Cr	Mn	Fe	Zn	As	Sr	Ba	Organic
Mass Percent	Leonardite	0.146	0.686	1.568	2.062	0.010	1.001	0.019	0.193	0.792	0.060	0.005	0.035	2.516	0.006	0.005	0.015	0.185	90.696
	Humic Acid	n/a	0.014	0.102	0.087	n/a	0.472	0.089	0.036	0.020	0.044	n/a	n/a	0.209	n/a	n/a	n/a	n/a	98.922

XRD Analysis

Figure 1 compares XRD profiles of the reference sample LFP/C as well as that of the LFP/G composite by humic acid. Both samples exhibit the typical pattern of olivine crystal structure (Pmnb group). Peaks locations and intensities for olivine-structured LFP are the last dataset on **Figure 1.** No impurities were found in both samples and the sharp peaks indicate a high degree of crystal purity. Also present in the LFP/G composite is a small peak around 2-theta 26.5°, which is the (002) peak of graphitized carbon. This peak is emphasized in the in-set on the right side of **Figure 1.** This result demonstrates humic acid tends to be graphitized during carbothermal reduction.

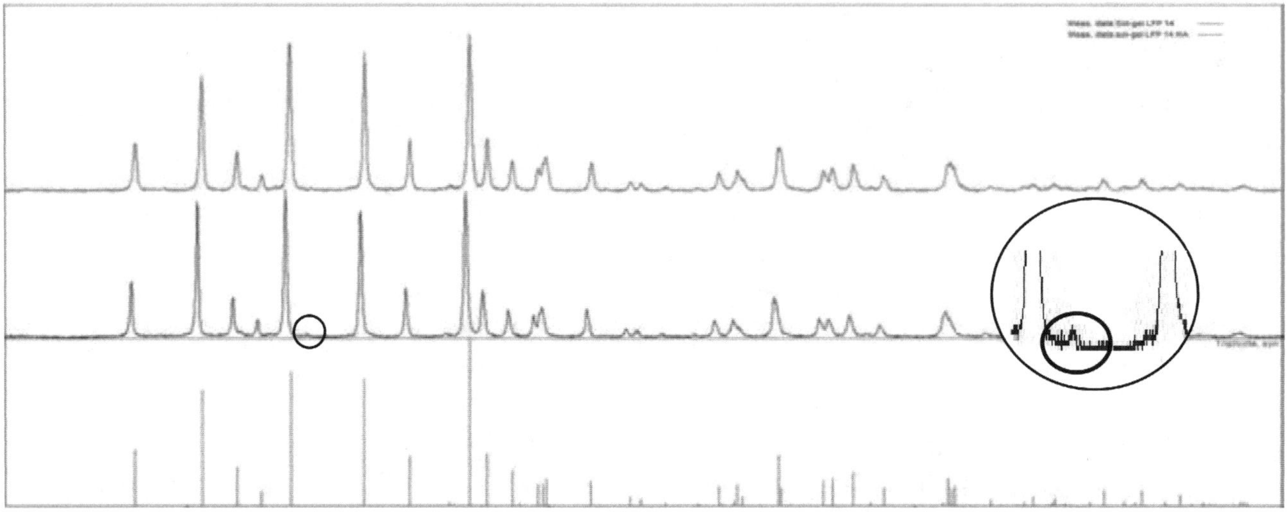

Figure 1. XRD Spectra of LFP/C reference sample (top), LFP/G composite (middle), and typical olivine-structured LFP (bottom). In-set area highlights the (002) peak of graphitized carbon.

Physical Characteristics

SEM images of the reference LFP/C sample as well as the LFP/G composite are shown below as **Figure 2. Figure 2a** depicts the reference LFP/C sample at 5000x magnification, with agglomerations of nano-sized particles. **Figure 2b** is an image of the LFP/G composite at 5000x magnification. It is apparent that LFP nanoparticles have formed on thin, graphene-like flakes. These thin flakes were not observed in the reference sample, indicating that they are a product of the humic acid. **Figure 2c** is an image of the thin flakes at 20,000x magnification. The nature of these plates corresponds strongly to highly ordered carbon in a form similar to few-layer graphene sheets.

Figure 2. SEM images of LFP samples. Reference LFP/C sample (A). LFP/G composite (B). High magnification of carbon plates (C).

Raman Analysis

Raman analysis was performed on samples after sintering to investigate the presence of highly ordered carbon or graphene. **Figure 3** compares the Raman spectra for the reference LFP/C material with that of the LFP/G composite sample. The reference sample has very similar peak intensities between the D- and G-bands, located at 1350 cm^{-1} and 1580 cm^{-1} respectively.

The I^D/I^G ratio for the reference sample is about 1.07, indicating an approximately equal amount of ordered and disordered carbon. The sample prepared with humic acid has a much greater G-band intensity with a I^D/I^G ratio equal to 0.528, indicating a much higher degree of ordered carbon.

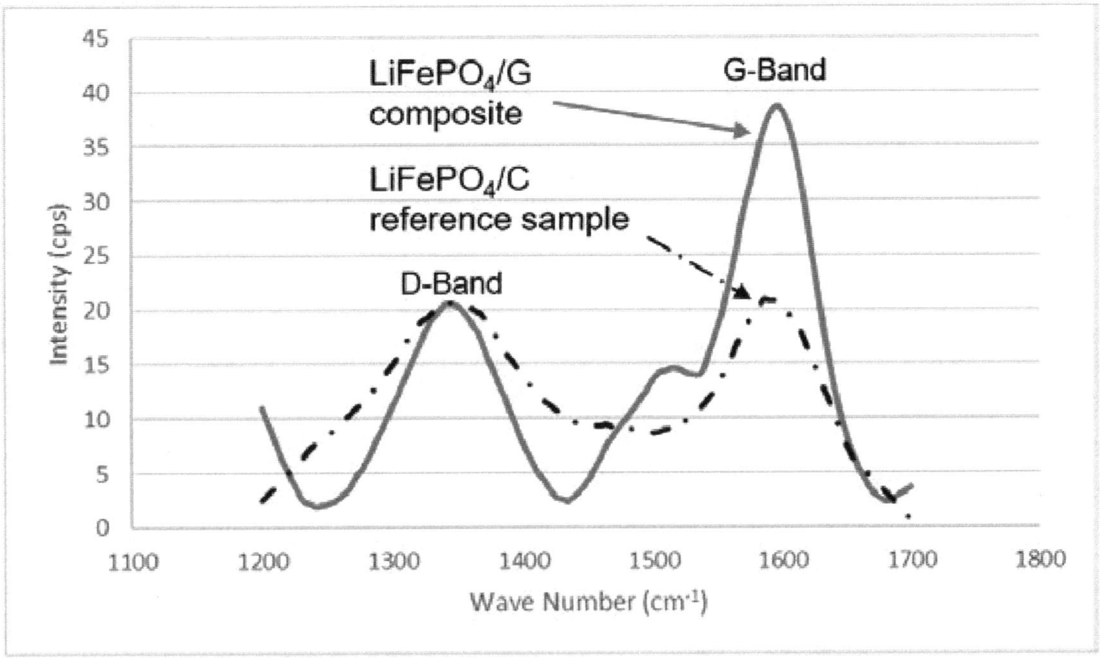

Figure 3. Raman spectra for the LFP/C reference sample and the LFP/G composite comparing ordered carbon (G-band) and disordered carbon (D-band).

Electrochemical Performance

Performance of the prepared coin cells was less than anticipated, likely due to the high amount of carbon in the samples. The reference sample prepared using dextrose had a carbon content measured at 6.91% wt. The sample prepared using humic acid had a carbon content of 10.49% wt.

Figure 4 is a comparison of charge/discharge voltage profiles for the reference sample and the LFP/G composite. The LFP/G composite profile displays a higher capacity than the reference sample, as well as some distortion at lower voltages around 2.7V. This could be due to the high theoretical capacity of graphene [16]. **Figure 5** compares the cycling capacity retention between the samples. Both samples are very similar, having specific capacities of about 100 mAh/g at 1C. As the LFP/G composite has a higher carbon content, it would be expected that the specific capacity would be greater than the reference sample if the carbon contents were normalized.

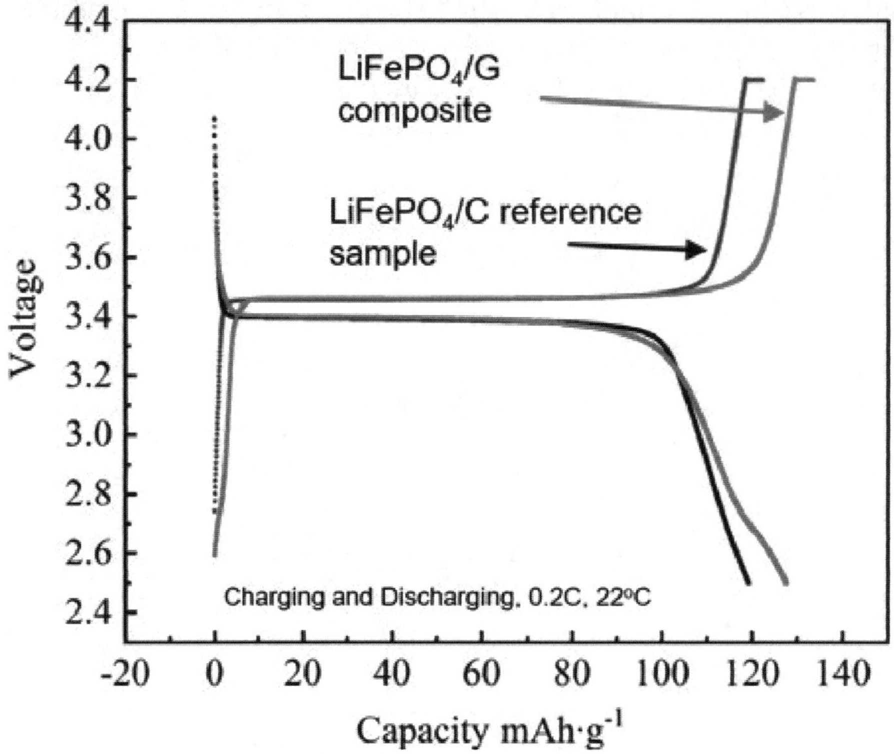

Figure 4. Charge/discharge voltage profiles for the LFP/C reference sample and the LFP/G composite.

Figure 5 compares charge/discharge rate capabilities of the two samples. The LFP/G composite performed slightly better than the reference sample at charge/discharge rates of 0.2C and 0.5C. The performance of both samples becomes virtually identical at a charge/discharge rate of 1C, which is attributed to the high carbon content in the LFP/G composite. The samples were tested at higher charge/discharge rates of 2C and 5C, however the results between the two samples were indistinguishable. This is likely due to the non-optimized carbon content in both samples, as well the difference in carbon content between them. If carbon content in the samples was optimized further, it would be expected that the LFP/G composite would significantly outperform the reference sample, as a sample containing graphene or few-layer graphite would ideally display high specific capacity at high charge/discharge rates.

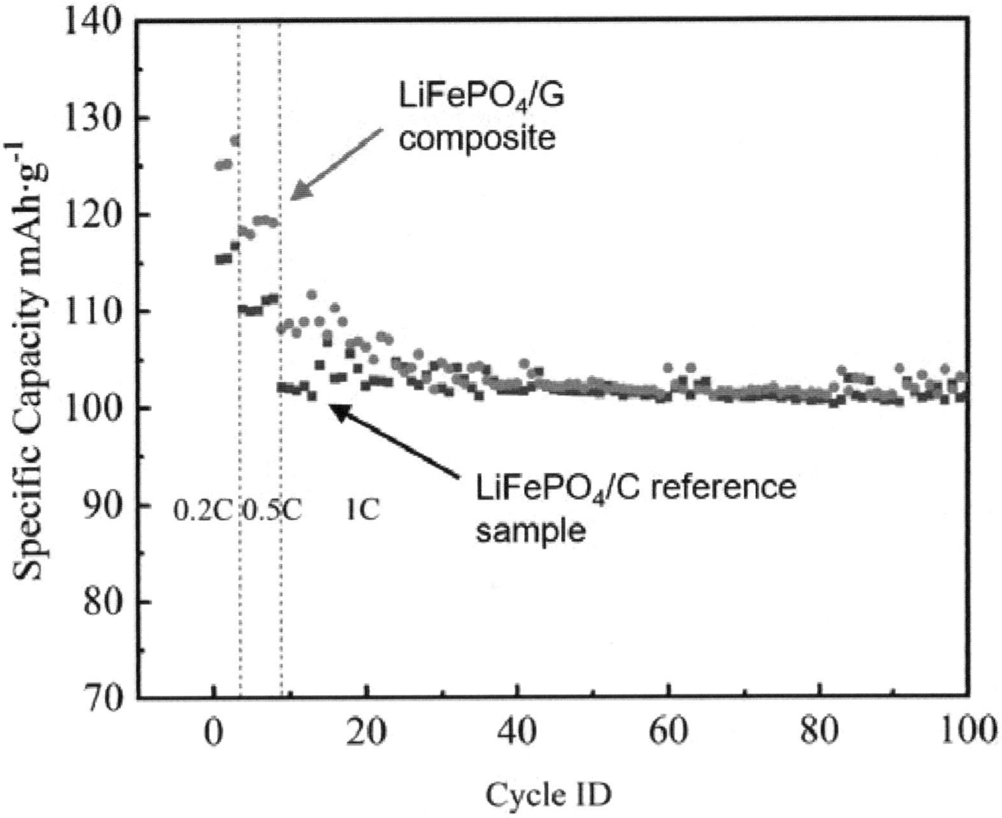

Figure 5. Charging/discharging rate capabilities of the LFP/C reference sample and the LFP/G composite. The rate is 0.2C for 4 cycles, 0.5C for 8 cycles, and then 1C for the remainder.

Conclusions

Humic acid was extracted from leonardite followed by iron and ash reduction to suitable levels for battery use. Lithium iron phosphate cathodes were prepared with humic acid and compared with those which were prepared using dextrose as a carbon source. Thin carbon sheets were observed under SEM in the LFP/G composite. This combined with the low I^D/I^G ratio for the composite are strong indicators that graphene is forming as a product of humic acid undergoing carbothermal reduction. Cathodes prepared with humic acid displayed higher electro-chemical performance, but in the future, the synthetic procedure for LFP/G need be optimized to further improve the electrochemical performance. In conclusion, humic acid has the potential in the manufacturing of environmentally benign cathode composite materials for Lithium ion batteries.

Acknowledgements

The authors gratefully acknowledge the financial support from the North Dakota Industrial Commission Renewable Energy Program (Project No. R-035-044) and Clean Republic LLC.

References

[1] J. Wang and X. Sun, "Understanding and recent development of carbon coating on LiFePO4 cathode materials for lithium-ion batteries," *Energy & Environmental Science,* vol. 5, no. 1, pp. 5163-5185, 2012.

[2] D. Jugovic and D. Uskokovic, "A review of recent developments in the synthesis procedures of lithium iron," *Journal of Power Sources,* vol. 190, no. 2, pp. 538-544, 2009.

[3] E.-S. M. Duraia, S. Niu, G. W. Beall and C. P. Rhodes, "Humic acid-derived graphene–SnO2 nanocomposites for high capacity lithium-ion battery anodes," *Journal of Materials Science: Materials in Electronics,* vol. 29, no. 10, pp. 8456-8464, May 2018.

[4] X. Zhou, F. Wang, Y. Zhu and Z. Liu, "Graphene modified LiFePO4 cathode materials for high power lithium ion batteries," *Journal of Materials Chemistry,* vol. 21, no. 10, pp. 3353-3358, 2011.

[5] R. Mo, Z. Lei, D. Rooney and K. Sun, "Facile synthesis of nanocrystalline LiFePO4/graphene composite ascathode material for high power lithium ion batteries," *Electrochimica Acta,* vol. 130, no. 6, pp. 594-599, 2014.

[6] "Sigma-Aldrich Graphene Oxide Powder, 15-20 sheets, 4-10% edge-oxidized," Sigma-Aldrich, [Online]. Available: https://www.sigmaaldrich.com/catalog/product/aldrich/796034?lang=en&region=US. [Accessed 27 August 2018].

[7] J. Barker, M. Y. Saidi and J. L. Swoyer, "Lithium Iron (II) Phospho-olivines Prepared by a Novel Carbothermal Reduction Method," *Electrochemical and Solid-State Letter,* vol. 6, no. 3, pp. A53-A55, 2003.

[8] C. Powell and G. W. Beal, "Graphene Oxide and Graphene from Low Grade Coal: Synthesis, Characterization, and Applications," *Current Opinion in Colloid & Interface Science,* vol. 20, no. 5-6, pp. 362-366, 2015.

[9] R. W. Youngs and C. M. Frost, "Humic Acids from Leonardite -- A Soil Conditioner and Organic Fertilizer," *Proceedings of the National Academy of Sciences of the United States of America,* vol. 17, no. 1, pp. 12-18, 1 January 1963.

[10] J. E. Gieseking, "Chemical Composition and Physical Properties of Humic Substances," in *Soil Components : Vol. 1: Organic Components*, Berlin, Heidelberg, Springer, 1975, pp. 1-211.

[11] L. Leonardite Products, "Leonardite Products," DAWA Solutions Group, 1 August 2018. [Online]. Available: http://leonarditeproducts.com/AboutUs. [Accessed 29 August 2018].

[12] G. W. Beal, "Method and System for Producing Graphene and Functionalized Graphene". Untited States of America Patent 8361430, 1 June 2012.

[13] N. N. Schwartz, "Production of Humic Acid". United States of America Patent 3398186, 23 December 1963.

[14] A. Piccolo, "Characteristics of Soil Humic Extracts Obtained by some Organic and Inorganic Solvents and Purified by HCl-HF Treatment," *Soil Science,* vol. 146, no. 6, pp. 418-426, 1988.

[15] D. L. Sparks, Ed., Methods of Soil Analysis Part 3—Chemical Methods, SSSA Book Series 5.3, Madison, Wisconsin: Soil Science Society of America, American Society of Agronomy, 1996, p. 1017.

[16] J. Zhu, R. Duan, S. Zhang, N. Jiang, Y. Zhang and J. Zhu, "The Application of Graphene in Lithium Ion Battery Electrode Materials," *Springer Plus,* vol. 3, no. 1, pp. 585-592, 2014.

531149 A Novel Dynamic Simulation Methodology for High Temperature Packed-Bed Thermal Energy Storage

Abstract:

A Novel Dynamic Simulation Methodology for High Temperature Packed-Bed Thermal Energy Storage

Thermal energy storage (TES) is becoming an increasingly attractive technology for improving power production efficiency and capacity. It has been shown that adding only 4 hours of storage to a solar hybrid steam injection gas turbine (STIG) cycle can increase the plant's capacity factor by up to 50%[1]. To maximize the benefits of thermal energy storage, it is necessary to develop storage systems capable of storage at similar temperatures as those used in the power cycle. Typical industry process temperatures range between 600–1000°C[2,3], while current applied TES systems consider 400–600°C to be their highest temperatures[4]. Demonstration of TES systems capable of storage at higher temperatures (in the 600-1000°C range) are necessary for continued progress. In addition, the ability to simulate behavior of such systems will accelerate their application to industry. Accurate prediction and simulation of thermal energy storage systems over a range of operating conditions and various storage media is obtained by reducing the three-dimensional storage system to two separate one-dimensional models, representing the thermal profile axially along the device in one, and radially within the individual storage media beads in the other. These models work harmoniously together to simulate the storage capabilities of the device. By modeling the individual storage media, the thermal profile within individual packing can be determined, and the energy density of the packing, and therefore the system as a whole, can be accurately predicted at any time during charging or discharging. In addition, configuration of the model in this manner allows for it be applicable to devices utilizing both sensible heat storage and latent heat storage using encapsulated phase change materials (PCM's). The simulation results are validated through development and operation of a matching experimental device. This device was constructed to be capable of using various types of storage media, while requiring little effort to change configurations to be suitable for sensible heat or latent heat storage materials. Experiments were conducted to demonstrate the flexibility of the device design, as well as to validate the applicability of the simulation model to inherently different types of systems.

531067 Effects of Co-Product Uses on Environmental and Economic Sustainability of Hydrocarbon Biofuel from One- and Two-Step Pyrolysis of Poplar

Abstract:

This study investigated the environmental and economic sustainability of liquid hydrocarbon biofuel production via fast pyrolysis of poplar biomass through two pathways: a one-step pathway that converted poplar via fast pyrolysis only, and a two-step pathway that includes a torrefaction step prior to fast pyrolysis (Kulas et al. 2018). Fast pyrolysis involves rapid thermal degradation of wood in the absence of air at a temperature of approximately 500-550°C with a short residence time of less than 2 seconds. The bio-oil product from pyrolysis is catalytically upgraded with hydrogen into hydrocarbon biofuel. Torrefaction, often referred to as mild pyrolysis, tends to enhance bio-oil properties by reducing water content, minimizing acidity, and reducing size reduction energy. The development of a two-stage fast pyrolysis process that involves a torrefaction pretreatment step prior to pyrolysis was investigated as an approach to minimize the energy consumption associated with size reduction and enhance the bio-oil quality.

Optimization of these fast pyrolysis-based biofuel processes were investigated through heat integration and alternative uses of the co-product biochar, which can be sold as an energy source to displace coal or natural gas, soil amendment or processed into activated carbon. The impacts of optimization through heat integration on the cost of hydrocarbon biofuel production as well as the environmental impacts were investigated through a techno-economic analysis (TEA) and life cycle assessment (LCA), respectively, with two-step and one-step processing compared to fossil fuels. The TEA indicates that a one-step heat integrated pathway with the production of activated carbon has a minimum selling price of \$3.23/gallon compared to \$5.16/gallon for a two-step heat integrated process with burning of the co-product biochar to displace coal. The LCA indicates that using the displacement analysis approach, a two-step heat integrated pathway had a global warming potential of -102 g CO2 equivalent/MJ biofuel compared to 16 CO2 equivalent/MJ biofuel for the heat integrated one-step pathway.

References

1. Kulas, D. Winjobi, O., Zhou, W., Shonnard, D.R. (2018), Effects of Co-product Uses on Environmental and Economic Sustainability of Hydrocarbon Biofuel from One- and Two-Step Pyrolysis of Poplar, *ACS Sustainable Chemistry & Engineering*, March 21, 2018, **DOI:** 10.1021/acssuschemeng.7b04390

536785 Optimizing Alkaline-Oxidative Pretreamtent of Hybrid Poplar to Maximize Lignin Co-Product Value

Abstract:

Optimizing Alkaline-Oxidative Pretreatment of Hybrid Poplar to Maximize Lignin Co-Product Value

Diversification of the product portfolio of lignocellulose-to-biofuels processes to include renewable, bio-based chemicals is widely recognized as an opportunity both to improve the economics of these processes and to buffer against market fluctuations. Thus, there is an obvious and compelling need to develop integrated processing approaches that enable multi-product biorefineries. In particular, there is an opportunity for technologies that can be integrated with conversion processes employing a biomass pretreatment/fractionation strategy, whereby it may be possible to utilize all or a fraction of the process-modified lignins for purposes other than as a solid fuel. Although the high heterogeneity of lignin and its low susceptibility to depolymerization creates a number of well-documented challenges complicating its conversion to higher-value products, process-derived lignin nevertheless represents an immense opportunity if they can be generated in a way that optimally couples with conversion strategies. We recently identified that a two-stage alkaline-oxidative pretreatment/delignification process is capable of yielding lignins from hardwoods that are highly susceptible to acidic depolymerization to monomers. In this work, we build on these compelling findings to develop an integrated processing approach through the optimization of this two-stage pretreatment to yield multiple products (lignocellulose-derived polysaccharides and lignin streams that are highly susceptible to depolymerization) with properties that are suited for applications across a diverse range of target markets and product values. Specifically, we will present results from the initial screening of alkaline pre-extraction conditions for 3 hybrid poplar lines over a range of times, temperatures (95-155°C), alkali loadings (50 to 150 mg NaOH/g biomass) to assess the impact of pre-extraction conditions on the yields and properties of the extracted lignins, as well as the impact of these processing conditions on integration with the second stage of the pretreatment and ultimate product yields, product value, and processing costs.

527302 Controlled Radical Polymerization of Lignin-Derived Bio-Oil for Melt-Spinnable Thermoplastic

Abstract:

Controlled radical polymerization of lignin-derived bio-oil for melt-spinnable thermoplastic

Wangda Qu[1], Yuerui Huang[2], Eric Cochran[2], Xianglan Bai[1]

[1] Department of Mechanical Engineering

[2] Department of Chemical and Biological Engineering

Lignin based polymers can be low-cost, green alternatives of petroleum-based polymers. In the present study, red oak lignin-derived bio-oil was used as the starting material to synthesize melt-spinnable thermoplastics. The lignin bio-oil was obtained by fast pyrolyzing red oak at 500 °C followed by water washing the crude bio-oil. The average Mw of the lignin bio-oil was 566 g/mol, and it consists of various phenolic monomers and oligomers. The NMR analysis indicated that there were three OH groups per unit in average. During the first-step of synthesis, the OH groups in the bio-oil was fully or partially methyacrylated. For the partly methyacrylated bio-oil, the remaining OH groups were either acetylated or left as is. In the following step, reversible addition fragmentation chain-transfer (RAFT) method was applied to polymerize the methyacrylated bio-oil. As a result, gelation was only observed with the fully methyacrylated bio-oil for thermosetting. Two other partially methyacrylated bio-oil with and without acetylation were successfully converted into thermoplastics. The Mw and polydispersity index of the polymer generated from the partially methylated bio-oil followed by acetylation were 16k g/mol and 1.53, respectively. The polymer also had a T_g of 130 ºC and T_d of 250 ºC, and the viscosity only increased from 376 and 878 Pa. s during isothermal rheology test at 210 ºC up to 10 min. The melt-spun fiber of this polymer was homogeneous in appearance and had a reduced brittleness compared to conventional lignin-based fibers. The yield of solid residue was 28% upon pyrolyzing to1000 ºC, implying that the fiber could be further developed into carbon fiber.

532086 Lignin Nanoparticles (LNPs) Fabrication through Tailored Lignin Reactivity By Innovative Sequential Organosolv Fragmentation Approach (iSOFA)

Abstract:

Sustainable biorefinery of lignocellulosic biomass to usable platform molecules for producing a broad and multifunctional array of value-added products are key to achieving a green biobased economy. Recently, lignin valorization has been considered as an essential process for the sustainability of biorefineries. Despite various efforts, lignin valorization is still hindered by the poor fractionation performance due to its inherent heterogeneity and low reactivity, which is highly sensitive to the fractionation approach employed. To address these challenges, four sequential organosolv fragmentation approach (SOFA) configurations were developed to deconstruct biomass, fractionate the lignin and tailor its chemistry and reactivity. In a biorefinery concept, carbohydrate output in each SOFA at different stages was taken into consideration. Upon the process, the glucose and xylose yield increased in the order: stage 3>stage 2>stage 1, suggesting the improved monomer sugar release from biomass at the later stage of SOFA.

The lignin with specific reactivity fractionated from each SOFA was subsequently used to fabricate lignin nanoparticles (LNPs) via self-assembly. Results showed that the effective diameter of LNPs from each SOFA followed the order stage 2 > stage 3 > stage 1, and the smallest effective diameter was 130 nm obtained from SOFA using ethanol plus sulfuric acid. The half width, polydisperity index (PDI), and zeta potential of these LNPs were less than 50 nm, 0.08, and -50 mV, respectively, suggesting the uniform distribution and good stability of LNPs. As observed, no specific aggradation occurred during long time storage, suggesting that the LNPs dispersion was very stable in pure water. These LNPs also presented good stability at a broad range of pH values and ionic strength. Taken together, LNPs produced from SOFA were more uniform and stable due to relatively small particles size, PDI, and low zeta potential. 2D HSQC and ^{31}P NMR analysis revealed that SOFA using ethanol plus sulfuric acid enriched G-type lignin, reduced S-type lignin, and thus decreased the S/G ratio. It decreased more Î²-O-4 and Î²â€“Î² linkage groups, indicating the de-polymerization of lignin. Most importantly, SOFA using ethanol plus sulfuric acid exposed more phenolic hydroxyl groups and less carboxylic hydroxyls groups, which was beneficial to the formation of uniformed and stable LNPs. As a result, high-quality LNPs with round shape, narrow size distribution, reduced polydispersity, and good stability have been fabricated through tailoring the lignin reactivity by SOFA. Fabrication of LNPs will open up a green and sustainable pathway for upgrading the lignin stream and thus contribute to the economic feasibility of biorefineries.

538618 Production of 2,5-Furan Dicarboxylic Acid (FDCA) in Ionic Liquid Media

Abstract:

2,5- furan dicarboxylic acid (FDCA) is a bio-based monomer for polyethylene furanoate (PEF), a viable replacement for fossil-based polyethylene terephthalate (PET). In this work, results from conversion of fructose to FDCA in a biphasic ionic liquid (IL)/Ï'-valerolactone (GVL) system, will be reported. Fructose is dehydrated to 5-hydroxymethylfurfural (5-HMF) in an acidic ionic liquid while simultaneously extracted into the immiscible GVL at high yields. The GVL environment also stabilizes the thermally-sensitive 5-HMF and facilitates catalytic oxidation of HMF to FDCA. This presentation will discuss results from catalyst screening studies. Reaction rate, product yield and catalyst reuse data will be presented.

538626 Converting Switchgrass into Cellulosic Sugars, Lipids, and Carotenoids

Abstract:

In this study, an acidified, aqueous DES comprising choline chloride: glycerol (ChCl:Gly) was used to fractionate switchgrass into three major streams under a relatively mild condition: cellulose-rich pulp, lignin, and xylose-rich liquor. The pulp showed good digestibility with about 89% glucose yield. The solvent can be recycled successfully and reused for at least four more pretreatment cycles while maintaining its pretreatment capability. The solvent recycling also improved the lignin recovery from the pretreatment liquor. Lignin recovered from different pretreatment cycles had the Î²-O-4 bonds preserved, and shared similar structures with native lignin. Using the pretreatment liquor as a substrate, the oleaginous yeast *Rhodotorula toruloides* produced 18.7 g/L biomass with lipid and carotenoid titers of 8.1 g/L and 15.0 mg/L, respectively. Overall, this study demonstrated a green process integrating chemical and biological methods toward full utilization of lignocellulosic biomass.

533566 Catalytic Hydrogenation of Furfural over Rumop: Probing Bimetallic andÂ Compositional Effects on Catalyst Performance

Abstract:

Catalytic Hydrogenation of Furfural over RuMoP: Probing Bimetallic and Compositional Effects on Catalyst Performance

Yolanda Bonita, Timothy P. Oâ€™Connell, and Jason C. Hicks*

Department of Chemical and Biomolecular Engineering, University of Notre Dame, 182 Fitzpatrick Hall, Notre Dame, IN 46556

The synthesis of furfuryl alcohol from furfural uses an environmentally toxic copper chromite catalyst in the process. As possible alternatives, Cu-based catalysts such as supported Cu, bimetallic Cu-Ni, Cu-Co, Cu-Fe, and Cu-Ru have been studied to replace copper chromite. However, none of these catalysts surpass the high selectivity exhibited by copper chromite for furfural hydrogenation to furfuryl alcohol.

We have recently discovered a variety of novel transition metal phosphides that are highly selective to furfuryl alcohol and active at near ambient temperature. Contrary to the literature that suggested group VIII metals are not selective to furfuryl alcohol due to the binding configuration, we have synthesized group VIII metal (i.e. Ru) phosphides that exhibit high selectivity to furfuryl alcohol, similar to that of copper chromite and higher than other Cu-based catalysts. Moreover, when comparing pseudo-first order rate constants, we see an increase of > 5 times in the site-normalized, initial rate and 99% selectivity for bimetallic RuMoP. This phenomena was also observed in our previous studies (i.e. hydrodeoxygenation reaction)[1-4], which showcases bimetallic character and potential control of the catalytic properties based on catalyst composition. Indeed, we have observed direct influence on the activity and selectivity by varying the composition of $Ru_xMo_{2-x}P$, where the measured pseudo-first order rate constant was enhanced for Ru-rich materials (x = 1.2). In this presentation, we will present a thorough examination of bimetallic RuMoP as a furfural hydrogenation catalyst through detailed characterization and kinetic experiments.

References

1. Rensel, D. J., Rouvimov S., Gin M. E., Hicks J. C. Highly selective bimetallic FeMoP catalyst for C-O bond cleavage of aryl ethers. *J. Catal.* Sep 2013;305:256-263.
2. Rensel, D. J., Kim J., Bonita Y., Hicks J. C. Investigating the multifunctional nature of bimetallic FeMoP catalysts using dehydration and hydrogenolysis reactions. *Appl. Catal. A.* Aug 25 2016;524:85-93.
3. Rensel, D. J., Kim J., Jain V., Bonita Y., Rai N., Hicks J. C. Composition-directed $Fe_xMo_{2-x}P$ bimetallic catalysts for hydrodeoxygenation reactions. *Catal. Sci. Technol.* 2017;7:1857-1867.
4. Bonita, Y., Hicks J. C. Periodic trends from metal substitution in bimetallic Mo-based phosphides for hydrodeoxygenation and hydrogenation reactions. *J. Phys. Chem. C.* 2018, ASAP.

532119 Etherification of 5-Hydroxymethylfurfural Using Zeolite Catalysts

Abstract:

Biomass is a promising feedstock for the production of modern chemicals. In this regard, polymers represent an attractive class of potential biobased materials, and it may be possible to synthesize a variety of polymeric materials starting from 5-hydroxymethylfurfural (HMF). In this study, we show that H-BEA zeolite is highly active and selective for the production of a variety of HMF ethers which would be suitable for subsequent polymerization.

Several catalysts were evaluated for their etherification activity, including: H-BEA and H-ZSM-5 zeolites, Amberlyst-15 (A-15), and Î³-Al_2O_3. These catalysts were chosen to evaluate a variety of active sites and pore structures. Catalysts possessing BrÃ,nsted acid sites (*i.e.*, H-ZSM-5 and H-BEA zeolites, and A-15) showed the highest selectivities to the ether product, with H-BEA achieving nearly quantitative selectivity. We hypothesize that the improved selectivity observed for H-BEA, compared with A-15, is due to confinement effects that inhibit both levulinate formation and biomolecular condensation (*i.e.*, resinification) reactions.

The rate of etherification is nearly constant for the series of ethanol, *n*-butanol, cyclohexanol, and phenol. Moreover, while the etherification rate saturates at high HMF concentrations, there is little variation in the rate across for alcohol concentrations ranging from 0.1 to 10 mol L^{-1}. These kinetics data, coupled with the invariance of the rate with varying alcohol data, suggest that the reaction proceeds through an oxocarbenium ion transition state that does not involve the etherifying alcohol.

532366 Mechanistic Insights into the Hydrogenolysis of Levoglucosanol over Bifunctional Platinum Silica-Alumina Catalysts in Tetrahydrofuran Solvent

Abstract:

The cellulose-derived platform molecule levoglucosenone (LGO) can be converted to a variety of high-value oxygenated chemicals. [1] LGO can be quantitatively hydrogenated to levoglucosanol (Lgol) over metal catalysts. [2] Herein, we report on the hydrogenolysis of Lgol to tetrahydrofurandimethanol (THFDM) and tetrahydropyran-2-methanol-5-hydroxyl (THP2M5H) over bifunctional platinum or palladium catalysts supported on silica-alumina in tetrahydrofuran solvent. [3,4] THFDM is an α,ω-diol and potential polymer precursor, while both THFDM and THP2M5H can be further upgraded to the commodity chemical 1,6-hexanediol.

A combination of experimental reaction kinetics studies and first principles calculations were used to investigate the mechanism and the roles of the metal and acid sites in this reaction. THFDM and THP2M5H are produced in parallel pathways, and the ketone intermediate THP2M5one is a precursor to THP2M5H. [3] The THFDM cis/trans ratio is independent of the Lgol threo/erythro ratio, suggesting that the mechanism passes through an acyclic intermediate which erases the stereochemistry of the Lgol reactant. ^{13}C radio-labeling was used to confirm the ring rearrangement forming THFDM. The catalyst undergoes moderate, reversible deactivation with time-on-stream in a continuous flow reactor. The reaction rates and product selectivities are comparable at 1.1% and 5.3% Pt loadings, indicating that the reaction rate is acid-limited at these metal loadings; the metal-catalyzed hydrogenation step is not rate-limiting (Figure 1). [4] The measured zero-order dependence in hydrogen indicates that the rate-limiting acid-catalyzed step precedes the fast hydrogenation step. As the Pt loading in Pt/SiAl catalysts is decreased, or when the bare SiAl support is separated from a Pt/SiO_2 catalyst in a dual-layer configuration, the selectivity towards identified products decreases (Figure 1). These results suggest that degradation reactions are avoided when the reactive intermediates formed over acid sites are rapidly hydrogenated over metal sites. First principles simulations were performed to investigate the energetics of the proposed reaction pathway consistent with the experimental results.

Figure 1. *Lgol consumption rate (left y-axis) and product selectivity (right y-axis) versus Pt loading for Pt/SiO_2-Al_2O_3 catalysts.*

These insights into the mechanism of Lgol hydrogenolysis provide a foundation for rationally designed processes to produce high-value chemicals from lignocellulosic biomass using bifunctional metal-acid catalysts.

References.

1. Cao, F.; Schwartz, T. J.; McClelland, D. J.; Krishna, S. H.; Dumesic, J. A.; Huber, G. W., Dehydration of cellulose to levoglucosenone using polar aprotic solvents. *Energy & Environmental Science* **2015,** *8* (6), 1808-1815.

2. He, J.; Huang, K.; Barnett, K. J.; Krishna, S. H.; Alonso, D. M.; Brentzel, Z. J.; Burt, S. P.; Walker, T.; Banholzer, W. F.; Maravelias, C. T.; Hermans, I.; Dumesic, J. A.; Huber, G. W., New catalytic strategies for [small alpha],[small omega]-diols production from lignocellulosic biomass. *Faraday Discussions* **2017,** *202* (0), 247-267.

3. Krishna, S. H.; McClelland, D. J.; Rashke, Q. A.; Dumesic, J. A.; Huber, G. W., Hydrogenation of levoglucosenone to renewable chemicals. *Green Chemistry* **2017,** *19* (5), 1278-1285.

4. Krishna, S. H.; Assary, R. S.; Rashke, Q. A.; Schmidt, Z. R.; Curtiss, L. A.; Dumesic, J. A.; Huber, G. W., Mechanistic Insights into the Hydrogenolysis of Levoglucosanol over Bifunctional Platinum Silica–Alumina Catalysts. *ACS Catalysis* **2018**, 3743-3753.

538548 Hydrothermal Stability of Chloromethyl Polystyrene Based Solid Acid Catalysts and Mechanism of Cellulose Hydrolysis

Abstract:

Cellulose is one of the major components of lignocellulosic biomass and the most abundant renewable polymer on earth. As such it could serve as a sustainable feedstock for the production of commercially important fuels and chemicals. However, cellulose conversion to smaller platform molecule is the major bottleneck for economically competitive utilization of lignocellulosic biomass.

The most common processes to depolymerize cellulose employ liquid acid or enzyme catalyzed hydrolysis to produce glucose. However, issues such as reactor corrosion, slow reaction rates, and inability to recycle limits the cost-effective conversion of cellulose. Alternatively, solid acids have been proposed as reusable catalysts for cellulose hydrolysis. Introduction of binding groups in the catalyst structure is hypothesized to increase the affinity between the solid acid and cellulose, brining the glycosidic bonds within contact with catalytic sites resulting in a higher hydrolysis rate. This hypothesis finds support in adsorption studies of small cellulosic molecules such as cellobiose and glucose. However, the mechanism of such catalytic activity has not been elucidated.

Chloromethyl polystyrene based bifunctional catalysts have showed high activity towards cellulose depolymerization. It is hypothesized the chlorine participates in hydrogen bonding with the hydroxyl groups in cellulose chains. However, the chloromethyl groups is rather label and could hydrolyze in hydrothermal conditions generating hydrochloric acid. Hydrochloric acid could also catalyze cellulose depolymerization, but its effect on the overall catalytic mechanism has not be evaluated.

In this work we studied partially sulfonated chloromethyl polystyrene catalyst for cellulose hydrolysis. Structural analysis of catalyst revealed that synthesis procedure results in little to no chlorine available on the external surface, but presence of hydroxyl groups, which can also participate in hydrogen bonding with cellulose. The catalyst showed high activity towards hydrolyzing cellulose resulting in levulinic acid and formic acid as major products. However, analysis of the solid and the liquid media revealed chloromethyl group hydrolysis and hydrochloric acid generation. This observation raised the question if the catalyst is indeed acting as a solid acid.

In order to elucidate the mechanism of hydrolysis we subjected the catalyst to hydrothermal reaction conditions in the absence of cellulose. Monitoring pH and chloride ion concentration, we developed a kinetic profile of hydrochloric acid generation. Cellulose hydrolysis kinetic models involve ion concentration as a preexponential factor on the expression for the reaction rate constant. We hypothesized that incorporating the time dependence of the acid concentration in the kinetic model of cellulose hydrolysis to glucose and subsequent dehydration to HMF, levulinic acid, and formic acid would be able to accurately predict experimental results. An acceptable agreement between model and experimental product yields pointed to the liquid acid acting as the primary catalyst, obscuring any catalytic activity, if any, of the solid catalyst. Further support was obtained by carrying out hydrolysis by a titrated solid acid catalyst, in which the sulfonic acid groups have been ion exchanged and, thus, rendered inactive; similar yields were obtained as the non-titrated catalyst. The results in this investigation highlight the need for systematic study of proposed solid acids for cellulose hydrolysis and consideration of potential release of homogeneous acids in the reaction media.

538054 Glycerol Transfer-Hydrogenation of Levulinic Acid Using Ru and Ir Carbene Organometallics Immobilized on Active Hydrotalcites

Abstract:

The mechanism and kinetics of levulinic acid (LA) hydrogenation to gamma-valerolactone (GVL) have been extensively studied over the last decade. Despite improvements in selectivity and the use of alternative hydrogen sources such as formic acid and secondary alcohols, economic production of GVL remains hindered by slow reaction rates and the high cost of current hydrogen sources. The use of glycerol, a waste byproduct in biodiesel synthesis, as a hydrogen donor could alleviate the economic challenges associated with GVL production. Herein we tested hydrogenation/deoxygenation of LA under dilute aqueous to neat conditions using Ru and Ir N-heterocyclic carbene (NHC) organometallic catalysts while also varying pH with potassium hydroxide. A total of 7 Ru and Ir homogeneous catalysts were screened for activity, the most active catalysts were then immobilized onto tunable hydrotalcites (HT). A 3:1 ratio of Mg:Al HT was used as a basis for all HT catalytic reactions while substitutions of Al with Fe, Ni, and Co were tested for improved catalytic activity. Products were identified using LCMS, HPLC, NMR, GCMS, and FTIR. The products formed and reaction rates were pH-dependent; acidic conditions favored ketal formation between LA and glycerol whereas alkaline conditions with an Ir(I) complex favored GVL and lactic acid formation at a turnover frequency (TOF) above 66,000 h^{-1} at 150ËšC for GVL. The Ir(I) catalyst achieved unprecedented reaction rates with glycerol, therefore, a comparison to reaction rates achieved with formic acid was demonstrated and showed negligible differences in rates and selectivity. Upon immobilizing the NHC catalysts on HT, the non-active Ru and Ir catalysts achieved activity over 10,000 h^{-1}. Under the same conditions the immobilized Ir(I)HT rate was reduced to 15,000 h^{-1} for GVL. The increase in activity from the other Ru and Ir(III) complexes tested are believed to be a result of the HT Lewis acid sites.

530511 Paired Electrocatalytic Hydrogenation and Oxidation of 5-Hydroxymethylfurfural for Efficient Production of Biorenewable Monomers

Abstract:

Electrochemical synthesis is a promising route for sustainable chemical production; however, its widespread application is hindered by economic limitations incurred from high capital and operating costs. In this presentation, we report highly efficient electrocatalytic conversion of 5-(hydroxymethyl)furfural (HMF) to biobased monomers by pairing HMF reduction and oxidation half reactions in a single electrochemical cell. HMF hydrogenation to 2,5-bis(hydroxymethyl)furan (BHMF) was achieved under mild electrolyte conditions and ambient temperature using a Ag/C cathode. Meanwhile, HMF oxidation to 2,5-furandicarboxylic acid (FDCA) with ~100% efficiency was facilitated under the same conditions by a homogeneous nitroxyl radical redox mediator, together with an inexpensive carbon felt anode. The selectivity and efficiency for Ag-catalyzed BHMF formation were sensitive to cathode potential, owing to competition from HMF hydrodimerization reactions and water reduction (hydrogen evolution). Moreover, the carbon support of Ag/C was active for HMF reduction and contributed to undesired dimer/oligomer formation at strongly cathodic potentials. As a result, high BHMF selectivity and efficiency were only achieved within a narrow potential range near â€“1.3 V. Fortunately, the selectivity of redox-mediated HMF oxidation was insensitive to anode potential, thus allowing HMF hydrogenation and oxidation half reactions to be performed together in a single cathode-potential-controlled cell. Electrocatalytic HMF conversion in a paired cell achieved high molar yields of BHMF and FDCA, and nearly doubled electron efficiency compared to the unpaired cells. Future research will be focused on rational design of an electrochemical flow reactor, in which the use of a membrane electrode assembly (MEA) can significantly lower the internal resistance and reduce the total energy demand.

525127 A Full FurfuralÂ Utilization over Ni/SiO_2 Catalysts

Abstract:

Furfural can be mass-produced from lignocellulose biomass and can be a platform chemical for producing valuable chemicals. In this study, we examine Ni/SiO_2 catalysts for the conversion of furfural under a hydrogen atmosphere. The reaction products include furfuryl alcohol (FA), furan, and propylene as major; butanal and other hydrogenated products, indicating the reaction routes involving side chain hydrogenation, decarbonylation, and ring opening. The reactivity and the product selectivity depend on the reaction temperature and the Ni particle size. The pretreatment by including calcination prior to hydrogen reduction leads to Ni/SiO_2â€"CR with large Ni particles (~ 15 nm) and a high selectivity to FA at below 200 °C. The Ni/SiO_2â€"R is hydrogen-pretreated without the calcination and it contains small Ni particles (~ 5 nm) and exhibits relatively high selectivity to furan. Moreover, the furan can be reacted seriously to produce propylene and CO by this Ni catalyst. The turnover frequency (TOF) of furfural conversion is 239 and 408 h^{-1}, respectively, on Ni/SiO_2-CR and Ni/SiO_2-R at 175 °C, when the former shows 100% selectivity to FA and the latter exhibits a selectivity of around 38% and 62%, to FA and furan, respectively. The results indicate that the furfural hydrotreating reaction over Ni catalysts is with structure sensitive and give a full furfural utilization, either to FA (a biofuel), or to propylene and CO (a chemical feedstock).

527041 Two Stage Stochastic Programming Modeling Approach for Energy Storage System Operation Under Uncertainty

Abstract:

Employing renewable energy sources has become an irreversible trend in the current energy systems of various sectors including process industries. Their decision-makers should cope with often occurring mismatches between varying energy supply and demand over multiple time periods. Considering the inherent variability of renewable output, it would be often witnessed that surplus energies from renewable sources is simply thrown away without being used. It is thus understandable to see the introduction of energy storage system (ESS) for the sake of the economic benefit of using the surplus again later. We can increase the penetration of renewable energies by installing the ESS. On the other hand, it is also right to point out that there is much to be explored in maximizing the potential of ESS with other entities in the energy system. In order to transform the expectation into reality, the corresponding decision-making framework should be rigorously designed. As an alternative, this paper investigates the impact of variation on the design and operation of ESS. Specifically, the design of ESS under uncertainty is transformed into a two-stage stochastic programming model. The stochastic model was formulated in terms of a mixed integer linear programming (MILP) problem. As illustrated in numerical examples, the proposed modeling approach allows us to gain the insight on renewable energy systems. More similar works are expected to establish the robust energy system in process industries.

527226 Theoretical and Experimental Investigation of the Microbial Degradation of Solitary Oil Microdroplets

Abstract:

Natural seeps and accidental releases of crude oil in the sea result in clouds of droplets that are carried away by underwater sea currents. The droplets may be created either at the sea surface during the breakup of an oil slick by sea waves, or at the seafloor during the atomization of live crude oil extruding at sufficiently high speed from a natural crack or a broken wellhead. Drifting droplet clouds disturb the established ecosystem dynamics and pose a high risk of toxic effects to many marine species. It is therefore of crucial importance to understand and quantify the factors that determine the retention of the dispersed droplets within the seawater column. Dissolution, biodegradation and aggregation with marine snow are the main physical processes that rule the fate of dispersed oil droplets in marine waters.

With regard to the biodegradation process, microbes have developed three fundamental strategies for accessing and assimilating oily substrates. Depending on their affinity for the oily phase and ability to proliferate in multicellular structures, microbes might either attach to the oil surface and directly uptake compounds from the oily phase, or grow suspended in the aqueous phase consuming solubilized oil, or form three-dimensional biofilms over the oil-water interface (Kapellos, 2017).

We have recently developed a compound particle model for the biodegradation of solitary oil microdroplets moving through a water column (Kapellos et al., 2018). The compound particle is of the core-shell type and consists of an oily core that is successively surrounded by a bioreactive skin of negligible thickness and another bioreactive shell of finite thickness. The bioreactive skin represents a thin layer of microbes that uptake oil directly from the oily phase, whereas the bioreactive shell represents a distinct biofilm phase. The new model accounts for all three modes of biodegradation: direct interfacial uptake, bioreaction in the bulk aqueous phase, and bioreaction in a biofilm formed around the droplet. A closed-form correlation has been established for the overall dissolution rate for a given particle configuration and two coupled ordinary differential equations have been derived for the evolution of the dimensions of the compound particle. Numerical analysis is currently used to extend the domain of validity of the model by relaxing key hypotheses. In this presentation, major findings of the theoretical analysis will be discussed in detail and in conjunction with preliminary results from the biodegradation of hexadecane droplets by Marinobacter sp. in a microfluidic setting.

Acknowledgements: This work has received funding from the European Unionâ€™s Horizon 2020 research and innovation programme under the Marie Sklodowska-Curie grant agreement No 741799 (project "OILY MICROCOSM").

References:

Kapellos, G.E. "Microbial strategies for oil biodegradation", In *Modeling of Microscale Transport in Biological Processes*; Becker, S.M., Ed.; Academic Press, pp. 19-39 (2017).

Kapellos, G.E., Paraskeva, C.A., Kalogerakis, N. and Doyle, P.S. "Theoretical insight into the biodegradation of solitary oil microdroplets moving through a water column", *Bioengineering*, 5:15, (2018).

530591 Modeling Faradaic Capacitive Deionization with Redox Active Porous Electrodes

Abstract:

Capacitive deionization (CDI) is an attractive technology for brackish water treatment, due to the low energy consumption at low source water salinity level compared to conventional water treatment technologies such as reverse osmosis or distillation.

It has been shown that the surface charge on the porous carbon electrodes has significant impact on CDI performance. By introducing immobilized chemical groups, the point of zero charge of carbon materials can be tuned, thus enhanced or inversed CDI performance could be obtained. Redox active electrodes have been investigated in CDI applications recently. The Faradaically induced chemical charges add to electronic charges within the electric double layers formed at the electrode surfaces to provide a much higher adsorption capacity than attained in purely capacitive processes, which is denoted as Faradaic capacitive deionization (FaCDI) here. The redox active ligand can be considered as variable chemical charges that depend on the cell voltage, as opposed to the fixed charges introduced by chemical modification.

Here, we developed a model which incorporates the effect of variable chemical charge due to general faradaic reactions of redox species immobilized on the electrode surfaces. The theory describes the variable chemical charge by the generalized Frumkin-Butler-Volmer equation across the stern layer as an extension of current electrosorption theory. As a first approximation, the model assumes the spacer channel is well-mixed with uniform electrosorption in each electrode. The model is able to predict the experimentally observed enhanced and inverted performance of CDI cells. Furthermore, the deionization performance of FaCDI cells is shown to be superior to that of CDI and enhanced-CDI systems with equilibrium adsorption capacities 50-100% higher than attained with CDI systems. By properly choosing the equilibrium potentials of the redox active material, FaCDI is able to achieve the same ion adsorption capacity at smaller voltages compared to conventional CDI, which indicates the promise of obtaining higher salt adsorption capacity with lower energy input through the use of FaCDI materials.

The model enables us to optimize the operation parameters of the process, such as flow rate, applied voltage as well as the design of CDI cells, such as the surface to volume ratio of the device, in order to obtain the optimal deionization performance.

532901 Finding the Ideal Feedstock for Biodiesel Production Via Generalized Kinetic Model for Transesterification and Saponification

Abstract:

The hunt for an assortment of energy sources is just about as old as humanity's need for energy. This search has been escalated by an expanding industrialization since the 19th century. The internal combustion engine is a prime example. Even though fuels obtained from fossil sources have been a prime choice since the onset of these engines, biofuels especially biodiesel have also found some interest. This interest has particularly gained momentum in the past few decades due to rapidly increasing energy demand and depletion in fossil fuel resources.

Biodiesel is a non-toxic, biodegradable, and environmentally benign fuel that can be used in diesel engines. The most commonly used method for the production of biodiesel is transesterification. It involves the reaction of oil with an alcohol in the presence of a catalyst to form fatty acid alkyl ester (biodiesel) and glycerol. Although, biodiesel is a promising alternative fuel, its production poses significant challenges due to its higher costs compared to that of petroleum-based diesel. Typically one of the major contributors to this issue is the cost of the feedstock, which accounts for approximately 60 to 80% of the total production cost.

Even though the literature is replete with the studies about modelling and optimization of transesterification [1, 2, 3, 4], very few researchers have focused their attention on the selection of feedstock for biodiesel production. Although, a few comparative studies have been reported in the literature [2, 5], all of them focus on static optimization, thereby neglecting the kinetics of biodiesel production. Therefore, in order to address these issues, we aim to achieve the following objectives:

1. Find optimal operating conditions of the reactor for biodiesel production using different vegetable oils as feedstock.
2. Develop meta-models depicting the effect of the different vegetable oils/their blends on the properties of the biodiesel produced using these feedstocks.
3. Use the developed meta-models to find the best possible feedstock for a particular country.

The purpose of this study is to find optimal feedstock (vegetable oil blend) for a particular country based on the standard property specifications and the availability of oils in that country. To this end, we use the generalized kinetic model, developed in our previous work [6] to find optimal conditions with 4,500 different blends of six major oils *viz.*, Jatropha, Linseed, Olive, Palm, Rapeseed, and Sunflower oil as feedstock. We employ a property based objective function (considering density, viscosity, cold point, pour point, cetane number, higher heating value, lower calorific value, and cold filter plugging point of biodiesel) to solve these optimization problems. We then employ correlations from the literature [3] to calculate properties of biodiesel formed using a particular feedstock and at corresponding optimal conditions. This is followed by developing meta-models with the computed properties of biodiesel as output variables and different characteristics of the feedstocks as the input variables. We use five different approaches to characterize the feedstock *viz.* a) average chain length and the unsaturation degree of the fatty acids that constitute the oil (adopted from [7], number of dimensions (N) = 2); b) average chain length, weighted unsaturation degree calculated based on the triglyceride composition of oil (proposed in this work, N = 2); c) triglyceride composition (proposed in this work, N = 34); d) weights on the considered oils for the different blends (proposed in this work, N = 6); and e) properties of oils, *i.e.* viscosity, density, heat capacity, enthalpy of vaporization, and vapour pressure calculated based on the triglyceride composition of oil (proposed in this work, N = 5). Finally, we use the best performing meta-models to obtain optimal feedstock for production of biodiesel in a particular country.

The predictions made using the developed meta-models are in accord with those made using the original models. We quantify this agreement, by computing the Percentage Average Absolute Error (PAAE) as follows,

PAAE = (Average Absolute Error/Average Response)x100

The error values are below 1% for most of the cases with the maximum value at 2.55%, which shows excellent prediction performance. Moreover, we use a normalized error metric to compare the performance of the meta-models developed using different data sets in this study. We observed that the meta-models developed using triglyceride composition and weight distribution of oils in the blend as input variables performed superior to the other meta-models. However, owing to the large number of dimensions in the former case, it is apt to use the latter. Finally, we used the meta-models to obtain optimal feedstock blend for a particular country. We analysed different scenarios based on the standard specification imposed on the biodiesel quality, availability and type of feedstock (*i.e.* edible or non-edible). We observed that the quality of biodiesel produced can change significantly based on the considered scenarios, thereby highlighting the importance of such a study. Besides, we also observed that it is possible to produce premium quality biodiesel even with just non-edible oils. Such a result can definitely bolster the efforts for commercialization of biodiesel production as the use of non-edible oils can significantly reduce the costs associated with the biodiesel production process.

REFERENCES

[1] V. Singh, K. Solanki, M. N. Gupta, Process optimization for biodiesel production, Recent patents on biotechnology 2 (2) (2008) 130{143.

[2] J.Yang, T. Astatkie, Q. S. He, A comparative study on the effect of unsaturation degree of camelina and canola oils on the optimization of bio-diesel production, Energy Reports 2 (2016) 211-217.

[3] S. Pinzi, D. Leiva, G. Arzamendi, L. Gandia, M. Dorado, Multiple response optimization of vegetable oils fatty acid composition to improve biodiesel physical properties, Bioresource technology 102 (15) (2011) 7280-7288.

[4] R. S. Kumar, K. Sureshkumar, R. Velraj, Optimization of biodiesel production

from manilkara zapota (l.) seed oil using taguchi method, Fuel 140 (2015) 90-96.

[5] T. Issariyakul, A. K. Dalai, Comparative kinetics of transesterication for biodiesel production from palm oil and mustard oil, The Canadian Journal of Chemical Engineering 90 (2) (2012) 342-350.

[6] P. Chhabra, S. Mosbach, M. Kraft, I. A. Karimi, A generalized kinetic model for transesterication and saponification (2017).

URL https://aiche.confex.com/aiche/2017/meetingapp.cgi/Paper/500597

[7] E. G. Giakoumis, A statistical investigation of biodiesel physical and chemical properties, and their correlation with the degree of unsaturation, Renewable Energy 50 (2013) 858-878.

530255 Accelerating the Generation of Coal Power Plant Property Models

Abstract:

Power plants are complex systems that are represented by first principles or empirical models. As power plants are developed with new technologies outside of well-studied thermodynamic regions, data-driven approaches have been used to model these complex systems [5]. For example, thermophysical properties or kinetic reaction models can be explained to an arbitrary level of complexity with an empirical model. The challenge is that these process models are not well understood, and it is not always intuitive what functional forms can be utilized to represent these systems. When optimizing proposed model systems, the efficiency of power plants relies on having accurate models. To maximize efficiency as well as to maintain safe operating conditions, it is essential that complex kinetic and thermophysical relationships are represented accurately, even though determining such a representation may be computationally expensive.

In this work we address the problem of speeding up the model generation process. In particular, we look at accelerating best subset selection algorithms with the use of GPU parallel computing. Best subset selection is a regression problem that seeks to determine a small set of features that best relates a set of inputs to an output [7]. There are many exact and heuristic approaches that can be used to solve the best subset selection problem, such as branch-and-bound [4], forward and backward selection [3], exhaustive search, the lasso [7], and mixed-integer optimization [6, 2, 1]. The proposed approach speeds up the model building process, allowing us to consider a larger feature space and generate more realistic models. The proposed methodology is compared against best subset selection techniques as well as the Automated Learning of Algebraic MOdels (ALAMO) methodology [2].

References

[1] D. Bertsimas, A. King, and R. Mazumder. Best subset selection via a modern optimization lens. The Annals of Statistics, 44:813–852, 2016.

[2] A. Cozad, N. V. Sahinidis, and D. C. Miller. Automatic learning of algebraic models for optimization. AIChE Journal, 60:2211–2227, 2014.

[3] M.A. Efroymson. Multiple regression analysis. Mathematical methods for digital computers, pages 191–203, 1960.

[4] C. Gatu and EJ Kontoghiorghes. Branch-and-bound algorithms for computing the best-subset regression models. Computational and Graphical Statistics, 15:139–156, 2006.

[5] S Lu and BW Hogg. Dynamic nonlinear modelling of power plant by physical principles and neural networks. International Journal of Electrical Power & Energy Systems, 22.

[6] A. Miller. Subset selection in regression. CRC Press, 2002.

[7] R. Tibshirani. Regression shrinkage and selection via the lasso. Royal Statistical Society, Methodological:267–288, 1996.

524382 Dynamic Optimisation of Water-Injection Wells Operation for Enhanced Oil Production from a Mature Oil and Gas Field

Abstract:

Field engineers in the oil and gas industry are constantly faced with the challenge of maintaining profitability amidst several operational constraints whose effects span a wide spectrum of timescales. Given the inevitably high environmental and financial stakes associated with the exploration and production of an oil and gas prospect, there is a strong incentive to enhance hydrocarbon recovery and production via systematic, model-based optimisation approaches [1-3]. In further response to this challenge, the application of sophisticated simulation methodologies to comprehensively ocapture the reservoir behaviour, multiphase flows (in wellbores and flowlines) [4-5] and gas-oil-water separation in the processing facilities is of exceptional industrial as well as academic interest. Liquid loading in gas wells and artificial lift design considerations, reservoir pressure maintenance via water injection, gas/water coning during production from deviated wells, pressure drop and liquid holdup of multiphase mixtures in highly deviated flowlines are some of the specific features of the problem which are frequently encountered and known to affect its complexity.

Simulation of these prevalent subsurface and surface phenomena does not always guarantee an accurate prediction of the onset of these problems, let alone the trouble-free operation of an existing petroleum field. In order to tackle the shortcomings and thus improve the fidelity of the currently employed models, it is necessary to also combine robust optimisation methods with these oil and gas CFD simulations [6-7]. This combination of simulation and optimisation algorithms increases the order of complexity due to model nonlinearity, non-convexity and the presence of discrete variables. There has been an increasing number of contributions in the application of optimisation techniques to field production, and this can be attributed to the advances in the development of specialised algorithms and accompanying computational power. To date, most of these attempts have focused on fields undergoing primary production through predominantly vertical wells [1-3].

This paper advances the state of the art in this field by formulating a production optimisation problem, via simultaneous consideration of production and injection wells with both vertical and deviated (inclined/horizontal) well geometries. A field undergoing secondary production (Enhanced Oil Recovery, EOR) is considered in this study. We use multiphase flow and reservoir simulation software in order to compute flowrates and pressure drops in wellbores and flowlines of the considered production network; extra measures are taken to ensure all flowlines are free from hydrate and wax at the prevalent temperature and pressure conditions. The problem is solved as a nonlinear program (NLP), comprising an economic objective function and several constraints to ensure operational feasibility. The adopted optimisation technique yields a reasonable increase in oil production compared to the usual and direct application of a black-box simulator. This also translates to an increased NPV with the current oil price, thus demonstrating the efficiency of the proposed method as a value addition tool.

REFERENCES

[1] Gunnerud, V., Conn, A. and Foss, B., 2013. Embedding structural information in simulation-based optimization. *Comput. Chem. Eng.* **53**: 35–43.
[2] Gerogiorgis, D.I., Georgiadis, M., Bowen, G., Pantelides, C.C. and Pistikopoulos, E.N., 2006. Dynamic oil and gas production optimization via explicit reservoir simulation. *Comput. Aided Chem. Eng.*, **21**: 179–184.
[3] Tavallali, M.S., Karimi, I.A., Teo, K.M., Baxendale, D. and Ayatollahi, S., 2013. Optimal producer well placement and production planning in an oil reservoir. *Comput. Chem. Eng.*, **55**: 109–125.
[4] Epelle, E.I. and Gerogiorgis, D.I., 2017. A multiparametric CFD analysis of multiphase annular flows for oil and gas drilling applications. *Comput. Chem. Eng.*, **106**: 645–661.
[5] Epelle, E.I. and Gerogiorgis, D.I., 2018. Transient and steady state analysis of drill cuttings transport phenomena under turbulent conditions. *Chem. Eng. Res. Des.* **131**: 520-544.
[6] Gunnerud, V. and Foss, B., 2010. Oil production optimization—A piecewise linear model, solved with two decomposition strategies. *Comput. Chem. Eng.*, **34**(11): 1803–1812.
[7] Kosmidis, V.D., Perkins, J.D. and Pistikopoulos, E.N., 2004. Optimization of well oil rate allocations in petroleum fields. *Ind. Eng. Chem. Res.* **43**(14): 3513–3527.

537394 Voluntary Oil Well Ignition As a Blowout Response: Analysis of Factors Influencing Viability

Abstract:

While most off-shore drilling use conventional steel rigs, for large reservoirs in shallow waters, artificial islands make an attractive alternative for offshore drilling because it offers more space, lower environmental impact, greater personnel safety, and lower cost. Drilling from an off-shore island may allow voluntary ignition as a blowout response if the oil contains sufficient gas. For voluntary ignition to be a viable blowout response, wellhead burning â€" that depends on a number of factors â€" needs to be nearly 100% efficient.

In this work, we investigate wellhead burning as a planned response to oilwell blowouts. The work presents a case study of proposed wellhead burning of 880 GOR, 27° API oil, should a blowout occur. The work takes into account some of the concerns regarding possible significant oil spill on Beaufort Sea as a result of incomplete combustion of the blowout effluent.

Burning of two-phase mixtures of oil and gas is a complex phenomenon that depends on the fraction of gas in the mixture, percentage of oil being entrained as liquid droplets in the gas stream, the size of the droplets, their residence time in the flame, among other parameters. Available models and correlations for the influence of these parameters on burning efficiency are examined in view of their relevance to wellhead burning.

This work will demonstrate that for some oil wells with high enough GOR, wellhead burning could be efficient enough for to be a reasonable response to a well blowout. In the studied case, a GOR of 880 combined with high total flowrate of the mixture, strongly implies nearly complete combustion of the well effluent during a blowout. The case study helped us formulate an algorithm to examine whether any particular well could be a possible candidate for wellhead burning as a planned response to a blowout. The approach suggests determining a wellâ€™s worst case discharge (WCD), estimating fluid velocities at well exit, estimating adiabatic flame temperature, and performing an energy balance to determine if all oil drops will evaporate.

AUTHOR INDEX

AUTHOR INDEX

AUTHOR INDEX

AUTHOR INDEX

AUTHOR INDEX

AUTHOR INDEX

AUTHOR INDEX

AUTHOR INDEX

AUTHOR INDEX

AUTHOR INDEX